COMPOSITE MATERIALS FOR INDUSTRY, ELECTRONICS, AND THE ENVIRONMENT

Research and Applications

COMPOSITE MATERIALS FOR INDUSTRY, ELECTRONICS, AND THE ENVIRONMENT

Research and Applications

Edited by

Omari V. Mukbaniani

Devrim Balköse

Heru Susanto

A. K. Haghi

AAP | APPLE ACADEMIC PRESS

Apple Academic Press Inc.
3333 Mistwell Crescent
Oakville, ON L6L 0A2 Canada

Apple Academic Press Inc.
1265 Goldenrod Circle NE
Palm Bay, Florida 32905 USA

Exclusive worldwide distribution by CRC Press, a member of Taylor & Francis Group

International Standard Book Number-13: 978-1-77188-740-3 (Hardcover)
International Standard Book Number-13: 978-0-42945-793-7 (eBook)

Library and Archives Canada Cataloguing in Publication

Title: Composite materials for industry, electronics, and the environment : research and applications / edited by Omari V. Mukbaniani, Devrim Balköse, Heru Susanto, A.K. Haghi.

Names: Mukbaniani, O. V. (Omar V.), editor. | Balköse, Devrim, editor. | Susanto, Heru, 1965- editor. | Haghi, A. K., editor.

Description: Includes bibliographical references and index.

Identifiers: Canadiana (print) 20190072474 | Canadiana (ebook) 20190072520 | ISBN 9781771887403 (hardcover) | ISBN 9780429457937 (eBook)

Subjects: LCSH: Composite materials.

Classification: LCC TA418.9.C6 C66 2019 | DDC 620.1/18—dc23

Library of Congress Cataloging-in-Publication Data

Names: Mukbaniani, O. V. (Omar V.), editor.

Title: Composite materials for industry, electronics, and the environment : research and applications / [edited by] Omari V. Mukbaniani, Devrim Balköse, Heru Susanto, A.K. Haghi.

Description: Toronto ; New Jersey : Apple Academic Press, [2019] | Series: Innovations in agricultural and biological engineering | Includes bibliographical references and index.

Identifiers: LCCN 2019007653 (print) | LCCN 2019009260 (ebook) | ISBN 9780429457937 (ebook) | ISBN 9781771887403 (hardcover : alk. paper)

Subjects: LCSH: Composite materials.

Classification: LCC TA418.9.C6 (ebook) | LCC TA418.9.C6 C5735 2019 (print) | DDC 620.1/18--dc23

LC record available at https://lccn.loc.gov/2019007653

Apple Academic Press also publishes its books in a variety of electronic formats. Some content that appears in print may not be available in electronic format. For information about Apple Academic Press products, visit our website at **www.appleacademicpress.com** and the CRC Press website at **www.crcpress.com**

ABOUT THE EDITORS

Omari V. Mukbaniani, DSc

Omari Vasilii Mukbaniani, DSc, is Professor and Chair of the Macromolecular Chemistry Department of Iv. Javakhishvili Tbilisi State University, Tbilisi, Georgia. He is also the Director of the Institute of Macromolecular Chemistry at Iv. Javakhishvili Tbilisi State University. He is a member of the Academy of Natural Sciences of Georgia. For several years he was a member of advisory board and editorial board of the Journal Proceedings of Iv. Javakhishvili Tbilisi State University (Chemical Series) and contributing editor of the journals *Polymer News, Polymers Research Journal,* and *Chemistry and Chemical Technology*. His research interests include polymer chemistry, polymeric materials, and chemistry of organosilicon compounds, as well as methods of precision synthesis to build block and the development of graft and comb-type structure. He also researches the mechanisms of reactions leading to these polymers and the synthesis of various types of functionalized silicon polymers, copolymers, and block copolymers.

Devrim Balköse, PhD

Devrim Balköse, PhD, is currently a retired faculty member in the Chemical Engineering Department at Izmir Institute of Technology, Izmir, Turkey. She graduated from the Middle East Technical University in Ankara, Turkey, with a degree in chemical engineering. She received her MS and PhD degrees from Ege University, Izmir, Turkey, in 1974 and 1977, respectively. She became Associate Professor in macromolecular chemistry in 1983 and Professor in process and reactor engineering in 1990. She worked as a research assistant, assistant professor, associate professor, and professor between 1970 and 2000 at Ege University. She was the Head of Chemical Engineering Department at Izmir Institute of Technology, Izmir, Turkey, between 2000 and 2009. She is now a retired faculty member in the same department. Her research interests are in polymer reaction engineering, polymer foams and films, adsorbent development, and moisture sorption. Her research projects are on nanosized zinc borate production, ZnO polymer composites, zinc borate lubricants, antistatic additives, and metal soaps.

Heru Susanto, PhD

Heru Susanto, PhD, is currently Head of the Information Department and researcher at the Indonesian Institute of Sciences, Computational Science & IT Governance Research Group. At present, he is also an Honorary Professor and Visiting Scholar at the Department of Information Management, College of Management, Tunghai University, Taichung, Taiwan. Dr. Susanto has worked as an IT professional in several roles, including Web Division Head of IT Strategic Management at Indomobil Group Corporation and Prince Muqrin Chair for Information Security Technologies at King Saud University. His research interests are in the areas of information security, IT governance, computational sciences, business process re-engineering, and e-marketing. Dr. Susanto received a BSc in Computer Science from Bogor Agricultural University, an MBA in Marketing Management from the School of Business and Management Indonesia, and an MSc in Information Systems from King Saud University, and a PhD in Information Security System from the University of Brunei and King Saud University.

A. K. Haghi, PhD

A. K. Haghi, PhD, is the author and editor of 165 books, as well as over 1000 published papers in various journals and conference proceedings. Dr. Haghi has received several grants, consulted for a number of major corporations, and is a frequent speaker to national and international audiences. Since 1983, he served as professor at several universities. He is currently Editor-in-Chief of the *International Journal of Chemoinformatics and Chemical Engineering* and *Polymers Research Journal* and on the editorial boards of many international journals. He is also a member of the Canadian Research and Development Center of Sciences and Cultures (CRDCSC), Montreal, Quebec, Canada. He holds a BSc in urban and environmental engineering from the University of North Carolina (USA), an MSc in mechanical engineering from North Carolina A&T State University (USA), a DEA in applied mechanics, acoustics and materials from the Université de Technologie de Compiègne (France), and a PhD in engineering sciences from Université de Franche-Comté (France).

CONTENTS

Contributors ..... *ix*

Abbreviations ..... *xiii*

Preface ..... *xvii*

SECTION I: Advanced Composites Systems 1

1. **Graphene-Based Fibers and Their Application in Advanced Composites System 3**
   Raghvendra Kumar Mishra

2. **Epoxy Composites Filled with Gypsum (Alabaster G-5): Possible Ways for Strengthening, Stabilization, and Structuration 25**
   D. L. Starokadomsky and A. A. Ishenko

3. **Structure and Properties of Polymer Composites Based on Statistical Ethylene–Propylene Copolymer and Mineral Fillers 45**
   N. T. Kahramanov, N. B. Arzumanova, I. V. Bayramova, and J. N. Gahramanly

4. **Mechanical and Vibrational Analysis of Microtubule Using Molecular Dynamic Simulations 57**
   M. Ghamami, M. H. Jafarabadi, and H. Nahvi

5. **Statically Dissipating and Luminescent Zinc Oxide/Linear Low-Density Polyethylene Composites 87**
   Filiz Özmıhçı Ömürlü and Devrim Balköse

SECTION II: Electronic and Ionic Composites 119

6. **Materials for Organic Transistor Applications 121**
   Andreea Irina Barzic and Razvan Florin Barzic

7. **Organic Electronics: For a Better Tomorrow 153**
   Ajith James Jose, Greta Mary Thomas, S. Parvathy, and Ann Treessa Wilson

8. **Basic Principle of Fluorescence and Design of Sensors for Ions 171**
   Ashish Kumar and Swapan Dey

9. **Superconductivity and Quantum Computing Via Magnetic Molecules** 201
Francisco Torrens and Gloria Castellano

SECTION III: Prospects and Challenges of Composites 211

10. **Novel Separation Processes, Water Science, and Composite Membranes: A Far-Reaching Review and a Vision for the Future** 213
Sukanchan Palit

11. **Genetic Algorithm, Fluidization Engineering, and the Futuristic Vision of Composites and Chemical Engineering** 235
Sukanchan Palit

12. **Computational Fluid Dynamics Techniques and Its Applications in Chemical Engineering, Petroleum Engineering, and the Vast Domain of Composite Science** 259
Sukanchan Palit

13. **Evolutionary Computation, Multiobjective Optimization, Simulated Annealing, and Composites and Polymer Science: A Short and Far-Reaching Review** 287
Sukanchan Palit

SECTION IV: Biobased Composites 317

14. **Passive Aerated Composting of Leaves and Predigested Office Papers** 319
A. Y. Zahrim, S. Sariah, R. Mariani, I. Azreen, Y. Zulkiflee, and A. S. Fazlin

15. **Taking into Account the Features of the Within-Field Variability of Composite Soil Fertility in Precision Agrotechnologies** 339
Rafail A. Afanas'ev, Genrietta E. Merzlaya, and Michail O. Smirnov

16. **Chemistry of Medicinal Plants, Foods, and Natural Products and Their Composites** 357
Laura María Solis-Salas, Luis Enrique Cobos-Puc, Cristóbal Noé Aguilar-González, Crystel Aleyvick Sierra Rivera, Anna Iliná, and Sonia Yesenia Silva Belmares

Index *383*

CONTRIBUTORS

Rafail A. Afanas'ev
Pryanishnikov All-Russian Scientific Research Institute of Agrochemistry, d. 31A, Pryanishnikova St., Moscow 127550, Russia

Cristóbal Noé Aguilar-González
Food Research Department, Faculty of Chemistry, Autonomous University of Coahuila, Blvd. Venustiano Carranza, Col. República Oriente, 25280, Saltillo, Coahuila, Mexico

N. B. Arzumanova
Institute of Polymer Materials of Azerbaijan National Academy of Sciences, Sumgait, Azerbaijan Republic

I. Azreen
Chemical Engineering Programme, Faculty of Engineering, Universiti Malaysia Sabah, Jalan UMS, 88400 Kota Kinabalu, Sabah, Malaysia

Devrim Balköse
Department of Chemical Engineering, İzmir Institute of Technology, Gulbahce, Urla, Izmir, Turkey

Andreea Irina Barzic
"Petru Poni" Institute of Macromolecular Chemistry, 41A Grigore Ghica Voda Alley, 700487 Iasi, Romania

Razvan Florin Barzic
Faculty of Mechanics, "Gheorghe Asachi" Technical University of Iasi, 43 Dimitrie Mangeron, 700050 Iasi, Romania

I. V. Bayramova
Institute of Polymer Materials of Azerbaijan National Academy of Sciences, Sumgait, Azerbaijan Republic

Gloria Castellano
Departamento de Ciencias Experimentales y Matemáticas, Facultad de Veterinaria y Ciencias Experimentales, Universidad Católica de Valencia San Vicente Mártir, Guillem de Castro-94, E-46001 València, Spain

Luis Enrique Cobos-Puc
Food Research Department, Faculty of Chemistry, Autonomous University of Coahuila, Blvd. Venustiano Carranza, Col. República Oriente, 25280, Saltillo, Coahuila, Mexico

Swapan Dey
Department of Applied Chemistry, Indian Institute of Technology (ISM), Dhanbad 826004, Jharkhand, India

A. S. Fazlin
Chemical Engineering Programme, Faculty of Engineering, Universiti Malaysia Sabah, Jalan UMS, 88400 Kota Kinabalu, Sabah, Malaysia

J. N. Gahramanly
Azerbaijan State University of Oil and Industry, Baku, Azerbaijan Republic

M. Ghamami
Department of Mechanical Engineering, Isfahan University of Technology, Isfahan, Iran

Anna Iliná
Food Research Department, Faculty of Chemistry, Autonomous University of Coahuila, Blvd. Venustiano Carranza, Col. República Oriente, 25280, Saltillo, Coahuila, Mexico

A. A. Ishenko
Cathedra of Machines for Metallurgy, Mariupol Priazovski State Technology University, Mariupol 87500, Ukraine

M. H. Jafarabadi
Department of Mechanical Engineering, Isfahan University of Technology, Isfahan, Iran

Ajith James Jose
Research and Postgraduate Department of Chemistry, St. Berchmans College, Changanacherry, Kerala 686101, India

N. T. Kahramanov
Institute of Polymer Materials of Azerbaijan National Academy of Sciences, Sumgait, Azerbaijan Republic

Ashish Kumar
Department of Applied Chemistry, Indian Institute of Technology (ISM), Dhanbad 826004, Jharkhand, India

R. Mariani
Chemical Engineering Programme, Faculty of Engineering, Universiti Malaysia Sabah, Jalan UMS, 88400 Kota Kinabalu, Sabah, Malaysia

Genrietta E. Merzlaya
Pryanishnikov All-Russian Scientific Research Institute of Agrochemistry, d. 31A, Pryanishnikova St., Moscow 127550, Russia

Raghvendra Kumar Mishra
Director, BSM Solar and Environmental Solution, A-348, Awas Vikas Colony, Sitapur, Unnao, Uttar Pradesh, India

H. Nahvi
Department of Mechanical Engineering, Isfahan University of Technology, Isfahan, Iran

Filiz Özmıhçı Ömürlü
Natural Anne Ekolojik Ürünler Çiğli İzmir, Turkey

Sukanchan Palit
Assistant Professor (Senior Scale), Department of Chemical Engineering, University of Petroleum and Energy Studies, Energy Acres, P.O. Bidholi via Premnagar, Dehradun 248007, Uttarakhand, India
43, Judges Bagan, P.O. Haridevpur, Kolkata 700082, India

S. Parvathy
Research and Postgraduate Department of Chemistry, St. Berchmans College, Changanacherry, Kerala 686101, India

S. Sariah
Chemical Engineering Programme, Faculty of Engineering, Universiti Malaysia Sabah, Jalan UMS, 88400 Kota Kinabalu, Sabah, Malaysia

Crystel Aleyvick Sierra Rivera
Food Research Department, Faculty of Chemistry, Autonomous University of Coahuila, Blvd. Venustiano Carranza, Col. República Oriente, 25280, Saltillo, Coahuila, Mexico

Sonia Yesenia Silva Belmares
Food Research Department, Faculty of Chemistry, Autonomous University of Coahuila, Blvd. Venustiano Carranza, Col. República Oriente, 25280, Saltillo, Coahuila, Mexico

Michail O. Smirnov
Pryanishnikov All-Russian Scientific Research Institute of Agrochemistry, d. 31A, Pryanishnikova St., Moscow 127550, Russia

Laura María Solis-Salas
Food Research Department, Faculty of Chemistry, Autonomous University of Coahuila, Blvd. Venustiano Carranza, Col. República Oriente, 25280, Saltillo, Coahuila, Mexico

D. L. Starokadomsky
Polymer Composite Laboratory, Chuiko Institute of Surface Chemistry, NAS of Ukraine, Kiev 03164, Ukraine

Greta Mary Thomas
Research and Postgraduate Department of Chemistry, St. Berchmans College, Changanacherry, Kerala 686101, India

Francisco Torrens
Institut Universitari de Ciència Molecular, Universitat de València, Edifici d'Instituts de Paterna, P.O. Box 22085, E-46071 València, Spain

Ann Treessa Wilson
Research and Postgraduate Department of Chemistry, St. Berchmans College, Changanacherry, Kerala 686101, India

A. Y. Zahrim
Chemical Engineering Programme, Faculty of Engineering, Universiti Malaysia Sabah, Jalan UMS, 88400 Kota Kinabalu, Sabah, Malaysia

Y. Zulkiflee
Chemical Engineering Programme, Faculty of Engineering, Universiti Malaysia Sabah, Jalan UMS, 88400 Kota Kinabalu, Sabah, Malaysia

ABBREVIATIONS

2D	two dimensional
3D	three dimensional
Alq3	tris(8-hydroxyquinoline) aluminum
AMP	antimicrobial peptide
AMPTES	aminopropyltriethoxysilane
anPOME	anaerobically digested palm oil mill effluent
AOP	advanced oxidation process
AP	aspartyl protease
a-Si:H	hydrogenated amorphous silicon
BCS	Bardeen–Cooper–Schrieffer
CES	central experiment station
CFD	computational fluid dynamics
co-PMSQ	copolymethylsilsesquoxane
CVD	chemical vapor deposition
DOS	4-dimethylamino-4′-(1-oxobutyl)-stilbene
DPPC	dipalmitoyl-L-α-phosphatidylcholine
D-PVA	dichromated polyvinyl alcohol
EMO	evolutionary multiobjective optimization
FCC	fluidized catalytic cracking
FCCU	fluidized catalytic cracking unit
FET	field-effect transistor
FFB	fresh fruit bunches
FMRP	fragile X mental retardation protein
FRET	fluorescence resonance energy transfer
FTIR	Fourier-transform infrared
GA	genetic algorithm
GI	germination index
GO	graphene oxide
GPS	global positioning receivers
GROMACS	Groningen machine for chemical simulations
GTR	general theory of relativity
HDPE	high-density polyethylene
HEP	high-energy physics

HOMO	highest occupied molecular orbital
ICT	intramolecular charge transfer
IGBT	insulated-gate bipolar transistor
I_{on}/I_{off}	on/off current ratio
IR	infrared
ISC	intersystem crossing
ITO	indium tin oxide
LCR	inductance–capacitance–resistance
LDPE	low-density polyethylene
LE	low-energy
LED	light-emitting diode
LINCS	linear constraint solver
LLDPE	linear low-density polyethylene
LUMO	lowest unoccupied molecular orbital
MEP	2*C*-methyl-D-erythritol-4-phosphate
MIM	metal–insulator–metal
MLCT	metal-to-ligand charge transfer
MOSA	multiobjective optimization algorithm
MOSFET	metal-oxide-semiconductor field-effect transistor
MT	microtubule
mZnO	micron-sized ZnO
NIR	near-infrared
NOM	natural organic matter
NP	nanoparticle
NPB	1,4-bis(1-naphthylphenylamino)biphenyl
nZnO	nano-sized ZnO
OFET	organic field-effect transistor
OLED	organic light-emitting diode
OM	organic matter
OTFT	organic thin-film transistor
PA	photopatternable acryl
PANI	polyaniline
PANI-CSA	polyaniline doped with camphorsulfonic acid
PARARI	Pryanishnikov All-Russian Agrochemistry Research Institute
PBD	2-(4-biphenyl)-5-(4-*t*-butylphenyl)-1,3,4-oxadiazole
PE	polyethylene
PEDOT	poly(3,4-ethylenedioxythiophene)

PEDOT:PSS	poly(3,4-ethylenedioxythiophene):poly(styrenesulfonate)
PEN	polyethylene naphthalate
PET	photo-induced electron transfer
PET	polyethylene therephthalate
PL	photoluminescence
PMMA	poly(methylmethacrylate)
PMSQ	polymethylsilsesquioxane
PMSQ-CN	co-PMSQ containing cyano group
POME	palm oil mill effluent
PSS	styrene sulfonic acid
PTMA	(2,2,6,6-tetramethyl piperidinyloxy-4-yl) methacrylate
PVA	polyvinyl alcohol
RBCG	resudue-based coarse graining
Rc	contact resistance
RFID	radio-frequency identification
R_{sheet}	sheet resistance
SBCG	shape-based coarse graining
SC	superconductivity
SSL	solid-state lighting
$S_{s\text{-}th}$	subthreshold slope
sZnO	submicron-sized ZnO
TEMPO	(2,2,6,6-tetramethylpiperidin-1-yl)oxyl
TFT	thin-field transistor
TOC	total organic carbon
TPBI	1,3,5-tris(*N*-phenylbenzimidizol-2-yl)benzene
TPD	*N,N'*-biphenyl-*N,N'*-bis(3-methylphenyl)-1, 1- biphenyl-4,4-diamine
UHMWPE	ultrahigh molecular weight polyethylene
UV	ultraviolet
UV–VIS	ultraviolet–visible
V_{on}	turn-on voltage
V_{th}	threshold voltage
XRD	X-ray diffraction

PREFACE

A composite is a combination of two or more chemically distinct and insoluble phases. The properties and performance of composites are far superior than those of the constituents.

Globally, composite technology and its applications have made tremendous progress during the last two decades.

Composites are one of the most widely used materials in industry, agriculture, engineering, and environment because of their adaptability to different situations and the relative ease of combination with other materials to serve specific purposes and exhibit desirable properties.

Moreover, the increasing demand for greener and biodegradable materials leading to the satisfaction of society requires a push toward the advancement of composite materials science.

The potential and application of high-performance composites has revolutionized engineering technology. In general, the research projects on production of composites and laminates are mostly based on the following premises: they are value-added products and they are cost-effective compared to traditional materials.

This book is divided into four sections.

Advanced composite systems are presented in Section 1.

Graphene-based fibers and their application in advanced composites system are discussed in Chapter 1.

Chapter 2 is concerned with the modification of epoxy resin Epoxy-520 with commercial gypsum G-5. Particles of gypsum were incorporated (10–75 wt%) into an epoxy-dian resin Epoxy-520. In Chapter 3, the influence of the multifunctional organic structurants and mineral fillers on the structural features and physicomechanical properties of polymer composites based on statistical ethylene–propylene copolymer is reviewed. In Chapter 4, microtubule polymers are studied as an alternative composites as a key engineering material. Chapter 5 includes the use of zinc oxide powder to improve the electrical conductivity and mechanical property of the polyethylene matrix composites.

In Section 2, different types of electronic and ionic composites are presented and discussed in detail. In this section, materials for organic transistor applications, along with organic electronics, are reviewed in Chapters 6 and 7.

The basic principle of the fluorescence emission process and designing of chemosensors are described in the next chapter. The last chapter presents superconductivity and quantum computing via magnetic molecules.

In Section 3, prospects and challenges of composites are reviewed and presented in detail.

In the last section of this book, different types of biobased composites and research opportunities are presented and introduced.

SECTION I
Advanced Composites Systems

CHAPTER 1

GRAPHENE-BASED FIBERS AND THEIR APPLICATION IN ADVANCED COMPOSITES SYSTEM

RAGHVENDRA KUMAR MISHRA*

BSM Solar and Environmental Solution, A-348, Awas Vikas Colony, Sitapur, Unnao, Uttar Pradesh, India

*E-mail: *raghvendramishra4489@gmail.com*

ABSTRACT

The accelerated development in our culture is directly connected with improvement in technology and also currently particularly with nanoscience as well as nanotechnologies. Nowadays, it is not possible to think about advancement without innovative materials; it has been observed that the world without varieties of useful gadgets, for example, mobile phones, foldable notebooks, electric cars, biosensors, airplanes, as well as various additional discoveries render our life with comfort and also convenience. Notably, a number of crucial developments for brand new prototypes and application of graphene have been shared recently, which is linked with energy storage, photovoltaic, theoretical knowledge of graphene, and synthesis of graphene and its derivatives' application in tissue engineering, bioimaging, drug delivery, and device designs. In this chapter, authors attempt to collocate the state-of-the-art development in the direction of new viewpoint the graphene and its nanofibers, applications of newly produced nanofibers based on graphene and its derivatives in the real worlds.

1.1 INTRODUCTION

The materials are a key factor to the technological innovation which makes our everyday life pleasant as well as fantastic practicality, for example, a variety of household products and also electronic gadgets such as mobile phones, laptops, excellent sensing gadgets, and so on.[1,2] The development of unique materials usually arrives with the most exciting and worthwhile era of scientific study in both theoretical as well as experimental aspects. Considering the cutting-edge material, it generally comes along with several tricky possibilities to reexamine old issues, along with problems are drawn that is one of the most desirable things to the scientific society.[3] Carbon is one of the most interesting components in the world. In addition to its characteristics, it consistently had a matter of intense awareness as well as vital investigation.[4–6] Several allotropic types of carbon are popular including fullerenes, carbon nanotubes, graphene, and so on. Nanoribbon is the key allotropes of the nanocarbon fraternity. Among these types of nanocarbon materials (fullerenes, carbon nanotubes graphene, as well as graphene nanoribbon), graphene is considered the most recently found member of the carbon family.[7] Therefore graphene is the newest allotrope of carbon found by Sir Andre Geim and coworkers. It is an entirely two-dimensional (2D) single layer, only one atom thick graphitic material, that is exceptionally strong as well as an outstanding conductor of electricity and heat. Due to these kinds of remarkable features such as electronic, thermal and magnetic features, and 2D structure, graphene has participated to necessary scientific studies in basic chemistry, physics, and materials science in addition to condensed matter physics.[8,9] Graphene pertains to a single-atom-layered aromatic carbon material; nevertheless, it is generally tricky to prevent double or even few layers throughout the synthesis. On account of the outstanding features, several researchers addressed graphene as a sensational form of carbon.[10] In this manner, graphene has emerged in form of a new expectation for the whole scientific society. It possesses extremely high inherent carrier mobility (200,000 cm^2/V^{-1} s^{-1}) and also displays the thermal conductivity many times higher than that of cupper. Moreover, it has the huge surface area which is possibly greater than single-walled carbon nanotube (2630 m^2 g^{-1}) and outstanding thermal and mechanical elastic properties.[11,12] The initial study into creating graphene and derivatives has primarily concentrated on oxidation of graphite. For instance, the most well-known method to create graphene in scalable quantity is the Hummers method which entails an oxidation–reduction process, in which the oxidation creates graphene oxide (GO) from graphite and consequently the reduction transforms the GO to graphene in the existence of a reluctant, for example, hydrazine or sodium borohydride.[13,14] In spite of this, currently, several strategies are being formulated for immediate as well as indirect manufacturing of graphene and its derivatives. It is vital to point out that use of functionalized as well as customized graphene is significantly higher than the pure graphene.[15] Figure 1.1 shows the scheme of the modified graphene over pristine graphene (pure graphene).

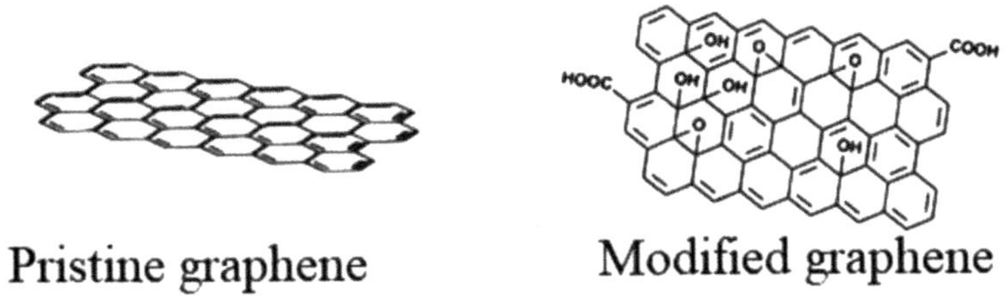

FIGURE 1.1 Schematic representation of the pristine graphene and modified graphene.

The real-time use of graphene and its derivatives has undoubtedly faced challenges such as scalable synthesis, storage, as well as appropriate processing. Consequently, desirable customization is extremely required on the surface of graphene.[16] Within this range, a number of groundbreaking efforts have been discussed which are primarily related to the inclusion of various functional moieties on the graphene surface. Among the variety of modification methods, stabilization and modification of graphene have drawn substantial attention.[17] Nevertheless, several of the commonly used modifications experience certain severe drawbacks. For example, several reduction techniques have been applied to complete the reduction of graphene from GO extracted functional active groups, although, it is not able to extract oxygen-containing moieties –OH, –COOH, and epoxy groups.[18,19] Thus, the reduction method provides O–H functional groups on graphene sheets that have a tendency to summate irreversibly. The irreversible accretion of graphene by means of π–π restacking will significantly encumber its production, storage, as well as distort its features.[20] As a result, stabilization as well as functionalization of graphene by means of modification of the material are required in order to prevent the unwanted aggregation.[21] The aim of this chapter is to summarize the synthesis, properties, and application of graphene and its nanofibers.

1.2 PREPARATION OF GRAPHENE AND ITS DERIVATIVE

Several steps have been developed for the manufacturing of graphene. However, each and every established strategy has its advantages and weaknesses. Among the range of methods, chemical exfoliation is the most typical as well as wieldy employed method; however, it is cumbersome, timeconsuming, and also includes utilization of very dangerous chemicals which additionally produce serious environmental difficulties.[22] Also, the structural defects are always established in the graphene sheet, which are straightly related to the specific features of this wonderful material. For the micromechanical cleavage (tape method introduced by Geim and coworker), in this instance, the quality of graphene is superb; nonetheless, the output is inadequate as well as highly time consuming.[23] Likewise, chemical vapor deposition (CVD) is truly one of the better techniques for ultrapure graphene synthesis. CVD technique for graphene manufacturing is costly as well as involves advanced equipment, so it is not extensively appropriate.[24] The electrochemical expansion method of graphene synthesis is cost effective, environmentally friendly, time effective, and possibly beneficial for the bulk manufacturing of graphene. Because of the experimental simplicity as well as simple filtration, this process of graphene synthesis has broadly been recognized in scientific society.[25,26] Lastly, Tour and coworkers have superbly opened up carbon nanotubes to create nanoribbons. However, the obstacle still remains with a high-yield technique which can exfoliate graphite effectively into solution-dispersible graphene sheets without having considerable chemical side effects as well as structural deterioration of the graphene sheets.[27,28] Thus, currently, the greatest issue for the scientists is to create a facile way for the generation of pure and single-layer graphene at commercial scale.[29] A summary of graphene synthesis methods is presented in Figure 1.2. Various functional derivatives of graphene are available based on the requirements in substantial scale by a variety of chemical reagents, which probably consists of an oxidation–reduction phenomenon.[30] The oxidation procedure of graphite sheets may help the exfoliation of GO sheets from the massive graphitic layers and the reduction method recovers the conductivity of GO by reestablishing its in-sheet conjugated arrangement. However, in several modifications of graphene sheets restoration of conductivity is impossible.[31–33]

1.3 CHARACTERIZATION OF GRAPHENE AND ITS DERIVATIVE

New technology or breakthrough of any innovative material, a chemical element, or a sophisticated molecular system involves a number of instrumental methods for its detection and filtration in order to balance its basic

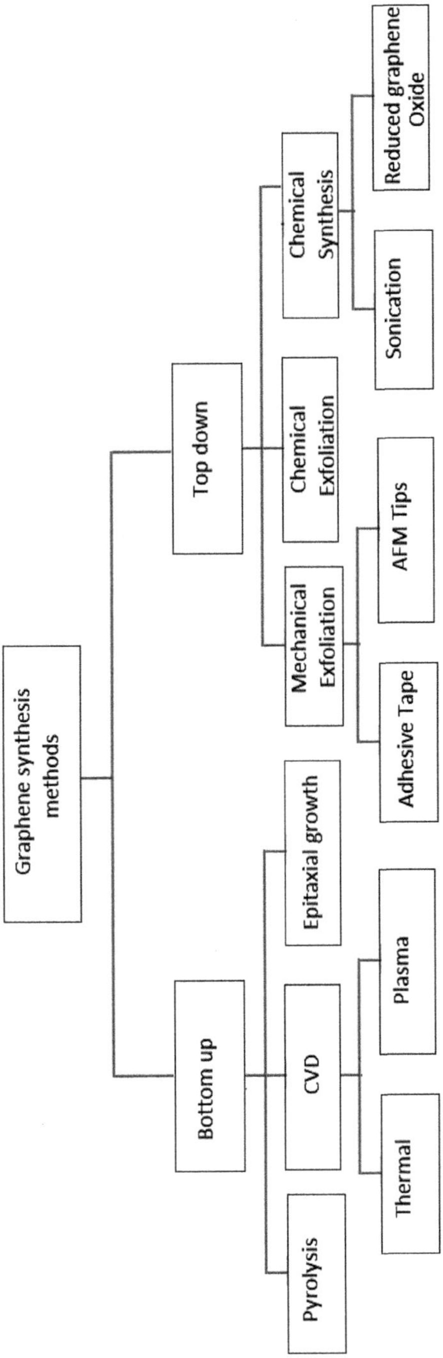

FIGURE 1.2 Methods of graphene synthesis.

features.[6,34–36] Among the several methodologies to characterize basic characteristics of samples (atoms, molecules, or even complex matters), spectroscopic approaches are the front line choice for the objective.[37,38] We learn about the building of atoms and molecules because of spectroscopy, which is based on interaction of electromagnetic radiation with matter. Various zones of the electromagnetic spectrum offer unique details due to such interactions. Therefore, spectroscopy is the study of the interaction between matter and radiated energy.[39,40] In addition, it was found via the analysis of visible light dispersed based on its wavelength through a prism. These days, the idea is extended tremendously to consist of any specific interaction with radiative energy in form of a function of its wavelength or frequency.[41,42] The spectroscopy techniques based on the quantum mechanics as well as Max Planck's justification of blackbody radiation, Albert Einstein's clarification of the photoelectric effect, and Niels Bohr's interpretation of atomic structure as well as spectra.[43,44] Spectroscopy is employed in physical and analytical chemistry due to the fact that atoms as well as molecules possess specific spectra. Because of this, these types of spectra are employed to identify, distinguish, as well as quantify details about the atoms and molecules. Spectroscopy can be utilized in astronomy and remote sensing. Among the distinct spectroscopic techniques, Fourier-transform infrared (FTIR) spectrometry is one of the most important instruments for organic chemists and also this approach has been widely employed by material chemists for recognition as well as characterization of nanomaterials such as carbon nanotubes, graphene, and their derivatives.[45,46] Chemical composition of carbon has constantly attracted enormous consideration of entire scientific society and in this manner graphene is fabulous. Graphene and related nanomaterials are synthesized via a range of techniques on the basis of practical application and it also consists of different types of functional activity groups together with structural units in order that after the finishing of synthesis FTIR spectrometry performs incredibly vital role in stabilizing features as well as recognition of these materials; due to the fact FTIR provides direct and extremely genuine details about which kinds of bonds, functional group, and structural units is found in sample.[47,48] On the basis of this theory, different kinds of bonds, also various functional groups, absorb infrared (IR) radiation of distinctive wavelengths. Thus, FTIR spectroscopy is a superb technique for the recognition of functional groups and it is usually employed to verify the purity of well-prepared samples.[49] Furthermore, it can be employed in the form of a primary indicator of functional groups which are present in an unidentified sample. A number of excellent articles have been published on different aspects of graphene and characterization of nanomaterials including graphene and graphene derivatives.[9,50] Figure 1.3 shows the structures of GO and graphene with IR active functional groups and double bonds highlighted by circles (left to right).

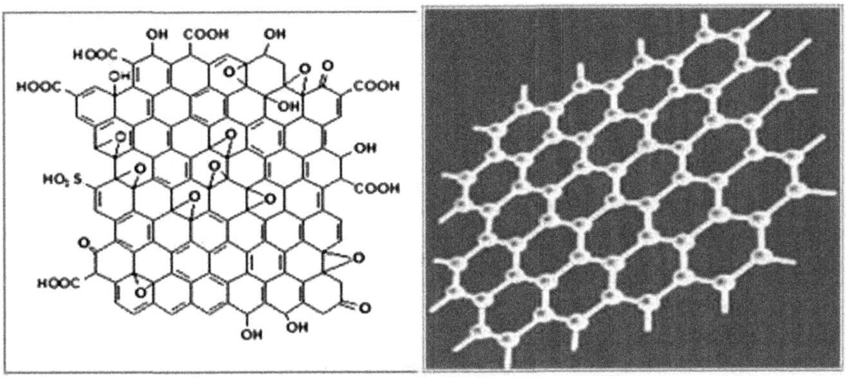

FIGURE 1.3 Structures of graphene oxide and graphene with IR active functional groups.

1.4 GRAPHENE AND ITS DERIVATIVE-BASED FIBERS

Graphene is a 2D crystalline layer with a monolayer of carbon atoms densely packed in an sp^2-bonded honeycomb lattice, and is viewed as an individual layer of the graphitic film in graphite. Graphene possesses outstanding mechanical, electrical, thermal, as well as optical properties, and translating these types of features to its constructed macrostructures is an essential concern for endorsing its application in sensible uses.[9,51,52] Graphene-based three-dimensional (3D) aerogels, 2D graphene-based membranes, and one-dimensional graphene-based fibers are being produced. As well as, graphene-based membranes maintain the present world ranking for the lightest material, with a density of 0.16 mg cm^{-3}, and have shown high-quality ability in the elimination of spilt oils.[15] Graphene-based membranes are generally created by infiltration or even CVD; in addition, it has a variety of applications in the vast field of energy storage. Wherein, one-dimensional graphene-based fibers have shown significantly better mechanical as well as electrical features in comparison with graphene-based 3D aerogels and also 2D graphene-based membranes.[53,54] One-dimensional graphene-based fibers are versatile as well as wearable.

Certain possible applications of one-dimensional graphene-based fibers have already been mentioned, such as conducting wires, fiber-shaped energy-storage gadgets, and worthwhile actuators.[17,55–57] In recent times, substantial endeavors have been generated on the marketing of one-dimensional graphene-based fiber from laboratory investigation to commercial applications. In the present section, we impart a state of the art on one-dimensional graphene-based fiber including their fabrication methods and the characteristics of fiber building blocks.[58,59]

1.4.1 FABRICATION METHODS OF GRAPHENE-BASED FIBERS

The leading methods of producing graphene-based fibers are influenced by those used to create traditional artificial fibers, which primarily consist of melt spinning as well as solution spinning.[60] Graphene has high-temperature stability; melting temperature of graphene is undoubtedly 4900 K that is actually greater than those of fullerene (4000 K) and also CNTs (4800 K).[61] Producing graphene fibers from graphene constructing blocks is a procedure of consistently aligning graphene sheets in a uniaxial direction along which the asymmetrical benefits of graphene are translated into macroscopic features.[62] Due to the unmeltable character of graphene, it is unfeasible to process graphene into nice fibers by the melt spinning procedures for polymeric nylon; melt spinning is actually not a choice for producing graphene-based fibers, although solution spinning is the suitable method. Hence, the generation of graphene-based fibers should apply the procedure of solution setup.[63] To manufacture graphene-based fibers in a constant fashion and to accomplish adequately top quality of the fiber is the primary consideration to enhance the capabilities as well as overall performance of fibers.[64] Consequently, there are certain requirements that should be satisfied including high-quantity manufacturing of solution-processable graphene, dispersibility of perfect graphene in typical solvents, consistent alignment of graphene, and the connection among graphene building blocks to maximize the performances.[19]

1.4.1.1 WET SPINNING METHODS FOR GRAPHENE-BASED FIBERS

The experiment pointed out the following fundamental procedures in accumulating wet-spun graphene fibers: first step—extruding into coagulation baths with the uniaxial flow; secondly, phase transition by coagulation with solvents swap; consequently, solidification by evaporating solvent; in addition to the final step, chemical or thermal reduction. By doing this, a GO dope is made by dispersing GO sheets into a steady water solution after injecting it into a coagulation bathtub to develop a gel-state fiber.[26,65,66] After coagulation for a particular time period, a GO fiber is received by the removal of the gel-state fiber after drying. A reduced GO fiber is additionally created by reducing the GO fiber as required. To make sure the consistent as well as the regular development of a GO gel-state fiber, the as-coagulated fiber must retain a specific mobility rate possibly by revolving the bath or drawing the fiber with a collection device. The most effective reduced GO fiber was constructed by spinning.[67,68] Nevertheless, it has restricted inaccurate control of the fiber motion speed, since it is dependent on the friction force between the fiber surface along with the coagulant, and the fiber circulation speed will not fluctuate proportionally to the bath revolving speed. Conversely, a collection device will provide as-synthesized fiber with a consistent drafting force along with a particular circulation speed. Hence, it gives a lot more benefits in creating fibers with precise draw ratio as well as favorable scalability.[69,70]

1.4.1.2 DRY SPINNING METHODS FOR GRAPHENE-BASED FIBERS

Dry spinning of one-dimensional graphene-based fibers additionally makes use of the GO dispersion in form of the spinning dope, although the coagulation bath is not applied. As an alternative, the GO dispersion is placed into in a tube and it is covered. It precipitated like a gel-state fiber by heating up or even by chemical reduction under substantial temperature.[47,62] Dry reduced GO fibers are received by extracting the solvent. GO dispersions were viewed as a colloid. High temperature brings the GO dispersant to transfer immediately, which destabilizes the equilibrium managed by the zeta potential, and therefore raises the possibility of GO sheet collision and precipitation. Alternatively, high temperature or chemical reduction are able to detach the oxygen-containing various active groups of GO, reducing the absolute zeta potential value of the GO dispersion.[31,62,71] Lastly, GO sheets precipitate and are put together into a fiber because of the insufficient electrostatic repulsion.[72]

1.4.1.3 ELECTROPHORETIC ASSEMBLY METHODS FOR GRAPHENE-BASED FIBERS

The electrophoretic event takes place in colloid solutions due to the fact that a charged particle will relocate under the effect of an electrical field.[73–75] A GO dispersion with the negatively charged GO sheets is known as a colloid. Therefore, an electrophoretic self-assembly technique was designed for graphene-based fibers' spinning.[75]

1.4.1.4 JET SPINNING METHODS FOR GRAPHENE-BASED FIBERS

Dry-jet wet spinning is an additional essential spinning way for traditional man-made fibers. Research has shown that PAN-based carbon fibers are generally spun with a high-concentration dope by means of this process, and the derived fibers presented much better mechanical features compared to the wet spinning process.[76,77] Moreover, dry-jet wet spinning methods are useful to spin fibers with a uniform structure. As a result, this process is furthermore favorable for graphene-based fibers synthesis.[63]

1.5 PROPERTIES OF GRAPHENE-BASED FIBERS

Owing to the excellent features of graphene fundamental blocks, it is anticipated that GF can have an excellent coincident combination of attributes in mechanical, electronic, as well as thermal features, by means of a magnificent arrangement of input on graphene fibers. Today, the majority of the investigation has accentuated the improvement of the mechanical and electrical capabilities of graphene-based fibers, and useful development has additionally been engineered in their thermal, actuating as well as other physical features.[55,62,78,79]

1.5.1 MECHANICAL PROPERTIES OF GRAPHENE-BASED FIBERS

The initial graphene fibers have a tensile strength of around 140 MPa at 5.8% elongation yield and experienced beneficial flexibility to render tight knots. The large discrepancy between the mechanical potency of graphene fibers along with the constituent graphene sheets has to be bridged. In the examination of the structural attributes of graphene fibers, three approaches from atomic to macroscopic scales are already suggested to update mechanical functionality[19,23,59,80]: (1) enhancing the alignment of graphene sheets along the fiber axis; (2) reducing the structural imperfections which include sheets boundaries, voids, as well as contaminants; and also (3) improving the interlayer interaction of the component graphene sheets by possibly covalent or alternatively noncovalent bonds. Controlled by the above tips, the mechanical potency of GF has risen to the GPa range in a stepwise design in just 3 years.[81] Drawing during the wet spinning method is being applied to improve the alignment of graphene sheets, and has consequently increased the mechanical functionality, particularly the ultimate strength of graphene fibers, 5 GPa. Drawing additionally offered the compactness of graphene laminates, thereby taking the elastic modulus to 47 GPa. The enhancement of the lateral size of graphene sheets has been proved to successfully reduce imperfect grain boundaries. To improve the interlayer interface on the atomic scale, authors developed hyperbranched molecules with multifunctional various active groups into interlayer galleries, and their biomimetic fibers with hydrogen bonding arrays show the mechanical strength of 550 MPa, which was improved to 650 MPa after chemical cross-linking.[59,80,82] In recent times, another research group has accomplished strength of 1.1–1.2 GPa by a combination of techniques to improve the alignment of graphene sheets along with the compactness of graphene fibers. These types of swift improvements defend the exciting concept of graphene fibers in the role of a meaningful high-performance carbonaceous fiber. The tensile mechanical functionality of graphene-based fibers depends mainly on the types of interactions such as chemical bond between the GO/reduced GO sheet and the doped component, the formation of hydrogen bonds in GO readily because groups are still present in reduced GO, van der Waals interaction, and so on.[16,83]

1.5.2 ELECTRICAL PROPERTIES OF GRAPHENE-BASED FIBERS

Without having doping element, the topmost electrical conductivity of graphene fibers at room temperature is 400 S cm^{-1}, similar to that of array-spun CNT fibers. In spite of this, the majority of one-dimensional graphene-based fibers are created from GO, an oxygenated graphene sheet with substantial electrical resistivity.[13,15] Consequently, the reducing treatment methods are of crucial significance in repairing the electrical conductivity of the GO sheet along with the resultant fiber. Researchers also have the option to furthermore enhance the conductivity by means of molecular doping to boost carrier density or even blending with other benchmark metals, for example, silver, copper and gold, alkali metals, and acids.[84,85]

1.5.3 THERMAL CONDUCTIVITY OF GRAPHENE-BASED FIBERS

The thermal conductivity of single-layer graphene was documented to be 5000 W m^{-1} K^{-1}. On the other hand, with a greater number of graphene layers, the free movement of each and every layer was restricted; in addition, the phonon-transport resistance is amplified. As a result, the thermal conductivity of multilayer graphene is lesser in comparison to the single-layer graphene.[50,86,87]

1.6 APPLICATION OF GRAPHENE-BASED FIBERS

Modern accomplishments in high-performance one-dimensional graphene-based fibers have gained analysis endeavors on their implementation in various field and gadgets. Possible applications in cables, energy-storage gadgets, worthwhile actuators, field emitters, and so forth, have been produced. The aim of this section is to provide an overview of the graphene-based fibers in various fields.[88]

1.6.1 GRAPHENE-BASED FIBERS FOR ACTUATORS

A variety of actuators of graphene fibers, for example, self-driven graphene-based fibers ducts, humidity actuators, torsional motors, as well as magnetic-driven springs are discussed. One-dimensional graphene-based fibers have confirmed suitable actuating functionality. For example, authors produced a voltage-sensitive reduced GO/PPy fiber. Research on a humidity-triggered GO fiber-based electronic switches as well as electric generators has been currently outlined.[17,19,89]

1.6.2 GRAPHENE-BASED FIBERS FOR FIELD EMITTERS

One-dimensional graphene-based fibers have significant electron-emitting pointers because of the significant width-to-thickness ratio, superior charger–carrier density, as well as good electrical conductivity, showing assurance for use in field emitters. The alignment of reduced GO sheets along the fiber axis along with the topological imperfections at the edges of such sheets that have been assumed to work as useful emission sites devote usually to the outstanding field-emission characteristics.[48,90,91]

1.6.3 GRAPHENE-BASED FIBERS FOR SPRINGS

One-dimensional graphene-based fibers were created in terms of springs with huge elongation as well as multicontrolled design. Spring-like structure of the one-dimensional graphene-based fibers by covering wet one-dimensional graphene-based fibers around cylindrical materials followed by thermally annealing the fiber to preserve the framework has been discussed. These graphene fibers-based springs revealed significant as well as completely recoverable elongation of 480% by mechanical deformation, as well as electrical static/magnetic fields have been additionally shown to be efficient stimuli to prompt this sensitive actuation tendency.[92]

1.6.4 GRAPHENE-BASED FIBERS FOR CATALYSIS

One-dimensional graphene-based fibers have likewise been proven to turn out to be useful in the catalytic reduction of oxygen. They discussed that nitrogen-doped reduced GO fibers experienced excellent electrocatalytic performance to oxygen in comparison with both nondoped reduced GO fibers as well as nitrogen-doped multiwalled CNTs, which provided the opportunity for one-dimensional graphene-based fibers in supplementary commercial applications.[93–95] Because of the suitable durability, outstanding conductivity, as well as the substantial surface area of such material, practical applications are presently concentrated on various perspectives.

1.6.5 GRAPHENE-BASED FIBERS FOR ENERGY STORAGE

Employing graphene-based films as well as frameworks in form of chemical supercapacitors for energy storage, graphene-based fibers are being viewed as a promising candidate for being valuable supercapacitor products which are adaptable which enables them to be knitted into the wearable gadget.[56] The capacitance of graphene fibers was evaluated, and the better result is found due to the extremely porous framework of the spun fibers. The inclusion of carbon nanotubes produced additional hierarchical pores in graphene fibers and consequently improved their capacitance in the instance of single-walled carbon nanotubes.[96,97] Additional elements including oxides (MnO_2 and Bi_2O_3) as well as conductive polymers (polypyrrole and polyaniline) additionally supported their pseudocapacitance to the overall performance of hybridized graphene fibers.[98]

1.7 CONCLUSION

Graphene fiber is high-class carbonaceous fiber with abundant functionalities. Its development considerably illustrates an innovative procedure to generate carbonaceous fibers simply from mineral graphite. The graphene fiber is covers efficiently by the thesis of solvated graphenes, the creation of liquid crystals, as well as wet spinning for neat or even composite graphene fibers. So far, a variety of graphene-based fibers configuration techniques have been formulated, such as wet spinning, dry spinning, dry-jet wet spinning, electrophoretic self-assembly, and so on. The fiber durability, as well as electrical conductivity, have attained 652 MPa and 416 S cm^{-1}, respectively. In spite of this, these types of activities are usually much lesser in comparison with those of carbon fibers and also CNT fibers, which can be related to several factors. To start with, the crystallinity of the beginning material such as GO must be enhanced. One-dimensional graphene-based fibers are usually more flexible as compared to carbon fiber and consequently possess a better thermal conductivity compared to those of both of the carbon fiber and also carbon nanotubes fibers; hence, one-dimensional graphene-based fibers also provide possible use in the field of thermal administration. An additional benefit for one-dimensional graphene-based fibers is the fact that the working expenditure is greatly reduced than that of carbon fiber or even carbon nanotubes fibers. An advanced enthusiasm for the growth and development of graphene fibers functions is also required.

KEYWORDS

- **graphene and its derivative**
- **nanofibers**
- **fabrication**
- **properties**
- **application**

REFERENCES

1. Riggs, C. New Gadgets on the Horizon: Part 2. *Libr. Hi Tech News*. **2006,** *23*, 16–18. DOI: 10.1108/07419050610668142.
2. Miluzzo, E.; Cornelius, C. T.; Ramaswamy, A.; Choudhury, T.; Liu, Z.; Campbell, A. T. Darwin Phones : The Evolution of Sensing and Inference on Mobile Phones. *Darwin* **2010,** *14*, 5–20. DOI: 10.1145/1814433.1814437.
3. Goh, P. S.; Ismail, A. F. Graphene-based Nanomaterial: The State-of-the-art Material for Cutting-edge Desalination Technology. *Desalination* **2015,** *356*, 115–128. DOI: 10.1016/j.desal.2014.10.001.
4. Jose Chirayil, C.; Abraham, J.; Kumar Mishra, R.; George, S. C.; Thomas, S. Instrumental Techniques for the Characterization of Nanoparticles. In *Thermal and Rheological Measurement Techniques for Nanomaterials Characterization*; Elsevier, 2017; pp 1–36. DOI: 10.1016/B978-0-323-46139-9.00001-3.
5. Mishra, R. K.; Cherusseri, J.; Bishnoi, A.; Thomas, S. Nuclear Magnetic Resonance Spectroscopy. In *Spectroscopic Methods for Nanomaterials Characterization*; Elsevier, 2017; pp 369–415. DOI: 10.1016/B978-0-323-46140-5.00013-3.
6. Mishra, R. K.; Cherusseri, J.; Allahyari, E.; Thomas, S.; Kalarikkal, N. Small-angle Light and X-ray Scattering in Nanosciences and Nanotechnology. In *Thermal and Rheological Measurement Techniques for Nanomaterials Characterization*; Elsevier, 2017; pp 233–269. DOI: 10.1016/B978-0-323-46139-9.00010-4.
7. Lim, H. E.; Miyata, Y.; Kitaura, R.; Nishimura, Y.; Nishimoto, Y.; Irle, S.; Warner, J. H.; Kataura, H.; Shinohara, H. Growth of Carbon Nanotubes Via Twisted Graphene Nanoribbons. *Nat. Commun.* **2013,** *4*. DOI: 10.1038/ncomms3548.
8. Novoselov, K. S.; Geim, A. K.; Morozov, S. V.; Jiang, D.; Katsnelson, M. I.; Grigorieva, I. V.; Dubonos, S. V.; Firsov, A. A. Two-dimensional Gas of Massless Dirac Fermions in Graphene. *Nature* **2005,** *438*, 197–200. DOI: 10.1038/nature04233.
9. Castro Neto, A. H.; Guinea, F.; Peres, N. M. R.; Novoselov, K. S.; Geim, A. K. The Electronic Properties of Graphene. *Rev. Mod. Phys.* **2009,** *81*, 109–162. DOI: 10.1103/RevModPhys.81.109.
10. Gao, W. The Chemistry of Graphene Oxide. In *Graphene Oxide Reduction Recipes, Spectroscopy, and Applications*; 2015; pp 61–95. DOI: 10.1007/978-3-319-15500-5_3.
11. Balandin, A. A.; Ghosh, S.; Bao, W.; Calizo, I.; Teweldebrhan, D.; Miao, F.; Lau, C. N. Superior Thermal Conductivity of Single-layer Graphene. *Nano Lett.* **2008,** *8*, 902–907. DOI: 10.1021/nl0731872.
12. Vashist, S. K. Advances in Graphene-based Sensors and Devices. *J. Nanomed. Nanotechnol.* **2012,** *4*. DOI: 10.4172/2157-7439.1000e127.
13. Marcano, D. C.; Kosynkin, D. V.; Berlin, J. M.; Sinitskii, A.; Sun, Z.; Slesarev, A.; Alemany, L. B.; Lu, W.; Tour, J. M. Improved Synthesis of Graphene Oxide. *ACS Nano.* **2010,** *4*, 4806–4814. DOI: 10.1021/nn1006368.
14. Dimiev, A. M.; Tour, J. M. Mechanism of Graphene Oxide Formation. *ACS Nano.* **2014,** *8*, 3060–3068. DOI: 10.1021/nn500606a.
15. Huang, X.; Qi, X.; Boey, F.; Zhang, H. Graphene-based Composites. *Chem. Soc. Rev.* **2012,** *41*, 666–686. DOI: 10.1039/c1cs15078b.
16. Karlický, F.; Kumara Ramanatha Datta, K.; Otyepka, M.; Zbořil, R. Halogenated Graphenes: Rapidly Growing Family of Graphene Derivatives. *ACS Nano.* **2013,** *7*, 6434–6464. DOI: 10.1021/nn4024027.

17. Varghese, S. S.; Lonkar, S.; Singh, K. K.; Swaminathan, S.; Abdala, A. Recent Advances in Graphene Based Gas Sensors. *Sens. Actuators B Chem.* **2015,** *218*, 160–183. DOI: 10.1016/j.snb.2015.04.062.
18. Voiry, D.; Yang, J.; Kupferberg, J.; Fullon, R.; Lee, C.; Jeong, H. Y.; Shin, H. S.; Chhowalla, M. High-quality Graphene Via Microwave Reduction of Solution-exfoliated Graphene Oxide. *Science* **2016,** *353* (80), 1413–1416. DOI: 10.1126/science.aah3398.
19. Compton, O. C.; Nguyen, S. T. Graphene Oxide, Highly Reduced Graphene Oxide, and Graphene: Versatile Building Blocks for Carbon-based Materials. *Small* **2010,** *6*, 711–723. DOI: 10.1002/smll.200901934.
20. Bagri, A.; Mattevi, C.; Acik, M.; Chabal, Y. J.; Chhowalla, M.; Shenoy, V. B. Structural Evolution During the Reduction of Chemically Derived Graphene Oxide. *Nat. Chem.* **2010,** *2*, 581–587. DOI: 10.1038/nchem.686.
21. Eigler, S.; Hirsch, A. Chemistry with Graphene and Graphene Oxide: Challenges for Synthetic Chemists. *Angew. Chem.* **2014,** *53*, 7720–7738. DOI: 10.1002/anie.201402780.
22. Bhuyan, M. S. A.; Uddin, M. N.; Islam, M. M.; Bipasha, F. A.; Hossain, S. S. Synthesis of Graphene. *Int. Nano Lett.* **2016,** *6*, 65–83. DOI: 10.1007/s40089-015-0176-1.
23. Banhart, F.; Kotakoski, J.; Krasheninnikov, A. V. Structural Defects in Graphene. *ACS Nano.* **2011,** *5*, 26–41. DOI: 10.1021/nn102598m.
24. Mattevi, C.; Kim, H.; Chhowalla, M. A Review of Chemical Vapour Deposition of Graphene on Copper. *J. Mater. Chem.* **2011,** *21*, 3324–3334. DOI: 10.1039/C0JM02126A.
25. Avouris, P.; Dimitrakopoulos, C. Graphene: Synthesis and Applications. *Mater. Today.* **2012,** *15*, 86–97. DOI: 10.1016/S1369-7021(12)70044-5.
26. Pei, S.; Cheng, H.-M. The Reduction of Graphene Oxide. *Carbon N. Y.* **2012,** *50*, 3210–3228. DOI: 10.1016/j.carbon.2011.11.010.
27. Kelly, K. F.; Billups, W. E. Synthesis of Soluble Graphite and Graphene. *Acc. Chem. Res.* **2013,** *46*, 4–13. DOI: 10.1021/ar300121q.
28. Lu, X.; Yu, M.; Huang, H.; Ruoff, R. S. Tailoring Graphite with the Goal of Achieving Single Sheets. *Nanotechnology* **1999,** *10*, 269–272. DOI: 10.1088/0957-4484/10/3/308.
29. Lee, J. H.; Lee, E. K.; Joo, W. J.; Jang, Y.; Kim, B. S.; Lim, J. Y.; Choi, S. H.; Ahn, S. J.; Ahn, J. R.; Park, M. H.; Yang, C. W.; Choi, B. L.; Hwang, S. W.; Whang, D. Wafer-scale Growth of Single-crystal Monolayer Graphene on Reusable Hydrogen-terminated Germanium. *Science* **2014,** *344* (80), 286–289. DOI: 10.1126/science.1252268.
30. Eda, G.; Chhowalla, M. Chemically Derived Graphene Oxide: Towards Large-area Thin-film Electronics and Optoelectronics. *Adv. Mater.* **2010,** *22*, 2392–2415. DOI: 10.1002/adma.200903689.
31. Su, C. Y.; Lu, A. Y.; Xu, Y.; Chen, F. R.; Khlobystov, A. N.; Li, L. J. High-quality Thin Graphene Films from Fast Electrochemical Exfoliation. *ACS Nano.* **2011,** *5*, 2332–2339. DOI: 10.1021/nn200025p.
32. Kaniyoor, A.; Baby, T. T.; Ramaprabhu, S. Graphene Synthesis Via Hydrogen Induced Low Temperature Exfoliation of Graphite Oxide. *J. Mater. Chem.* **2010,** *20*, 8467. DOI: 10.1039/c0jm01876g.
33. Zhao, J.; Pei, S.; Ren, W.; Gao, L.; Cheng, H. M. Efficient Preparation of Large-area Graphene Oxide Sheets for Transparent Conductive Films. *ACS Nano.* **2010,** *4*, 5245–5252. DOI: 10.1021/nn1015506.

34. Thomas, S.; Thomas, R.; Zachariah, A. K.; Mishra, R. K. *Thermal and Rheological Measurement Techniques for Nanomaterials Characterization*, 1st ed.; Elsevier, 2017. DOI: 10.1016/B978-0-323-46139-9.01001-X.
35. Mishra, R. K.; Zachariah, A. K. Thomas, S. Energy-dispersive X-ray Spectroscopy Techniques for Nanomaterial. In *Microscopy Methods in Nanomaterials Characterization*; Elsevier, 2017; pp 383–405. DOI: 10.1016/B978-0-323-46141-2.00012-2.
36. Jayanarayanan, K.; Rasana, N.; Mishra, R. K. Dynamic Mechanical Thermal Analysis of Polymer Nanocomposites. In *Thermal and Rheological Measurement Techniques for Nanomaterials Characterization*; Elsevier, 2017; pp 123–157. DOI: 10.1016/B978-0-323-46139-9.00006-2.
37. Thomas, S.; Thomas, R.; Zachariah, A. K.; Mishra, R. K. Microscopy Methods in Nanomaterials Characterization. In *Microscopy Methods in Nanomaterials Characterization*, 1st ed.; Thomas, S., Thomas, R., Zachariah, A. K., Mishra, R. K., Eds.; Elsevier, 2017; p 432. DOI: 10.1016/B978-0-323-46141-2.01001-4.
38. Thomas, S.; Thomas, R.; Zachariah, A. K.; Mishra, R. K. *Spectroscopic Methods for Nanomaterials Characterization*, 1st ed.; Elsevier, 2017. DOI: 10.1016/B978-0-323-46140-5.01001-3.
39. Maria, C. Application of FTIR Spectroscopy in Environmental Studies. In *Advanced Aspects of Spectroscopy*; 2012. DOI: 10.5772/48331.
40. Pérez-Juste, I.; Faza, O. N. Interaction of Radiation with Matter. In *Structure Elucidation in Organic Chemistry: The Search for the Right Tools*; 2015; pp 1–26. DOI: 10.1002/9783527664610.ch1.
41. Wan Ismail, W. Z.; Goldys, E. M.; Dawes, J. M. Extended Emission Wavelength of Random Dye Lasers by Exploiting Radiative and Non-radiative Energy Transfer. *Appl. Phys. B Lasers Opt.* **2016,** *122*, 1–9. DOI: 10.1007/s00340-016-6321-3.
42. Ni, X.; Kildishev, A. V.; Shalaev, V. M. Metasurface Holograms for Visible Light. *Nat. Commun.* **2013,** *4*. DOI: 10.1038/ncomms3807.
43. Steinberg, A.; Kwiat, P.; Chiao, R. Quantum Optical Tests of the Foundations of Physics. In *Springer Handbook of Atomic, Molecular, and Optical Physics*; Springer: New York, NY, 2006; pp 1185–1213. DOI: 10.1007/978-0-387-26308-3.
44. Niaz, M.; Klassen, S.; McMillan, B.; Metz, D. Reconstruction of the History of the Photoelectric Effect and Its Implications for General Physics Textbooks. *Sci. Educ.* **2010,** *94*, 903–931. DOI: 10.1002/sce.20389.
45. Berthomieu, C.; Hienerwadel, R. Fourier Transform Infrared (FTIR) Spectroscopy. *Photosynth. Res.* **2009,** *101*, 157–170. DOI: 10.1007/s11120-009-9439-x.
46. Casadio, F.; Leona, M.; Lombardi, J. R.; Van Duyne, R. Identification of Organic Colorants in Fibers, Paints, and Glazes by Surface Enhanced Raman Spectroscopy. *Acc. Chem. Res.* **2010,** *43*, 782–791. DOI: 10.1021/ar100019q.
47. Stankovich, S.; Dikin, D. A.; Dommett, G. H. B.; Kohlhaas, K. M.; Zimney, E. J.; Stach, E. A.; Piner, R. D.; Nguyen, S. T.; Ruoff, R. S. Graphene-based Composite Materials. *Nature* **2006,** *442*, 282–286. DOI: 10.1038/nature04969.
48. Chowdhury, S.; Balasubramanian, R. Recent Advances in the Use of Graphene-family Nanoadsorbents for Removal of Toxic Pollutants from Wastewater. *Adv. Colloid Interface Sci.* **2014,** *204*, 35–56. DOI: 10.1016/j.cis.2013.12.005.
49. Romanos, J.; Beckner, M.; Stalla, D.; Tekeei, A.; Suppes, G.; Jalisatgi, S.; Lee, M.; Hawthorne, F.; Robertson, J. D.; Firlej, L.; Kuchta, B.; Wexler, C.; Yu, P.; Pfeifer, P.

Infrared Study of Boron-carbon Chemical Bonds in Boron-doped Activated Carbon. *Carbon N. Y.* **2013,** *54*, 208–214. DOI: 10.1016/j.carbon.2012.11.031.
50. Ferrari, A. C.; Meyer, J. C.; Scardaci, V.; Casiraghi, C.; Lazzeri, M.; Mauri, F.; Piscanec, S.; Jiang, D.; Novoselov, K. S.; Roth, S.; Geim, A. K. Raman Spectrum of Graphene and Graphene Layers. *Phys. Rev. Lett.* **2006,** *97*. DOI: 10.1103/PhysRevLett.97.187401.
51. Li, X.; Zhu, Y.; Cai, W.; Borysiak, M.; Han, B.; Chen, D.; Piner, R. D.; Colomba, L.; Ruoff, R. S. Transfer of Large-area Graphene Films for High-performance Transparent Conductive Electrodes. *Nano Lett.* **2009,** *9*, 4359–4363. DOI: 10.1021/nl902623y.
52. Choi, W.; Lahiri, I.; Seelaboyina, R.; Kang, Y. S. Synthesis of Graphene and Its Applications: A Review. *Crit. Rev. Solid State Mater. Sci.* **2010,** *35*, 52–71. DOI: 10.1080/10408430903505036.
53. Singh, V.; Joung, D.; Zhai, L.; Das, S.; Khondaker, S. I.; Seal, S. Graphene Based Materials: Past, Present and Future. *Prog. Mater. Sci.* **2011,** *56*, 1178–1271. DOI: 10.1016/j.pmatsci.2011.03.003.
54. Wang, H.; Yuan, X.; Zeng, G.; Wu, Y.; Liu, Y.; Jiang, Q.; Gu, S. Three Dimensional Graphene Based Materials: Synthesis and Applications from Energy Storage and Conversion to Electrochemical Sensor and Environmental Remediation. *Adv. Colloid Interface Sci.* **2015,** *221*, 41–59. DOI: 10.1016/j.cis.2015.04.005.
55. Huang, X.; Yin, Z.; Wu, S.; Qi, X.; He, Q.; Zhang, Q.; Yan, Q.; Boey, F.; Zhang, H. Graphene-based Materials: Synthesis, Characterization, Properties, and Applications. *Small* **2011,** *7*, 1876–1902. DOI: 10.1002/smll.201002009.
56. Ke, Q.; Wang, J. Graphene-based Materials for Supercapacitor Electrodes: A Review. *J. Mater.* **2016,** *2*, 37–54. DOI: 10.1016/j.jmat.2016.01.001.
57. Zheng, S.; Wu, Z. S.; Wang, S.; Xiao, H.; Zhou, F.; Sun, C.; Bao, X.; Cheng, H. M. Graphene-based Materials for High-voltage and High-energy Asymmetric Supercapacitors. *Energy Storage Mater.* **2017,** *6*, 70–97. DOI: 10.1016/j.ensm.2016.10.003.
58. Chang, H.; Wu, H. Graphene-based Nanocomposites: Preparation, Functionalization, and Energy and Environmental Applications. *Energy Environ. Sci.* **2013,** *6*, 3483. DOI: 10.1039/c3ee42518e.
59. Kim, K. S.; Zhao, Y.; Jang, H.; Lee, S. Y.; Kim, J. M.; Kim, K. S.; Ahn, J. H.; Kim, P.; Choi, J. Y.; Hong, B. H. Large-scale Pattern Growth of Graphene Films for Stretchable Transparent Electrodes. *Nature* **2009,** *457*, 706–710. DOI: 10.1038/nature07719.
60. Raghvendra, K. M.; Sravanthi, L. Fabrication Techniques of Micro/Nano Fibres Based Nonwoven Composites: A Review. *Mod. Chem. Appl.* **2017,** *5*, 206. DOI: 10.4172/2329-6798.1000206.
61. Geim, A. K.; Novoselov, K. S. The Rise of Graphene. *Nat. Mater.* **2007,** *6*, 183–191. DOI: 10.1038/nmat1849.
62. Xu, Z.; Gao, C. Graphene Chiral Liquid Crystals and Macroscopic Assembled Fibres. *Nat. Commun.* **2011,** *2*. DOI: 10.1038/ncomms1583.
63. Seyedin, S.; Romano, M. S.; Minett, A. I.; Razal, J. M. Towards the Knittability of Graphene Oxide Fibres. *Sci. Rep.* **2015,** *5*. DOI: 10.1038/srep14946.
64. Bunsell, A. R. *Handbook of Tensile Properties of Textile and Technical Fibres*; 2009. DOI: 10.1533/9781845696801.
65. Dikin, D. A.; Stankovich, S.; Zimney, E. J.; Piner, R. D.; Dommett, G. H.; Evmenenko, G.; Nguyen, S. T.; Ruoff, R. S. Preparation and Characterization of Graphene Oxide Paper. *Nature* **2007,** *448*, 457–460. DOI: 10.1038/nature06016.

66. Hu, Y. C.; Hsu, W. L.; Wang, Y. T.; Ho, C. T.; Chang, P. Z. Enhance the Pyroelectricity of Polyvinylidene Fluoride by Graphene-oxide Doping. *Sensors (Switzerland)* **2014,** *14*, 6877–6890. DOI: 10.3390/s140406877.
67. Chong, J. Y.; Aba, N. F. D.; Wang, B.; Mattevi, C.; Li, K. UV-enhanced Sacrificial Layer Stabilised Graphene Oxide Hollow Fibre Membranes for Nanofiltration. *Sci. Rep.* **2015,** *5*. DOI: 10.1038/srep15799.
68. Arbelaiz, A.; Fernández, B.; Cantero, G.; Llano-Ponte, R.; Valea, A.; Mondragon, I. Mechanical Properties of Flax Fibre/Polypropylene Composites. Influence of Fibre/Matrix Modification and Glass Fibre Hybridization. *Compos. Part A Appl. Sci. Manuf.* **2005,** *36*, 1637–1644. DOI: 10.1016/j.compositesa.2005.03.021.
69. Jia, Y.; Yan, W.; Liu, H.-Y. Carbon Fibre Pullout Under the Influence of Residual Thermal Stresses in Polymer Matrix Composites. *Comput. Mater. Sci.* **2012,** *62*, 79–86. DOI: 10.1016/j.commatsci.2012.05.019.
70. Pfrang, A.; Hüttinger, K. J.; Schimmel, T. Adhesion Imaging of Carbon-fibre-reinforced Materials in the Pulsed Force Mode of the Atomic Force Microscope. *Surf. Interface Anal.* **2002,** *33*, 96–99. DOI: 10.1002/sia.1170.
71. Pei, S.; Zhao, J.; Du, J.; Ren, W.; Cheng, H. M. Direct Reduction of Graphene Oxide Films into Highly Conductive and Flexible Graphene Films by Hydrohalic Acids. *Carbon N. Y.* **2010,** *48*, 4466–4474. DOI: 10.1016/j.carbon.2010.08.006.
72. Parades, J. I.; Villar-Rodil, S.; Martínez-Alonso, A.; Tascón, J. M. D. Graphene Oxide Dispersions in Organic Solvents. *Langmuir* **2008,** *24*, 10560–10564. DOI: 10.1021/la801744a.
73. Stankovich, S.; Dikin, D. A.; Piner, R. D.; Kohlhaas, K. A.; Kleinhammes, A.; Jia, Y.; Wu, Y.; Nguyen, S. B. T.; Ruoff, R. S. Synthesis of Graphene-based Nanosheets Via Chemical Reduction of Exfoliated Graphite Oxide. *Carbon N. Y.* **2007,** *45*, 1558–1565. DOI: 10.1016/j.carbon.2007.02.034.
74. Pham, V. H.; Dang, T. T.; Hur, S. H.; Kim, E. J.; Chung, J. S. Highly Conductive Poly(methyl methacrylate) (PMMA)-reduced Graphene Oxide Composite Prepared by Self-assembly of PMMA Latex and Graphene Oxide Through Electrostatic Interaction. *ACS Appl. Mater. Interfaces* **2012,** *4*, 2630–2636. DOI: 10.1021/am300297j.
75. Chavez-Valdez, A.; Shaffer, M. S. P.; Boccaccini, A. R. Applications of Graphene Electrophoretic Deposition. A Review. *J. Phys. Chem. B.* **2013,** *117*, 1502–1515. DOI: 10.1021/jp3064917.
76. Pomfret, S. J.; Adams, P. N.; Comfort, N. P.; Monkman, A. P. Electrical and Mechanical Properties of Polyaniline Fibres Produced by a One-step Wet Spinning Process. *Polymer (Guildf).* **2000,** *41*, 2265–2269. DOI: 10.1016/S0032-3861(99)00365-1.
77. Arafat, M. T.; Tronci, G.; Yin, J.; Wood, D. J.; Russell, S. J. Biomimetic Wet-stable Fibres Via Wet Spinning and Diacid-based Crosslinking of Collagen Triple Helices. *Polym. (United Kingdom).* **2015,** *77*, 102–112. DOI: 10.1016/j.polymer.2015.09.037.
78. Kuilla, T.; Bhadra, S.; Yao, D. H.; Kim, N. H.; Bose, S.; Lee, J. H. Recent Advances in Graphene Based Polymer Composites. *Prog. Polym. Sci.* **2010,** *35*, 1350–1375. DOI: 10.1016/j.progpolymsci.2010.07.005.
79. Hu, Y.; Sun, X. Synthesis and Biomedical Applications of Graphene: Present and Future Trends. *Adv. Graphene Sci.* **2013,** 55–75. DOI: 10.5772/55666.
80. Xu, Z.; Sun, H.; Zhao, X.; Gao, C. Ultrastrong Fibers Assembled from Giant Graphene Oxide Sheets. *Adv. Mater.* **2013,** *25*, 188–193. DOI: 10.1002/adma.201203448.

81. Kuila, T.; Bose, S.; Mishra, A. K.; Khanra, P.; Kim, N. H.; Lee, J. H. Chemical Functionalization of Graphene and Its Applications. *Prog. Mater. Sci.* **2012,** *57*, 1061–1105. DOI: 10.1016/j.pmatsci.2012.03.002.
82. Cranford, S. W.; Buehler, M. J. Mechanical Properties of Graphyne. *Carbon N. Y.* **2011,** *49*, 4111–4121. DOI: 10.1016/j.carbon.2011.05.024.
83. Zhu, Y.; Murali, S.; Cai, W.; Li, X.; Suk, J. W.; Potts, J. R.; Ruoff, R. S. Graphene and Graphene Oxide: Synthesis, Properties, and Applications. *Adv. Mater.* **2010,** *22*, 3906–3924. DOI: 10.1002/adma.201001068.
84. Kato, R.; Tsugawa, K.; Okigawa, Y.; Ishihara, M.; Yamada, T.; Hasegawa, M. Bilayer Graphene Synthesis by Plasma Treatment of Copper Foils Without Using a Carbon-containing Gas. *Carbon N. Y.* **2014,** *77*, 823–828. DOI: 10.1016/j.carbon.2014.05.087.
85. Anwar, Z.; Kausar, A.; Rafique, I.; Muhammad, B. Advances in Epoxy/Graphene Nanoplatelet Composite with Enhanced Physical Properties: A Review. *Polym. Plast. Technol. Eng.* **2016,** *55*, 643–662. DOI: 10.1080/03602559.2015.1098695.
86. Ghosh, S.; Bao, W.; Nika, D. L.; Subrina, S.; Pokatilov, E. P.; Lau, C. N.; Balandin, A. A. Dimensional Crossover of Thermal Transport in Few-layer Graphene. *Nat. Mater.* **2010,** *9*, 555–558. DOI: 10.1038/nmat2753.
87. Xu, X.; Pereira, L. F. C.; Wang, Y.; Wu, J.; Zhang, K.; Zhao, X.; Bae, S.; Tinh Bui, C.; Xie, R.; Thong, J. T. L.; Hong, B. H.; Loh, K. P.; Donadio, D.; Li, B.; Özyilmaz, B. Length-dependent Thermal Conductivity in Suspended Single-layer Graphene. *Nat. Commun.* **2014,** *5*. DOI: 10.1038/ncomms4689.
88. Meng, F.; Lu, W.; Li, Q.; Byun, J. H.; Oh, Y.; Chou, T. W. Graphene-based Fibers: A Review. *Adv. Mater.* **2015,** *27*, 5113–5131. DOI: 10.1002/adma.201501126.
89. Shao, Y.; Wang, J.; Wu, H.; Liu, J.; Aksay, I. A.; Lin, Y. Graphene Based Electrochemical Sensors and Biosensors: A Review. *Electroanalysis* **2010,** *22*, 1027–1036. DOI: 10.1002/elan.200900571.
90. Kim, J.; Kim, F.; Huang, J. Seeing Graphene-based Sheets. *Mater. Today.* **2010,** *13*, 28–38. DOI: 10.1016/S1369-7021(10)70031-6.
91. Ju, L.; Shi, Z.; Nair, N.; Lv, Y.; Jin, C.; Velasco, J.; Ojeda-Aristizabal, C.; Bechtel, H. A.; Martin, M. C.; Zettl, A.; Analytis, J.; Wang, F. Topological Valley Transport at Bilayer Graphene Domain Walls. *Nature* **2015,** *520*, 650–655. DOI: 10.1038/nature14364.
92. Cabot, J. M.; Duffy, E.; Currivan, S.; Ruland, A.; Jalili, R.; Mozer, A. J.; Innis, P. C.; Wallace, G. G.; Breadmore, M.; Paull, B. Characterisation of Graphene Fibres and Graphene Coated Fibres Using Capacitively Coupled Contactless Conductivity Detector. *Analyst* **2016,** *141*, 2774–2782. DOI: 10.1039/C5AN02534F.
93. Patel, M.; Feng, W.; Savaram, K.; Khoshi, M. R.; Huang, R.; Sun, J.; Rabie, E.; Flach, C.; Mendelsohn, R.; Garfunkel, E.; He, H. Microwave Enabled One-pot, One-step Fabrication and Nitrogen Doping of Holey Graphene Oxide for Catalytic Applications. *Small* **2015,** *11*, 3358–3368. DOI: 10.1002/smll.201403402.
94. Bag, S.; Roy, K.; Gopinath, C. S.; Raj, C. R. Facile Single-step Synthesis of Nitrogen-doped Reduced Graphene Oxide-Mn_3O_4 Hybrid Functional Material for the Electrocatalytic Reduction of Oxygen. *ACS Appl. Mater. Interfaces* **2014,** *6*, 2692–2699. DOI: 10.1021/am405213z.
95. Yu, D.; Goh, K.; Wang, H.; Wei, L.; Jiang, W.; Zhang, Q.; Dai, L.; Chen, Y. Scalable Synthesis of Hierarchically Structured Carbon Nanotube-graphene Fibres for Capacitive Energy Storage. *Nat. Nanotechnol.* **2014,** *9*, 555–562. DOI: 10.1038/nnano.2014.93.

96. Buglione, L.; Pumera, M. Graphene/Carbon Nanotube Composites Not Exhibiting Synergic Effect for Supercapacitors: The Resulting Capacitance Being Average of Capacitance of Individual Components. *Electrochem. Commun.* **2012,** *17*, 45–47. DOI: 10.1016/j.elecom.2012.01.018.
97. Qian, H.; Bismarck, A.; Greenhalgh, E. S.; Shaffer, M. S. P. Carbon Nanotube Grafted Carbon Fibres: A Study of Wetting and Fibre Fragmentation. *Compos. Part A Appl. Sci. Manuf.* **2010,** *41* (9), 1107–1114. DOI: 10.1016/j.compositesa.2010.04.004.
98. Yu, G.; Hu, L.; Liu, N.; Wang, H.; Vosgueritchian, M.; Yang, Y.; Cui, Y.; Bao, Z. Enhancing the Supercapacitor Performance of Graphene/MnO_2 Nanostructured Electrodes by Conductive Wrapping. *Nano Lett.* **2011,** *11*, 4438–4442. DOI: 10.1021/nl2026635.

CHAPTER 2

EPOXY COMPOSITES FILLED WITH GYPSUM (ALABASTER G-5): POSSIBLE WAYS FOR STRENGTHENING, STABILIZATION, AND STRUCTURATION

D. L. STAROKADOMSKY[1,*] and A. A. ISHENKO[2]

[1]*Department of Composites, Chuiko Institute of Surface Chemistry, NAS of Ukraine, Kiev 03164, Ukraine*

[2]*Cathedra of Machines for Metallurgy, Mariupol Priazovski State Technology University, Mariupol 87500, Ukraine*

[*]*Corresponding author. E-mail: km17@ua.fm*

ABSTRACT

New developments in the synthesis of polymer composites have enabled the processing of exciting new, cheap, and biocompatible fillers for epoxy composites. This chapter is concerned with the modification of epoxy resin, Epoxy-520 with commercial gypsum G-5. Particles of gypsum were incorporated (10–75 wt%) into an epoxy-dian resin Epoxy-520. It formed additional filler structures in composite, which is related to the products of crystallization of gypsum under the action of the water admixtures and atmospheric water.

It is shown that the introduction of the gypsum "alabaster G-5" in epoxy-dian resin allows us to obtain the composites with acceptable compressive strength, resistance to water and solvents, improved abrasion resistance, and greatly improved (1.3–1.7 times that of control resin) adhesion to steel.

The fire-resistance of composites increases (1.5–2 times) due to the presence of a gypsum content, due to domination of inflammable inorganic phase and reduction of polymer phase content.

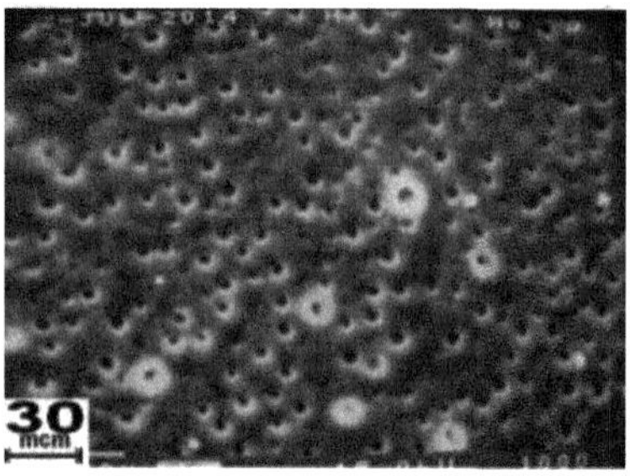

Epox. nonfilled

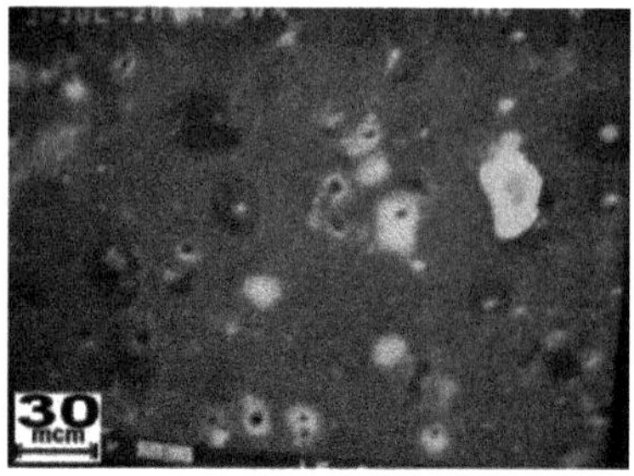

Epox. with 60 wt% of gypsum

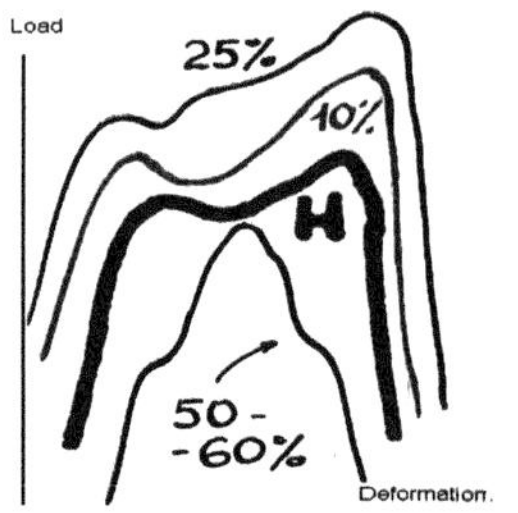

Diagrams "compression load—deformation" for 0% (H) and 10–60 wt% of gypsum

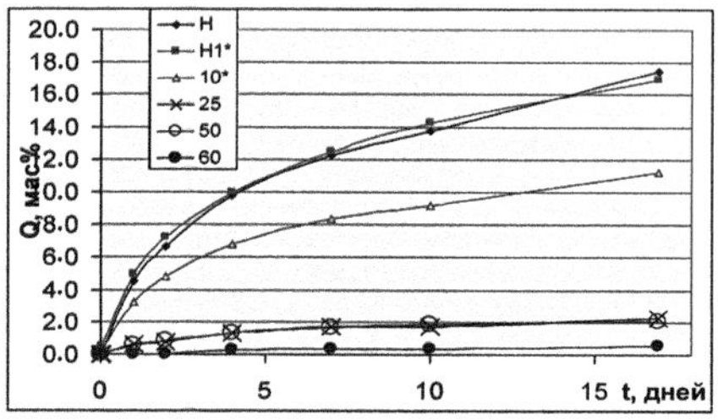

Swelling in ink-solvent of composites with 0–60 wt% of gypsum

Resistance to a physically aggressive organic environment (acetone, ink-solvent) after filling will increase significantly in all cases. However, chemical resistance in nitric acid was not observed. Microscopy shows a substantial compaction of the composite with increase in the filling, which is due to reducing the number and size of the pores. This is confirmed by the sharp decrease in swelling of the composite in the ink-solvent with increasing filler content. The results allow us to propose a method of producing a hardened and cheapened polyepoxide composites for industrial and municipal applications.

2.1 INTRODUCTION

Epoxy-olygomers (Figs. 2.1 and 2.2) were created in 1891 by A. N. Dianine (Russia).[1] Since 1940–1950, they became the universal resins for industrial, service, and remont specialties. Now, the market of most popular GMA-based dian-resin is estimated by millions of dollars (Fig. 2.3).

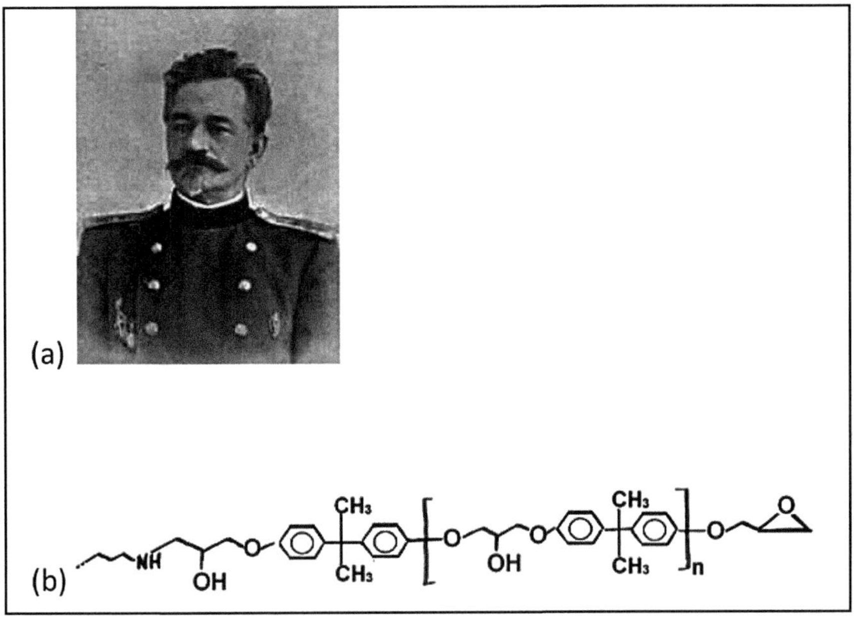

FIGURE 2.1 (a) Professor, Doctor of Chemistry Alexander Dyanine (1851–1918). (b) The chemical formula of epoxy resin.

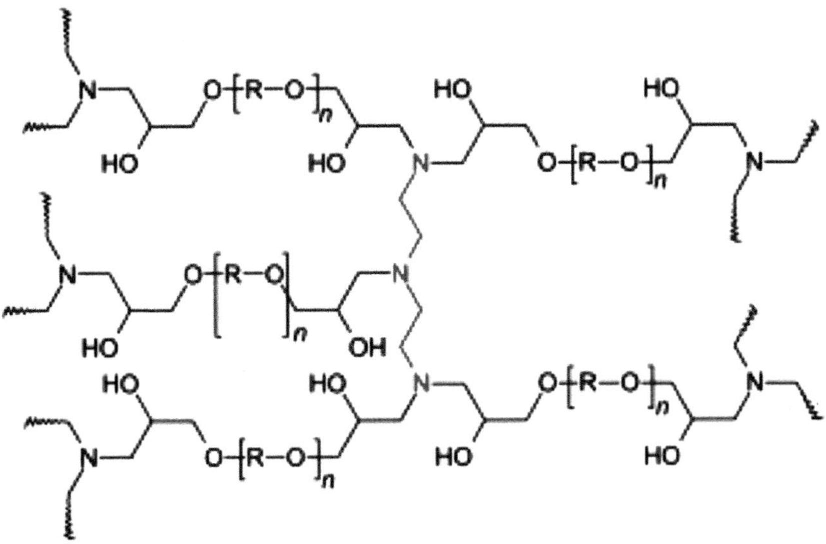

FIGURE 2.2 Typical epoxy-polymer [Wikipedia].

Gypsum is a cheap and universal material that has good compatibility with epoxy resins.[1–8] The polymer–gypsum and polymer–cement composites are popular in building[5] and industrial devices,[2–7] shipping/fishing,[8] etc., that leads to research in this field.[1–8] Epoxides are used as additives to building compounds, as epoxy polymer-cement concrete, as a compound at par with acrylic latex[3] and PVA-substances.[6] The gypsum (plaster, alabaster) polymer materials also are a nominant object of research.[1–5,7]

Our first data[1] showed that the introduction of gypsum into epoxy is capable of producing composites with enhanced adhesion and acceptable strength indices. According to Ref. [2], the content of cement or gypsum in polymer concrete is 60–90 wt%, and the most promising ratio "filler:resin" is 1:1. In Ref. [2], a water–epoxy emulsion was used; gypsum and cement are capable of self-curing under the action of water.

It is also known[8] that when using an epoxy with gypsum or cement, the surface of this boundary layer increases sharply and the polymerization proceeds faster than without the filler. In addition, heat removal from this boundary layer is limited and self-heating of the mixture is observed.

2.2 MATERIALS AND METHODS

The complex of composite characteristics have been investigated: visual (transparency), adhesion strength, resistance to aggressive environments (swelling and decomposition), thermal stability, etc.

The gypsum fillers of various trademarks were used for composite preparation. There are gypsum "Alebaster G-5" for building works (Ukrainian TM Gypsovyk, Stolarow, Polymin; italiano stomatologic ModelGips; Turkish decorative Saten Gips) that was taken as filler of epoxy-resin Epoxy-520 (Czech production). This epoxy-resin was filled by 50 wt% of one of gypsum fillers. After 1 month, compositions were hardened by polyethylene-polyamine in a ratio hardener:resin 1:5.

For swelling, the tablets of template (lens with $d = 1$ cm, $h = 0.1$–0.2 cm) was immersed in ink-solvent Inkwin (ether-acetate), acetone, and concentrated nitric acid. The changes in weight of templates calculated in % versus his initial mass.

For strength, different methods of sets were used next set of tests (according soviet/Russian industrial standards GOST, Fig. 2.4):

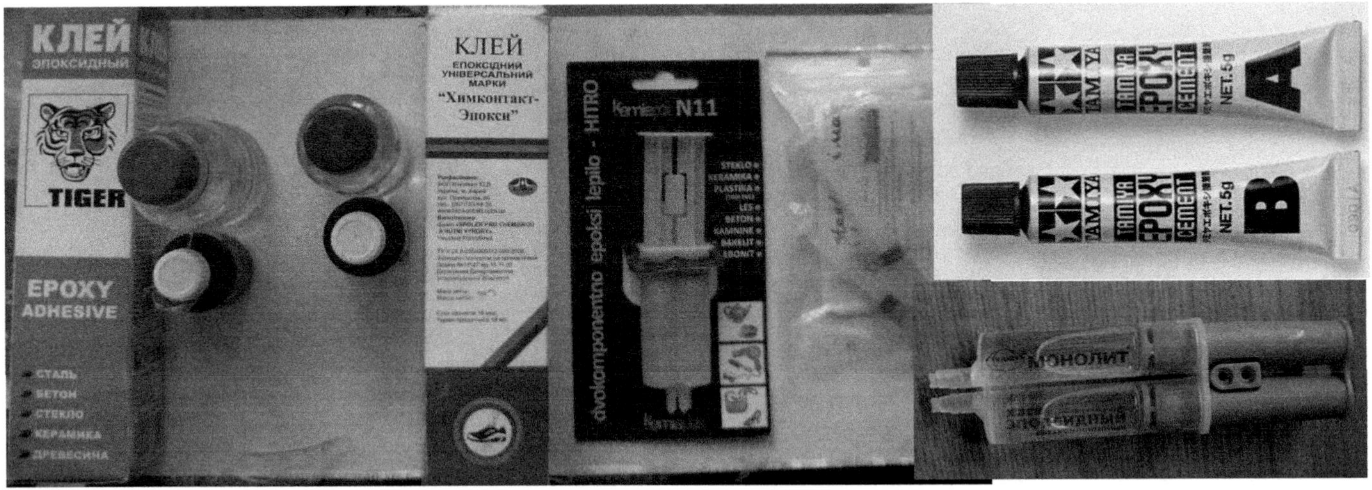

FIGURE 2.3 Epoxy-olygomers (in ex-USSR markets) with hardener in commercial tare.

FIGURE 2.4 Visualization of test methods.

- *Adhesion strength at tearing* of glued steel cylinders (area of glued surface 5 cm^2)
- *Compressive strength* (cylinders with $d = 6.7$ mm, $h = 10.5 \pm 1$ mm)
- *Tensile strength* (plates 60 × 10 × 1.8 mm)

For statistical verification of data, the 5–10 identical templates (according to Student's criteria) of each composite were tested. The result average F was calculated after excluding of minimal result ($F = (F_1 + F_2 + \ldots F_{i-1})/(i-1)$, where i—number of templates).

2.3 MORPHOLOGY OF INVESTIGATED EPOXY-COMPOSITES

2.3.1 MORPHOLOGY OF EPOXY–GYPSUM (10–75 wt%) COMPOSITES

After addition of gypsum, obtained epoxy–gypsum compositions become nontransparent, brown, and viscous suspension (Fig. 2.5). But its characteristic of hardening does not change.

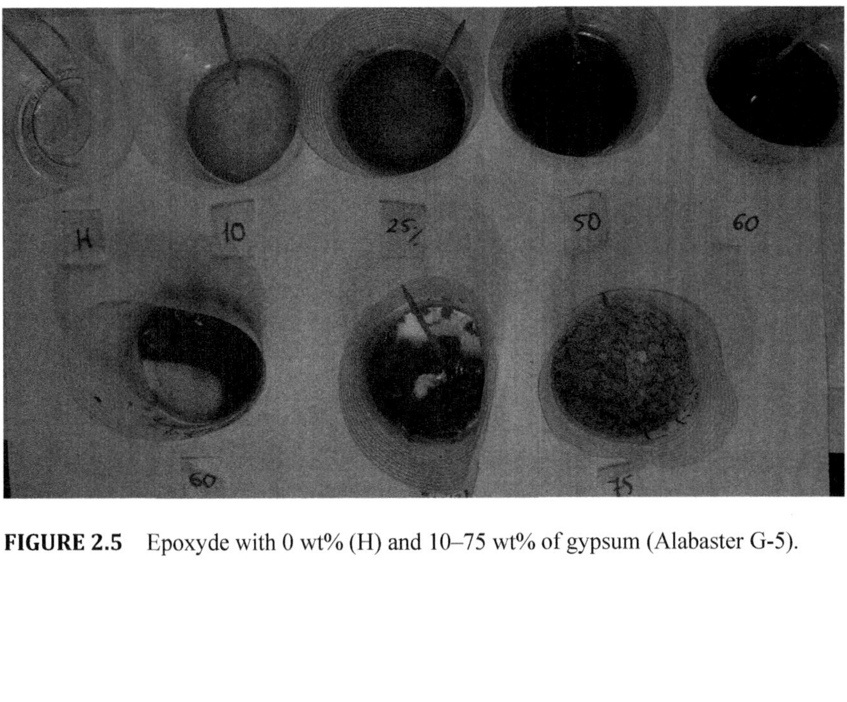

FIGURE 2.5 Epoxyde with 0 wt% (H) and 10–75 wt% of gypsum (Alabaster G-5).

After hardening, unfilled polymer have porous structure with some pure domain/spherulites structures (Fig. 2.6). It is evident from Figure 2.6 that the unfilled composition (H) is characterized by an approximately homogeneous distribution of a large number of pores and bubbles. As a rule (Fig. 2.6, photos 3 and 4), no noticeable formations in a homogeneous porous structure are visible. But along with the pores, it is also possible to detect rare domain structures of the polymer.

After filling up to 25 wt%, the composite structure does not change. In some cases, filler particles are visible on the images. Correspondingly, the X-ray analysis of the distribution along Ca atoms shows the aggregative ("constellations") distribution of the filler particles (Fig. 2.7, photo 7). A significant decrease in porosity occurs after filling 50 wt% (Fig. 2.8). At the same time, the filler particles become visible (light, nonspherical, in Fig. 2.8). Even more pronounced are these changes with an increase for filling up to 60 wt% (Fig. 2.9). With a detailed magnification (Fig. 2.9), the inhomogeneities or the aggregates evidently caused by the presence of the filler are clearly visible. X-ray analysis of calcium atoms shows a relatively homogeneous distribution of the filler, but with the presence of its dense clusters, which are reflected by seals on SEM images (Fig. 2.9, last two images).

2.4 STRENGTH TESTS OF EPOXY-GYPSUM (10–60 *wt%*) COMPOSITES

From Table 2.1 it follows: as the content of gypsum increases, the adhesion to steel increases substantially. At high concentrations (50% and 60% by weight), there is an almost twofold (1/7 times!) increase in adhesive strength. This can be explained by:

- optimizing the structure of the composite with gypsum,
- gypsum's own adhesion to steel, and
- elimination of surface-adsorbed water on the steel by gypsum clusters.

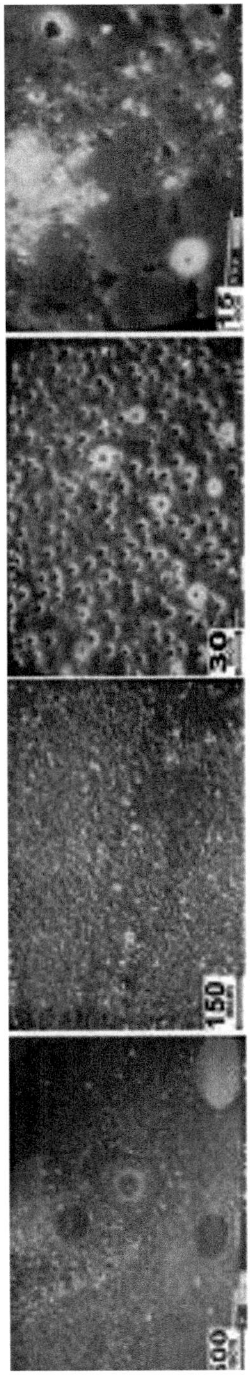

1 2 3 4

FIGURE 2.6 SEM images of hardened unfilled (0 wt% gypsum) polymer.

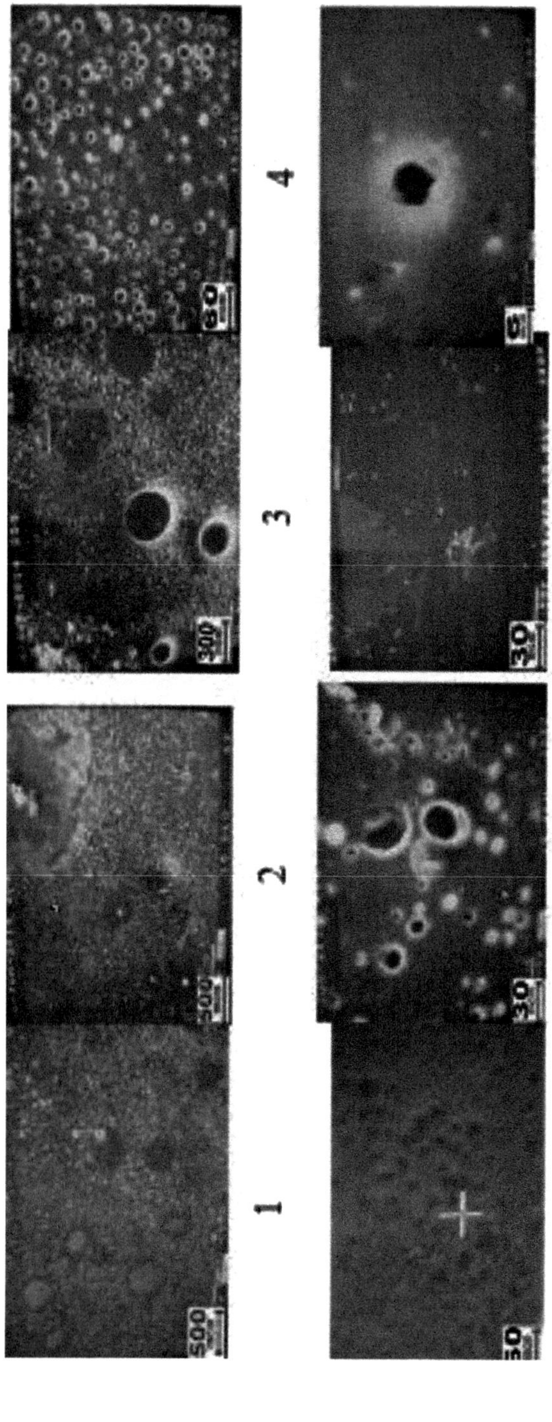

FIGURE 2.7 SEM images for epoxy with 25 wt% of gypsum.

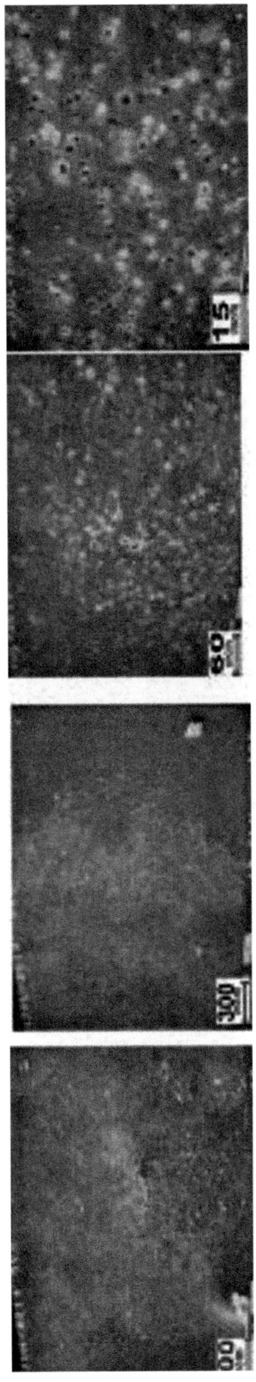

FIGURE 2.8 SEM images of composites with 50 wt% of gypsum.

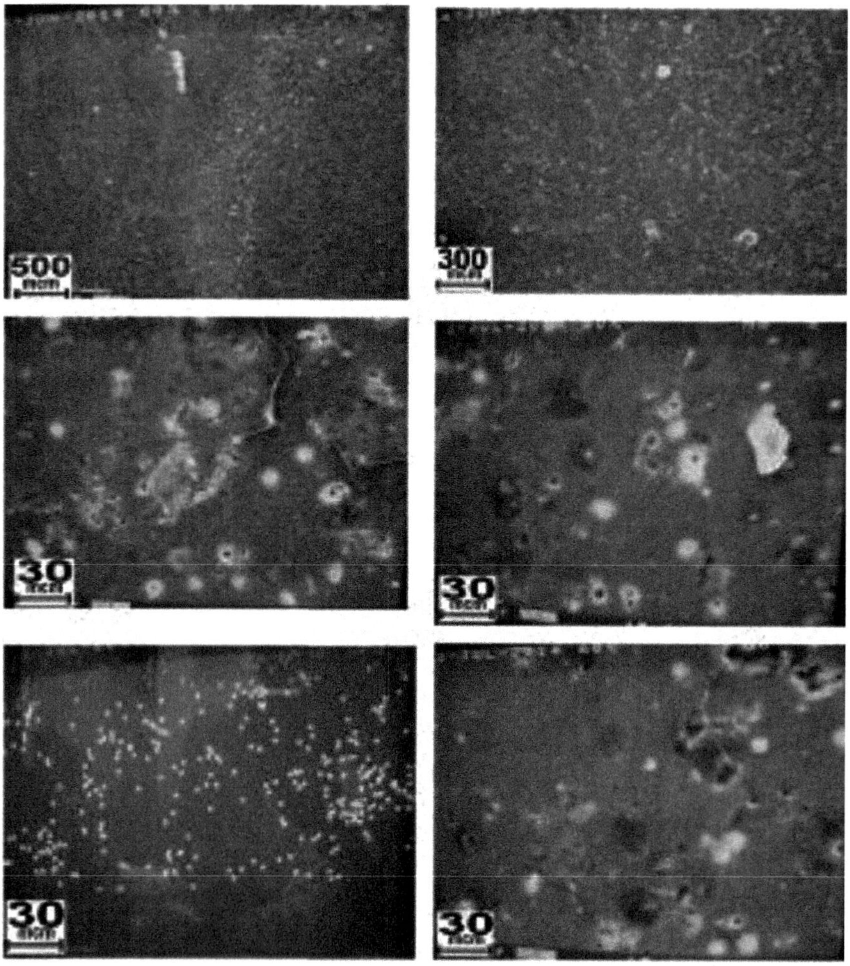

FIGURE 2.9 SEM images of composites with 60 wt% of gypsum.

TABLE 2.1 Adhesion (Loading of Tear Destruction, kgF) to Steel for Composites with Gypsum.

	0 wt%	25 wt%	50 wt%	60 wt%
Loading, kgF	460	540	780	780
% to Unfill. (Unfill. = 100%)	100%	125%	170%	170%
Loading-maximal value, kgF	480	670	920	850
% to max. value for Unfill. (max. Unfill. = 100%)	100%	140%	192%	177%

After filling, most part of physico-mechanic parameters improves (Tables 2.1 and 2.2). It is the result of structure optimization and reaction of filler with water clusters.

TABLE 2.2 Compression Strength and Modulux for Composites with Gypsum TM "Gypsovyk."

	0% (Nonfilled)	10 wt%	25 wt%	50 wt%	60 wt%
Loading, kgF	400 (100%)	400 (100%)	395 (99%)	370 (93%)	385 (96%)
Modulus, kg/cm^2	9600	1020	1000	1020	1720

After filling, a character of destruction changes, reflecting the transition of composites from plastic (for unfilled samples) to an elastic state. Thus, unfilled samples are first compressed (barreled), and upon reaching the plasticity limit, they are destroyed with the formation of a number of longitudinal cracks. The finally destroyed specimen is a strongly bent and distorted cylinder with a multitude of cracks and chips (Fig. 2.10).

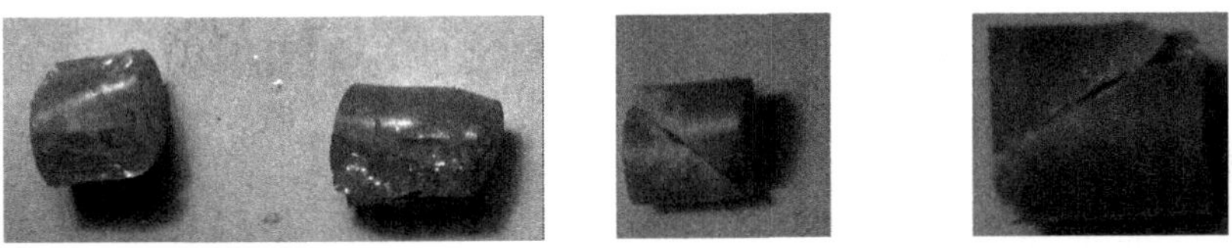

Nonfilled epoxy

50 wt% of gypsum-1 (TM Gypsovyk) and 2 (TM Stolarow)

FIGURE 2.10 Samples destructed by compression test (by axis of cylindrical template) of 50 wt%-filled composites.

After filling with gypsum and chalk, the destruction occurs through Chernov-Luders diagonal (or the longitudinal fracture, for composites with chalk, Fig. 2.11). This shows a qualitative change in the structure of the composite to an elastic nonplastic, with a change of the modulus of elasticity. Thus, despite the practical invariability of load and compression failure, 50% filling with gypsum and cement transfers polyepoxide to a new nonplastic composite state.

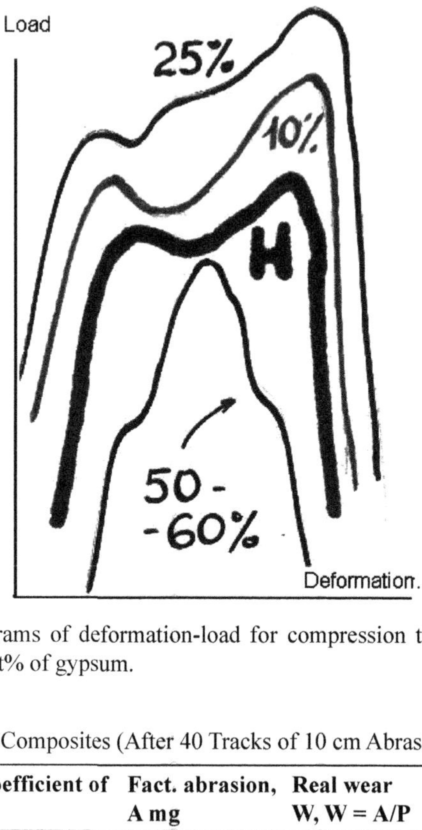

FIGURE 2.11 Diagrams of deformation-load for compression test of samples with 0 wt% (H) and 10–60 wt% of gypsum.

TABLE 2.3 Wear of Composites (After 40 Tracks of 10 cm Abrasive Paper P180).

Composite and its coefficient of density P = ρ/ρ_H	Fact. abrasion, A mg	Real wear W, W = A/P	Effect, in % to nonfilled (H)
Nonfilled (H)	**81**	**81**	100% (Comparison standard)
50% gypsum-1 (TM Gypsovyk), $P = 1.5$	97	65	80% (Better by 20%)
C 50% gypsum-2 (TM Stolarow), $P = 1.5$	90	60	74% (Better by 26%)
C 50% gypsum-3 (TM Polymin), $P = 1.5$	90	60	74% (Better by 26%)

2.5 THERMO- AND CHEMI-RESISTANCE

2.5.1 EPOXY-GYPSUM (10–75 wt%) COMPOSITES

Resistance of epoxy composite to aggressive ink-solvent and to acetone is obviously growing after filling. Swelling of epoxy composite in ink-solvent decreases about 1.5–2 times for 10 wt% and in 7–20 times for 25–60 wt% (Fig. 2.12). For acetone, a decrease in swelling about 1.5–2 times takes place. In both cases, the most effective were 60 wt%-filled composite (Figs. 2.12 and 2.13).

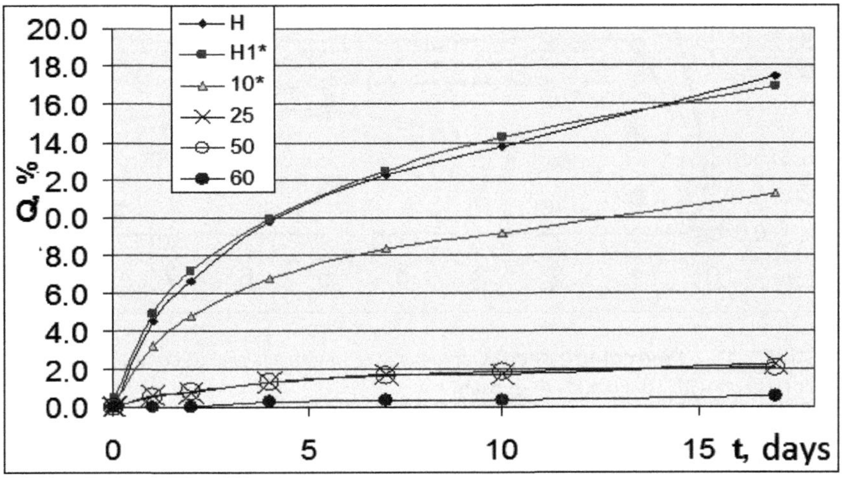

FIGURE 2.12 (See color insert.) Swelling in ink-solvent (TM Inkwin) of composites without filler (H and H1) and with 10–60 wt% of gypsum.

This is evidently a consequence of the compaction of the composite structure after filling and the reduction of porosity (which also is shown microscopically, see SEM images in Figs. 2.6–2.9).

In concentrated nitric acid (a model of a strongly acidic/oxidizing aggressive medium), the resistance of composites after filling with gypsum is markedly reduced. The lifetime (the time before the transition to a viscous or foamed state) of an unfilled polymer is in 2–5 times magnitude higher than for any gypsum-filled (Table 2.4). The author does not explain this pattern from chemical positions, taking into account the theoretical inertness of calcium sulfate (gypsum) with respect to nitric acid. Most likely, gypsum changes the structure of the composite in the direction of less stability of the polymer network to liquid-phase (but not thermal, see below) oxidation.

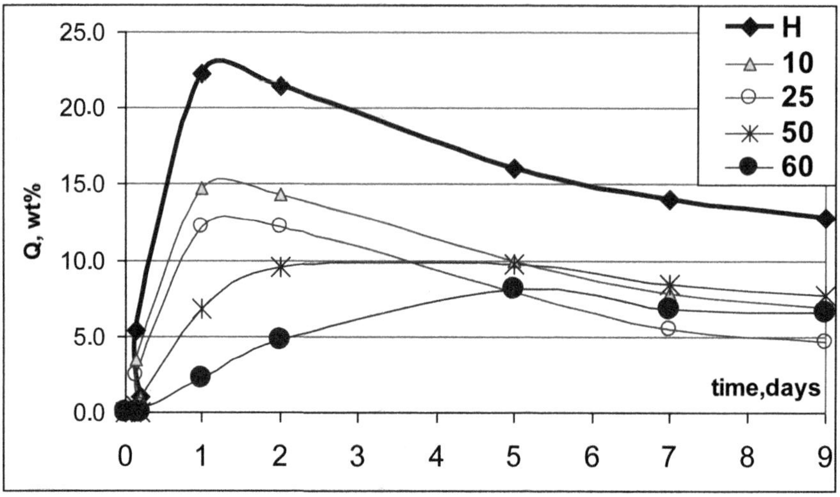

FIGURE 2.13 (See color insert.) Curves of swelling in acetone of composites without filler (H) and with 10–60 wt% of gypsum.

TABLE 2.4 Lifetime (Days) of Composites in Concentrated HNO_3, at 26°C.

H	H1	10	25	50	60
>6	>6	1	1	3	2

TABLE 2.5 The Lifetime of Samples from 50% of Different Gypsums in Acidic Acid at 15°C.

Sample	H1	Gypsovyk	Polymin	Modelgips	Saten Gips	Stolarow
Time, days	17	16	16	15	15	15
Character of destruction	Spread/wipe	Foam	Foam	Foam	Foam	Foam

Note: Designations: H1, H2, and H3 are unfilled samples of a different mass; G, C, P, It, S—gypsum—respectively, TM Gypsovyk, Stolarow, Polimin, ItalmodelGips (Italy), (Turkey).

2.5.2 RESISTANCE TO FLAME/BURNING

As can be seen from Table 2.6, that filling significantly increases (2–2.5 times) the composite resistance to the open fire. It is important to note that for most gypsum and cement filling, even after 5 s, the composition is incapable of self-inflaming—this requires even more time at the fire (up to 7 s). Unlike water-based fillers, chalk does not give such a significant effect, increasing only the time of self-ignition (Table 2.6).

TABLE 2.6 Resistance to Open Flames.

Sample	H1	H2	GypsG	Cement	GypsP	ItalGyps	SatenG	GypsC	Chalk
Time, s	2	2	5.5	3.5	5	5	3.5	5	4
Character	Inflames	Inflames	*Not Infl.*	*Not Infl.*	*Not Infl.*	Inflames	*Not Infl.*	*Not Infl.*	Inflames
	Time of self-ignition for fire-resistant samples after removal of fire								
Time, s	–	–	6–7	5–6	6–7	–	5	6–7	–
Character			Inflames	Inflames	Inflames		Inflames	Inflames	

Note: H1 and H2 are unfilled samples of different mass.
GypsC, Ctolarov; GypsG, gypsum Gypsovyk; GypsP, Polymin; Inflames, self-ignited after removal of fire; ItalGyps, modelgyps for stomatology; NotInfl, does not ignited (self-extinguishing) after removal of fire; SatenG, SatenGypsum.

Unfilled templates destruct completely at 200–250°C, but filled templates are resistant up to 300°C (Fig. 2.14). Filling increases in 2–3 times at time of inflammation (Table 2.6) and resistance to open fire.

2.6 CONCLUSIONS

1. The filling of epoxy resin with various gypsum substances leads to the formation of viscous-flow composites—as a rule gray-green, in some cases white or yellow (decorative gypsum) colors. The technology (time, % of hardener, etc.) for their curing is identical to that for unfilled resin.
2. Microscopy reveals the existence of ordered dendrite structures in composites with gypsum. This is evidently due to the ability to cure of gypsum under action of water traces.

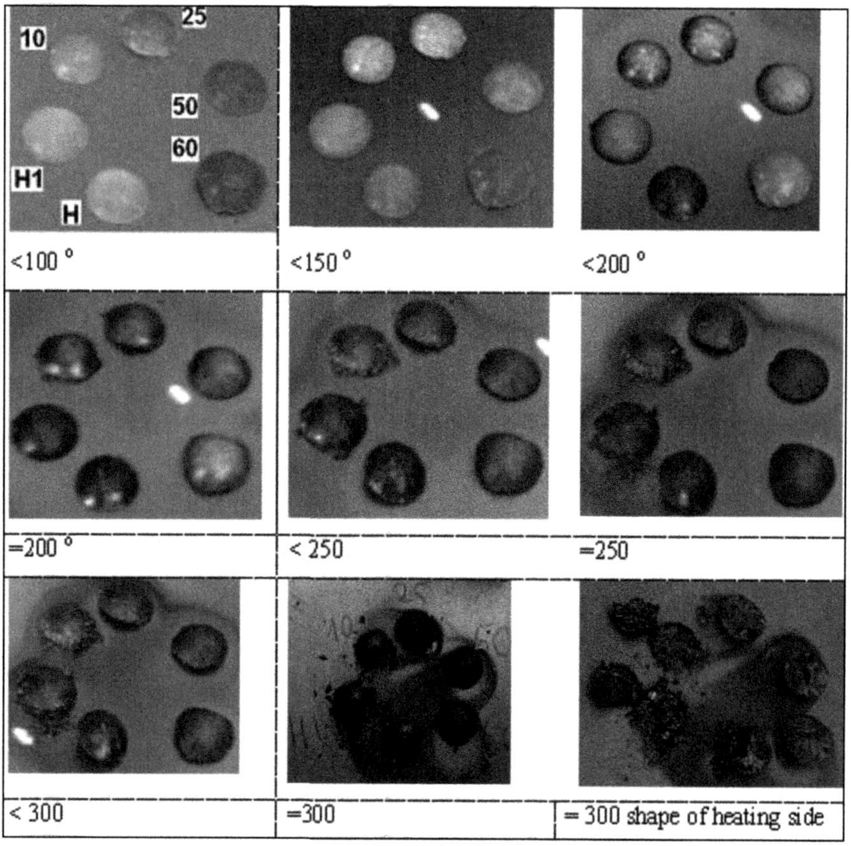

FIGURE 2.14 Visual changes in thermo-treatment of composite samples without filler (H and H1) and with 10–60 wt% of gypsum.

3. Filling can lead to hardening of composites. There is considerable potential for increasing the adhesion strength to steel. The resistance to abrasion after filling with all the investigated dispersions increased by 10–20%.
4. Resistance to aggressive liquids after filling changes ambiguously. Resistance to physically aggressive solvent (acetone) cardinally improves—there is no destruction of the composite and the degree of swelling decreases. A significant decrease of washing-out by acetone (unhardened) composite phase indicates an increase in degree of polymer conversion after filling. Resistance in a strongly acidic/oxidizing medium (lifetime in concentrated nitric acid)

somewhat decreases after filling (from 17 to 15–16 days), and decomposition occurs by foaming (for unfilled—by spreading).

5. The filling of composites with gypsum (50 wt%) leads to a significant increase in fire resistance. The burning-time of the composition increases by 2–2.5 times—from 2 to 5 s, and even more (6–7 s) for self-ignition state.

Thus, filling with gypsum makes it possible to obtain composites with acceptable or enhanced strength parameters, which can significantly increase fire resistance and resistance to physical swelling.

KEYWORDS

- **epoxy composite**
- **gypsum**
- **filling**
- **strength**
- **adhesion**
- **swelling**
- **resistance**
- **microstructure**

REFERENCES

1. Starokadomsky, D. A Long Life of Epoxy Dyan Resin. *Nauka Zhizn ("Science and Life", in Russian)*, **2018,** *N1*, 66–71.
2. Samatadze, A. I. Forming of Structure and Properties of Polymer Composites from Emulsions of Thermoreactive Olygomers with Gypsum or Cement. Dissertation Ph.D. Speciality 05.17.06 (Technology and recycling of polymers & composites). Moscow, 2011, Edited by INHS RAN (in Russian).
3. Colak, A. Characteristics of Acrylic Latex-modified and Partially Epoxy-impregnated Gypsum. *Cement Concrete Res.* **2001,** *31*, 1539–1547.
4. Colak, A. Physical and Mechanical Properties of Polymer—Plaster Composites. *Mater. Lett.* **2006,** *60*, 1977–1982.
5. Arijan, M.; Sobolev, K. The Optimization of Gypsum-based Composite Material. *Cement Concrete Res.* **2002,** *32*, 1725–1728.

6. Hazimmah, S. D. S.; Cheng, H. T. Engineering Properties of Epoxy Polymer Cement Concrete Reinforced with Glass Fibers. *J. Eng. Appl. Sci.* **2011,** *6*(3), 191–199.
7. Najim, T. S.; Al-Zubaidy, A. A.; Yassin, S. A. *Physical And Mechanical Properties of Polymer-Gypsum Composite*; 2016; pp 1–12. https://oud.academia.edu/suhadyasin, www.academia.edu/.../Physical_and_mechanica.
8. Starokadomsky, D. Physical and Mechanical Properties and Micro/Nanostructure of Epoxy Composites Filled with Gypsum, Chalk and Cement. *Compos. Nanostruct.* **2018,** *N1* (37), 39–51.

CHAPTER 3

STRUCTURE AND PROPERTIES OF POLYMER COMPOSITES BASED ON STATISTICAL ETHYLENE–PROPYLENE COPOLYMER AND MINERAL FILLERS

N. T. KAHRAMANOV[1,*], N. B. ARZUMANOVA[1], I. V. BAYRAMOVA[1], and J. N. GAHRAMANLY[2]

[1]*Institute of Polymer Materials of Azerbaijan National Academy of Sciences, Sumgait, Azerbaijan Republic*

[2]*Azerbaijan State University of Oil and Industry, Baku, Azerbaijan Republic*

[*]*Corresponding author. E-mail: najaf1946@rambler.ru*

ABSTRACT

The influence of the multi-functional organic structure-forming agent and mineral fillers on the structural features and physico-mechanical properties of polymer composites based on statistical ethylene–propylene copolymer is reviewed. The principal possibility of improvement of strength characteristics and fluidity of polymer composites in the process of joint use of mineral fillers and ingredients is shown.

3.1 INTRODUCTION

In recent years, scientists worldwide have all increasingly been attracted to researches aimed at studying the influence of the structure and composition of mineral fillers on the process of formation of crystalline permolecular organization in polymer materials.[1–7] Unique ability of finely divided particles of minerals to affect the mechanism of occurrence and growth of heterogeneous crystal nucleus in the polymer matrix has been detected in the process of mixing the polymers with the minerals. The latter fact has an essential impact on improving the properties of polymer composites filled with minerals.

The following are widely applied in order to implement a set of measures aimed at changing the mechanical properties of the original polymer matrix:

- processing additives that improve processing conditions;
- additives that modify the mechanical properties (structurants, plasticizers, reinforcing fillers, etc.).

There with, the efficiency of simultaneous use of number of additives has the combined action on the properties of the polymer composite. The mechanism of the existence of the components in the boundary areas of interphase area is predetermined by the physical and physicochemical forms of interaction existing in the additive-macrochain filler system.[7–10]

There are almost no systematic studies aimed at establishing the effect of multifunctional organic structurants on the structural characteristics and properties of filled polymer composites. The urgency of this problem is that the introduction of the minimum concentrations of organic structurants contributes not only to improving the deformation–strength characteristics of polymer composites but also significantly facilitates the miscibility of the blend components and, as a consequence, their processability.[11–12]

We have repeatedly confirmed in our works[8–12] that the use of organic structurants significantly affects the formation of finely divided spherulitic structures that contribute to the improvement of the technological compatibility of limited-compatible polymers,[11] increasing the strength properties, melt flow index and processability of composite materials.[12]

Investigations on the effects of finely divided particles of the organic structurants on mechanism of formation of permolecular structures in polymer composites and processes occurring in the interphase area and boundary areas of polymer–filler system still remain open. The occurrence of certain structures in the filled polymers and the consequential impact of filler on the regularity of the changing of their properties are one of the most important criteria predetermining the degree of "strengthening" of the polymer base. The increase in the elasticity coefficient and strength of the samples by dispersing the filler is considered as a form of "strengthening" of polymer composites.

Taking into account the complexity and the insufficient illumination of this problem in the literature, the purpose of this chapter is to show how essential the role of multifunctional structurants in the regularities of changes of main physico-mechanical and rheological properties of filled polymer composites is.

3.2 EXPERIMENTAL

The South Korean Industrial statistical ethylene–propylene copolymer (REP) (ethylene content 2–4%), trade name RP2400, has been used as the polymer base. The properties of the polymer are given below:

Ultimate tensile stress—28.5 MPa
Flexural modulus—975 MPa
Tensile strain—600%
Melt flow index—0.36 g/10 min
Vicat softening point—125°C
Melting point—138°C

Cement, chalk, and silica flour have been used as the fillers of polymer.

Cement: The main components of this building material are the binding materials of inorganic origin. The particle size varies in the range of 107–400 nm.

Silica flour: Typical chemical composition of the silica flour is: SiO_2 (99.46%), Fe_2O_3 (0.048%), Al_2O_3 (0.21%), TiO_2 (0.027%), CaO (0.021%). Bulk density—653 kg/m^3. The average particle size—50–100 nm.

Chalk: The base of the chemical composition of the chalk is calcium carbonate with a small amount of magnesium carbonate, but generally there is also noncarbonate part, mainly metal oxides. The average particle size—200–400 nm.

Alizarin and zinc stearate have been used as organic structurants.

Zinc stearate $(C_{17}H_{35}COO)_2Zn$: —White amorphous powder, melting point—403 K, used concentration 0.3–1.0 wt%, used as the lubricant agent in the processing of polymers by injection molding and extrusion.

Alizarin $(C_{14}H_8O_4)$: 1, 2-dihydroxy anthraquinone, colorant red crystals with molecular mass—240.2, melting point—562 K. Below is the structural formula of alizarin:

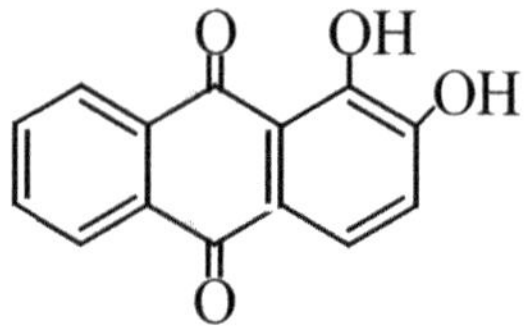

The polymer compositions were prepared through the process of mechanochemical modification (hot rolling) at temperature of 463 K, the rolling time—10 min.

For carrying out physico-mechanical testing of the polymer composites they are subjected to pressing at temperature of 473 K. Samples were punched from these to determine flexural modulus, ultimate tensile stress, and tensile strain of filled composites. Ultimate tensile stress and tensile strain were determined in accordance with the GOST 11262-80 (State Standard). The flexural modulus was determined in accordance with the GOST 4648-71 (State Standard).

Melt index of composites was determined on the IIRT device at temperature of 190°C and load of 5 kg.

The crystallization temperature was determined on IIRT device fitted for dilatometric measurements.[13]

3.3 RESULTS AND DISCUSSION

Taking into account the complexity and insufficient study of this problem, it seemed interesting to carry out a phased approach to the study of regularity of changes in the structure and properties of polymer composites, depending on the type of filler and structurant. It should be noted that often in the literature zinc stearate is only lubricant agent, which improves the processing of polymer composites, but do not reveal the main reasons for the improvement of the quality of the products obtained with the participation of this ingredient. At the same time, alizarin is known as a colorant and therefore practically is not perceived as a structurant in polymer composites.

For a more complete interpretation of detected regularities turn to the results of experimental studies, shown in Table 3.1. As can be seen from this table, the introduction of such fillers such as cement, silica flour, and chalk (without ingredients) contributes to continuous growth of flexural modulus. As regards ultimate tensile stress, the maximum value is accounted for by the samples with 5–10 wt% filler content. Further increase in the concentration of these fillers leads to a natural decrease in ultimate tensile stress, tensile strain, and melt index of samples. Reducing the strength of filled composites during uniaxial tension is the evidence to complex processes in the interspherulite area. We do not exclude that in small concentrations of the filler particles (5–10 wt%), the latter predominantly involved in the formation of heterogeneous nucleation.[14] A further increase in the concentration of filler helps in the process of crystallization from the melt and the crystal growth of particles in excess pushed in interspherulite amorphous area. If the crystallinity of the polymer base is approximately 50%, the amount of filler in the amorphous field is doubled. In other words, if the polymer is introduced 20 wt% filler, its concentration in the amorphous field of semicrystalline REP is approximately 40 wt%. Accumulating in this space, the filler particles reduce conformational mobility "continuous chains," increase the stiffness of the amorphous field, which immediately affects the deterioration of strength and tensile strain of polymer composites.

Attention shall be paid to the fact that the separate introduction of 1 wt% alizarin and zinc stearate into the REP leads to serious changes in its qualitative characteristics. For example, in this case, it provided a significant increase in all its strength characteristics, tensile strain, and melt flow index (MFI) of REP. It is clear that such significant improvement in the quality characteristics of the REP can be clearly interpreted based on the peculiarities of the formation of fine spherulitic permolecular structures.

The important aspect is that the simultaneous introduction of alizarin and zinc stearate into the composition of REP leads to even more improvement of final properties of polymer composites. It is precisely this feature of joint participation of alizarin and zinc stearate toward improving the quality of REP that can be regarded as a "synergetic effect." Therefore, in our further investigations in all the filled composites of REP, these ingredients were introduced simultaneously in the amount of 1 wt% alizarin and 1 wt% zinc stearate.

According to the data given in Table 3.1, introduction of the described ingredients into the composition of filled composites leads to a substantial improvement of their properties. Pairwise comparison of samples of filled composites with or without ingredients can establish that the maximum increase in the flexural modulus is 22–25%, ultimate tensile stress 15–21%, and MFI 4.0–4.8 times. There is a reason to

TABLE 3.1 The Composition and the Physico-mechanical Properties of the Filled Composites Based on REP Containing 1 wt% Alizarin and 1 wt% Zinc Stearate.

No.	Composition of polymer compound	Ultimate tensile stress, MPa	Flexural modulus, MPa	Tensile strain, %	MFI, g/10 min
1	REP	28.5	975	600	0.36
2	REP + 1% alizarin	29.8	1001	750	1.56
3	REP + 1% zinc stearate	29.1	988	680	2.98
4	REP + alizarin + zinc stearate	30.4	1006	780	3.11
5	REP + 5% cement	31.3	1010	560	0.44
6	REP + 5% cement + alizarin + zinc stearate	32.2	1028	640	1.94
7	REP + 10% cement	31.5	1025	320	0.35
8	REP + 10% cement + alizarin + zinc stearate	wz.6	1176	510	1.98
9	REP + 20% cement	24	1075	85	0.25
10	REP + 20% cement + alizarin + zinc stearate	32.3	1198	250	1.85
11	REP + 30% cement	22.8	1080	25	Do not flow
12	REP + 30% cement + alizarin + zinc stearate	24.6	1215	125	0.95
13	REP + 5% silica flour	29.8	1015	240	0.35
14	REP + 5% silica flour + alizarin + zinc stearate	31.4	1022	525	2.04
15	REP + 10% silica flour	27	1020	180	0.3
16	REP + 10% silica flour + alizarin + zinc stearate	30.8	1132	295	1.84
17	REP + 20% silica flour	22.6	1040	70	0.11

TABLE 3.1 *(Continued)*

No.	Composition of polymer compound	Ultimate tensile stress, MPa	Flexural modulus, MPa	Tensile strain, %	MFI, g/10 min
18	REP + 20% silica flour + alizarin + zinc stearate	27.3	1155	110	1.92
19	REP + 30% silica flour	20.8	1050	35	Do not flow
20	REP + 30% silica flour + alizarin + zinc stearate	24.5	1175	70	0.73
21	REP + 5% chalk	27.1	980	495	0.37
22	REP + 5% chalk + alizarin + zinc stearate	30.3	1035	210	2.01
23	REP + 10% chalk	25.8	1000	315	0.2
24	REP + 10% chalk + alizarin + zinc stearate	27.9	1121	145	1.72
25	REP + 20% chalk	19.8	1010	95	0.12
26	REP + 20% chalk + alizarin + zinc stearate	24.7	1123	105	1.45
27	REP + 30% chalk	18.2	1015	15	Do not flow
28	REP + 30% chalk + alizarin + zinc stearate	22.6	1155	65	0.69

assume that the smaller spherulite size, the fewer defects in crystalline formations and the more likely uniformly dispersing of the filler particles in interspherulite area and generally throughout the volume of the polymer matrix.

On the other hand, we do not exclude that alizarin and zinc stearate may to some extent contribute to laying the grain of surface of filler particles, thereby creating favorable conditions for the aggregate flow of the melt and supporting maximum wall sliding in evaluating of MFI of samples. In view of nature of the filler and polymer, and mutual dispersion, the process of choosing the optimal ratio of components in the polymer matrix may be associated with certain difficulties. In several cases this circumstance is explained by a variety of approaches to the interpretation of the observed regularities.[13,14] Depending on the type and concentration of filler, the latter has a significant structuring influence on polymer base both in solid and in viscous flow state.

For example, Table 3.2 presents the results of investigations on the influence of type and concentration of the filler and organic structurants on the crystallization temperature of the polymer composites. Analyzing the data contained in this table, it can be stated that the used organic structurants have a strong impact on the value of crystallization temperature of composites. Particularly, there is strong influence of alizarin. Joint use of discussed organic structurants once again has confirmed the existence of "synergism" effect, which was expressed in the increase of crystallization temperature of composites 5–7°C. It is characteristic that filler itself (except chalk) exhibits a structure-forming effect expressed in increase of the temperature of crystallization of polymer composites. As shown from this table, the most intensive crystallization process occurs in those samples in which cement and silica flour are used as a filler. Changing the crystallization temperature of the polymer matrix is only possible when foreign finely dispersed particle of organic or mineral origin is capable of forming heterogeneous crystallization nucleus.

The filled polymer composites may have two types of crystallization nucleus: homogeneous and heterogeneous. In our view, homogeneous nucleus is the primary microcrystalline formation which emerged as a result of thermo fluctuation changes in the melt of polymer at the level of oriented segments of macrochains. In the process of further cooling these micro-oriented areas grow into larger crystalline formations—spherulites. Heterogeneous crystallization nucleus is formed with the participation of solid foreign particles, which can orient macrosegments of the polymer matrix on their surface. It is obvious that during polymer processing by extrusion or injection molding, the cooling process of products is accompanied, at first, by formation of heterogeneous crystallization nucleus, and then homogeneous crystallization nucleus.[13,14]

TABLE 3.2 Influence of Structurants and Mineral Fillers on Heat Resistance and Crystallization Temperature of REP.

No.	Composition of polymer compound	Chilling point, °C	Vicat softening point, °C
1	REP	141	125
2	REP + 1% alizarin	144	125
3	REP + 1% zinc stearate	142	125
4	REP + alizarin + zinc stearate	146	125
5	REP + 5% cement	142	127
6	REP + 5% cement + alizarin + zinc stearate	147	128
7	REP + 10% cement	143	129
8	REP + 10% cement + alizarin + zinc stearate	147	130
9	REP + 5% silica flour	143	127
10	REP + 5% silica flour + alizarin + zinc stearate	148	129
11	REP + 5% chalk	141	125
12	REP + 5% chalk + alizarin + zinc stearate	142	126

3.4 CONCLUSION

Thus, on the basis of the foregoing, it can be concluded that the use of structurants with various mineral fillers has a positive impact on the improvement of physico-mechanical, thermal, and rheological properties of filled polymer composites. It becomes obvious that the use of mineral fillers and structurants allows solving a number of problems related to improving the technological capabilities of processing of filled polymer composites. As a matter of fact, new polymer composites with improved operating and processing characteristics were generated in the process of modification of REP.

KEYWORDS

- **ultimate tensile stress**
- **flexural modulus**
- **interfacial area**
- **continuous chains**
- **spherulites**
- **crystallinity**
- **crystallization**
- **filler**

REFERENCES

1. Berlin, A. A.; Volfson, S. A.; Oshman, V. Q. *Principles of Creation Composite Materials;* Chemistry Publ.: Moscow, 1990, p 240.
2. Lipatov, Y. S. *Physical Chemistry of Filled Polymers;* Chemistry Publ.: Moscow, 1977, p 304.
3. Keleznev, V. N.; Shershenov, V. A. *Chemistry and Physics of Polymers;* High School Publ.: Moscow, 1988, p 312.
4. Osama, A. X.; Osipchik, V. S.; Petuxova, A. V.; Kravchenko, T. P.; Kovalenko, V. A. Modification of Filled Polypropylene. *Plastic Mass* **2009,** *1*, 43–46.
5. Osipchik, V. S.; Nesterenkova, A. I. Talc Filled Compositions Based on Polypropylene. *Plastic Mass* **2007,** *6*, 44–46.
6. Ermakov, S. N.; Kerber, M. L.; Kravchenko, T. P. Chemical Modification and Mixing of Polymers During Reactive Extrusion. *Plastic Mass* **2007,** *10*, 32–41.
7. Pesetsky, S. S.; Boqdanovich, S. P. Nanocomposites Obtained by Dispersing Clay in Polymer Melts. In *Polymer Composites and Tribology*; International Scientific and Technical Conference: Gomel, 2015, pp 5.
8. Kahramanov, N. T. The Mechanism of Modifying the Permolecular Structure of Polyolefins by Grafting Acrylic Monomers. *Macromol. Comp.* **1990,** *32A*(11), C.2399–2403.
9. Kahramanov, N. T.; Baladzhanova, G. M.; Shahmaliyev, A. M. Investigation of Sorption Kinetics of Graft Copolymers on the Filler Surface. *Macromol. Comp.* **1991,** *32B*(5), 325–329.
10. Kahramanov, N. T.; Kahramanly, Y. N.; Faradzhev, G. M. The Properties of Filled Crystalline Polymers. *Azerbaijan Chem. J.* **2007,** *2*, 135–141.
11. Kahramanov, N. T.; Meyralieva, N. A.; Kahramanly, Y. N. The Technological Parameters of Processing of PP Compositions Filled by Natural Zeolite. *Plastic Mass* **2011,** *1*, 57–59.

12. Kahramanov, N. T.; Hajiyeva, R. S.; Kuliev, A. M.; Kahramanly, Y. N. The Influence of Different Ingredients on the Properties of the Polymer Mixtures Based on Polyamide and Polyurethane. *Plastic Mass* **2013,** (12), 9–13.
13. Kahramanov, N. T.; Dyachkovsky, F. S.; Buniyat-zade, A. A. Volumetric Properties and Crystallization of Polymerization-filled Polyethylene. In *Compilation IX. Synthesis of Polymerization-filled Polyolefins*; Institute of Chemical Physics, Academy of Sciences of the USSR: Chernogolovka, 1982; pp 130.
14. Kahramanov, N. T.; Arzumanova, N. B. The Problematic Questions of Mechano-chemical Synthesis of Polymer Compositions During Their Processing. *Int. Sci. Inst. Educatio Novosibirsk* **2015,** *3*(10), C.147–C.148.

CHAPTER 4

MECHANICAL AND VIBRATIONAL ANALYSIS OF MICROTUBULE USING MOLECULAR DYNAMIC SIMULATIONS

M. GHAMAMI*, M. H. JAFARABADI, and H. NAHVI

Department of Mechanical Engineering, Isfahan University of Technology, Isfahan, Iran

**Corresponding author. E-mail: mghamami@yahoo.com*

ABSTRACT

Microtubule (MT) polymer is one of the most abundant proteins in the body and one of the three main components of cells framework. This protein is responsible for bearing different loads on the cells and providing the necessary force to alter the cell proliferation, spontaneous migration, and the division of a cell. Also, MT plays a key role in internal transportation in the cells. Therefore, studying the vibrational behavior and mechanical properties of MTs is a significant issue which has attracted many experimental and theoretical researches toward itself. In this research, the study of mechanical and vibrational behavior of MTs is performed using molecular dynamics simulation and exact coarse-graining methods.

4.1 INTRODUCTION

Cells are the smallest unit of living organisms that define their physical behavior can facilitate understanding the biological processes in living existences. The mechanical properties of the cells are significantly determined by the cytoplasm, which consists of three main parts, protein microtubule (MTs), intermediate filaments, and actin. MTs play a major role in many intracellular operations. They are actually hard, hollow, and cylindrical strands of eukaryotic cellular cytoskeletons. MTs are significantly responsible for the cells shape and their strength.[1] Inside the cell, MTs act as pathways for the motion of the protein engine and play an important role in a wide range of functions, such as cell division, cellular mobility, and intracellular transport. The biological functions of the MTs depend to a large extent on their mechanical properties. Because most of these roles relied on mechanical and dynamical properties of MT, it is substantial to determine mechanical behavior of these nanostructures. Over the past few decades, many efforts have been made to identify these biological nanostructures. However, with the advent of new and empirical techniques such as high-resolution electron microscopes, our understanding of mechanical behavior of MTs has improved, but there are still unknown patterns that should be studied. Recently, the analysis of the dynamical behavior of the MTs has been considered and it has been shown that a systematic study of protein MTs vibrations can greatly increase the understanding of its biological functions.

These nanostructures are only found in eukaryotic cells. They are one of the most abundant proteins in the body and one of the three main components of the cell's framework. Also, the MT plays a key role in intracellular transport.[2] In Figure 4.1a, an example of a fluorescent photo taken from this nanostructure is shown. In Figure 4.1b, kinesin and MT are shown. Kinesin as a protein motor which benefits from MTs as its railroads for transportation inside the cells. Motor proteins such as kinesin and dynein are responsible for displacement of many elements like food, drug, and viruses in cells.

MTs are microscopic tubes made of a kind of protein called tubulin. Tubulins are globular proteins produced through peptide bonds of amino acids which make polymeric chains of proteins, shown in Figure 4.2.

MT contains two kinds of tubulins, α and β and assembles from dimers of α- and β-tubulin that are shown in Figure 4.3.[3]

4.1.1 MATHEMATICAL MODELING OF PHYSICAL SYSTEMS

Mathematical models can be applied to study natural phenomena. With incrementing the system complexity, more advanced models are required to determine the complex nature of the events. Models in the computational community are based on two basics: Newtonian classical mechanics and quantum mechanics. In general terms, these two models study a range of materials from a macroscale to a subatomic scale. In Newtonian

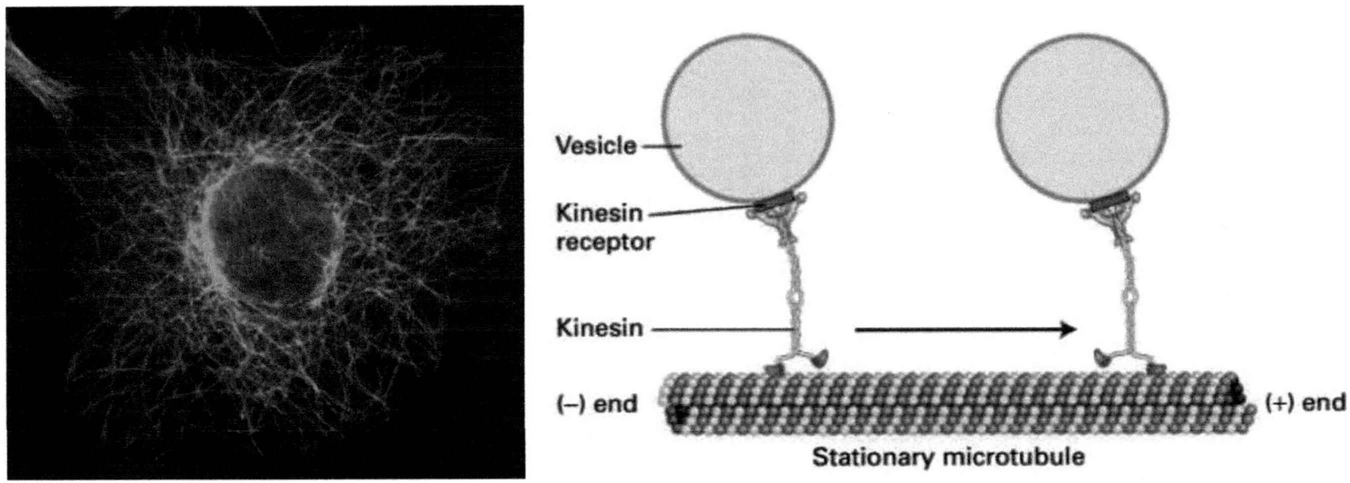

(a) (b)

FIGURE 4.1 (a) Fluorescence image of the accumulation of MTs and (b) kinesins motion on MT.

Source: Adapted from Ref. [2].

classical mechanics, the Newtonian motion equations relate to the various components of the material being dominant. Considering the accuracy of this model in examining the phenomena in nature, the methods based on this model are very popular. According to the application of this method, different types of models are presented to study phenomena in different scales. These are:

1. Continuum mechanics
2. Dislocations dynamics
3. Molecular methods

In the study of macroscale phenomena, we can assume a material as a cloth. In this way, we can obtain the equations governing the body by writing the Newton equations for a volumetric element of an object and integrating it over the entire object. In spite of the large application of this model in studying phenomena in large scale, the classical interconnected model is basically incapable of expressing a bunch of phenomena that arise from the atomic structure of the body. These include:

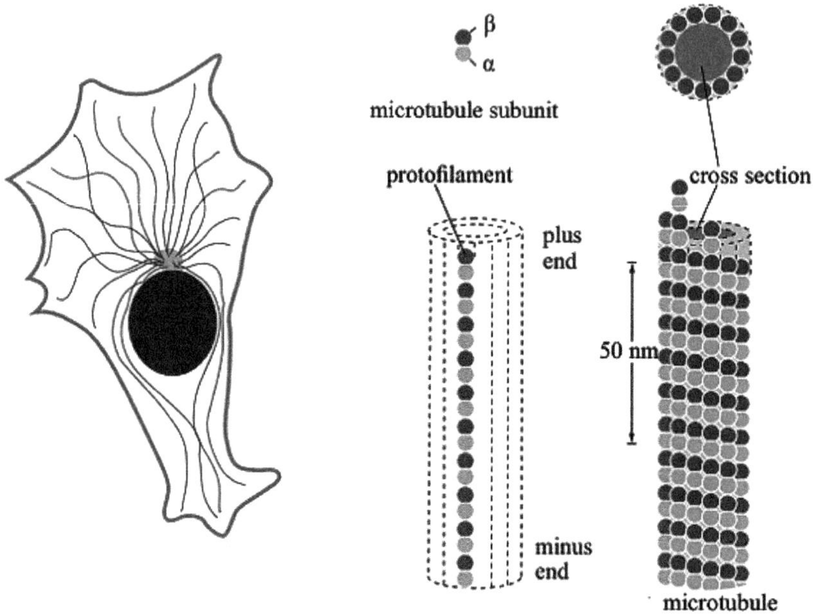

FIGURE 4.2 Substructure and dimension of MTs in a neural cell.

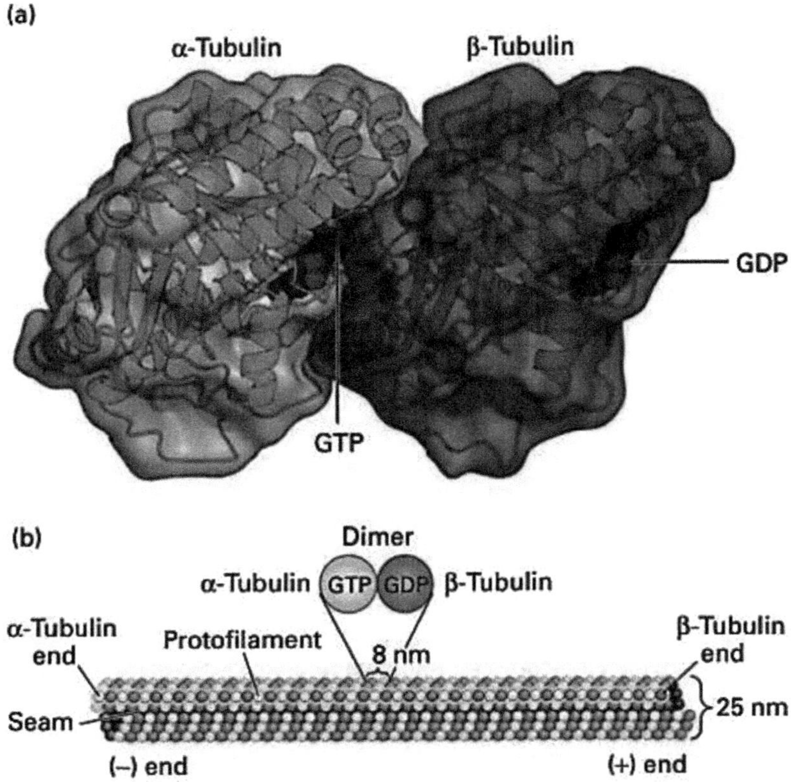

FIGURE 4.3 (See color insert.) Details of MT's substructure, α- and β-tubulin.

1. A pattern of dislocation in creep and fatigue
2. Heterogeneity of plastic deformation
3. Impact of size, geometry, and tension status on submission characteristics
4. Dependence of material resistance on a microscale as much as MTs and their orientation in the matrix of matter
5. The nature of the dependence of the resistance of MTs due to the links in their structure

In molecular models, all atoms are modeled individually and their relationship is expressed by atomic potentials. These models are very capable of examining very fine nanoscale structures. These models are divided into two categories: statistical and deterministic. The Monte Carlo method is a group of statistical methods. In this method, the motion equation is not solved, but the energy relationship with the potential location of atoms is investigated. The determinants in which direct motion equations are solved have different types of molecular dynamics that have a special place in studies.

In models based on quantum mechanics, the presence of finer beings is considered from atoms such as electrons, protons, and neutrons. In fact, intermediate atomic power depends on the status and characteristics of subatomic objects. The fundamental equation in quantum mechanics is the Schrödinger equation, which solves this differential equation into a special value problem. By finding the equation above, the density function for determining the probability of the presence of electrons around the atomic nucleus is achievable. Therefore, solving the Schrödinger equation provides valuable information on the structure of electronics, conductivity, electrical resistance, geometric characteristics of atomic bonds, and energy in the molecular structure. Numerical approximations are available to solve this equation, but these methods are very costly and their use is for noneconomic large molecules. An extremely important application of quantum-based methods is to offer a detailed picture of the interatomic interactions which appear as atomic potentials.

As the dynamic behavior of MTs is investigated on a nanoscale, classical mechanics of continuum environments cannot predict the dynamic behavior of these nanostructures. On the other hand, laboratory methods are costly and complex. Interestingly, new methods such as molecular dynamics theory are more accurate and more efficient. A molecular dynamics method is a form of computer simulation in which atoms and molecules are allowed to interact with the known laws of physics for a period of time, and follow the motion of the atoms. Since molecular systems generally contain a large number of particles, it is not possible to obtain the characteristics of complex systems analytically. Simulation of molecular dynamics solves this problem by employing the computational method. This method creates a link between experiment and theory and is considered as a virtual experiment. Molecular dynamics method investigates the relationships between the structure of molecules, the motion of molecules, and intermolecular interactions, and is a regular procedure. Its laws and theories come from mathematics, physics, and chemistry, and employ algorithms from computer science and information theory. This method was first used in theoretical physics in the 1950s, but today it is used more in material science and molecular biology. The molecular dynamics anticipates the "real" dynamic behavior of the system, which can be used to calculate the system average properties, in a specified time duration. By applying Newton's equations of motion, a set of successive atomic states is obtained. Molecular dynamics is a determinant method, in the sense that the state of the system can be predicted when the current state is known.

4.1.2 LITERATURE REVIEW AND PREVIOUS WORKS

In this section, a review of the history of research on the study of microscopic behavior is presented to help it set up a logical link between the research data of the past and the present study and make a theoretical framework for research. So far, a few studies have been done on the mechanical properties of the multiring model using molecular dynamics simulation, but the results of these studies will be initially investigated.

Among the researches carried out by molecular dynamic simulation, the research most closely related to the subject of this study is the study by Wells et al.,[3] which is the closest work to determine the mechanical properties of the MT using all-atom method and modeling of one loop of MT, and applying the force field Chemistry at Harvard Macromolecular Mechanics (CHARMM) 27. Another noticeable work on this nanostructure using simulating molecular dynamics was the study of kinesin interacting with tubulin dimer through all-atom approach by Zheng and Chakraborty.[4]

One of the most important tools used in this field is the continuum mechanics. Modified continuum modeling of nanostructures has drawn our attention. Modified continuum models not only benefit from the computational efficiency of classical continuum method but also at the same time produce accurate results in comparing with atomistic models. Nanoscale phenomena such as size-effects and surface energy effects which are a weakness of classical continuum modeling, could be investigated using modified continuum models. Most of these mathematical models are easily available and unlike atomistic models, they seldom require the use of professional software or hardware.[5–10]

Among the researches carried out in continuum mechanics field, Civalek et al.[11] investigated bending and lateral vibration of MT using Euler theory and Eringen nonlocal theory. In this study, MT is modeled like a clamped-free nanobeam as the only boundary condition considered. By extracting the Euler beam relations and applying nonlocal correction, coefficients aim to the expression of the nanoscale behavior. Provided relations are solved by numerical methods. The nonlocal parameter has been exerted selectively and no optimization has been done for it. Wang et al.[12] in another study, considering the MTs as an orthotropic cylindrical shell, studied the vibrational behavior of MT. In the research, dimensional correction methods were not applied. Ghorbanpour et al.[13] scrutinized the vibration behavior of the MT due to the kinesin protein engine movement upon it. As one of the strengths of the work, the cellular environment of the MT is taken into account. The MT is located in the cytoplasm within the cell's environment, which has significant effects on its mechanical behavior. The biofluid is modeled like a Pasternak elastic medium in relations, and the mathematical model is derived using the modified coupling stress theory for an orthotropic nanobeam. Considering the elastic property of the viscoelastic environment and regardless of its viscosity, the biofluid environment has not been accurately modeled. Kinesin and dynein have been modeled as nanoparticles, and the equations of motion have been obtained using the Euler theory, couple stress theory, and the Hamilton principle. By analytical solving of the obtained equations, the maximal microtubule deflection pertaining to the first mode shape due to movement of kinesins, with different speeds, are extracted. The weakness of this research is lack of using a precise method to calibrate the results and optimizing parameters and coefficients for dimensional correction. A remarkable research on the nonlinear vibration of MT is done by Shen,[14] the research applied nonlocal shell theory for a cylindrical orthotropic model of MT. The parameters are estimated at various temperatures, and thus the frequencies and shape of the modules are estimated for the nonlinear model. This study was based on another study by the same author[15] in which the mechanical properties of MT at various temperatures were extracted by a shear deformable nonlocal model. Another considerable work on the vibration of MT is investigated by Xiang et al.[16] using the mesh-free computational framework and modeling MT in the elastic medium. In this research, various scale coefficients and different numbers for the network have been used. The first to sixth frequencies are determined with various coefficients and number of nodes considered are plotted for different lengths. Applying finite element method has led to solving equations for all types of boundary conditions, and this results in an expanded and relatively comprehensive research. This research used modified continuum mechanic methods that essentially, reduces the accuracy of results.

Motamedi et al.[17] in another research investigate on nanobiomechanical properties of MT using both molecular dynamics simulation and multi-scale modelings method. By drawing the force-displacement diagram for tubulin, they could find a way to use a multiscale modeling technique with combining molecular dynamics and finite element method. Gao et al.[18] studied the effects of small-scale on the mechanical behavior of the MT using nonlocal theory. Gorbanpour et al.[19] studied the vibrational behavior of MTs, with considering biofluid around the MT, using the modified couple stress theory. The MT was modeled using orthopedic beam and the surface effects of fluid on MT vibrations were investigated; for this purpose, various correction coefficients were applied. Considering the biofluid effects and surface impacts, are the strength of work and relying on a method based on the continuum mechanics without optimizing the coefficient using a more exact method, is one of the weak points of this research. Benninajjaryan et al.[20] researched the size effect and elastic foundation on MT vibration using a cylindrical shell couple stress model.

Mokhtari et al.[21] in a research, investigated the MT-free vibration using the elastic cylindrical shell model and the stress and strain gradient theory. In this research, a thin-shell model has been exerted which is considered to be a weak point and is also limited to a simple Navier problem-solving as the other shortcomings of this research. The strength of that is considering the values consistent with previous studies results, for the dynamic coefficient of l_d and the static size coefficient l_s, for the strain gradient theory. The results of their research are in suitable agreement with other works in continuum mechanics. Another research was conducted by Zeverdejani et al.[22] who applied linear and nonlinear Euler model and strain gradient theory. Following this, the equations of motion have been extracted, and considering two simple–simple and clamped–clamped boundary conditions, these equations have been solved. The effects of the size and temperature and the geometry of MT on its natural frequencies are investigated too. The same mentioned researchers, in the other study[23] used the strain gradient theory to investigate the free vibrations of MT that is modeled by the cylindrical shell, and by extraction and solving equations, and taking into account the cytoplasm environment in the elastic form, frequencies are obtained for various radiuses. Also, MT frequencies of different thicknesses have been investigated.

Another remarkable research in this area was carried out by Farajpour et al.[24] They studied how to smart control MT using the piezoelectric cylindrical shell. Since the discovery of harmful and biodegradable effects of carbon nanotubes in recent years, piezoelectric has been considered as a substitute for nanoelectromechanical topics. Therefore, in the study, for the purpose of smart control of the vibration and buckling of the MT, a piezoelectric nanoshell modeling has been suggested. The cytoplasm surrounding is modeled as a Pasternak foundation. For modeling of nanoshells using Kirchhoff-love assumptions accompanied with Eringen's nonlocal theory and Hamilton principle. Corresponding relations are extracted and solved numerically using differential quadrature method. The strength of this research is a comprehensive study of the subject and the use of accurate theories based on continuum mechanics. Of course, like other earlier studies in this field, research developed only using modified methods of continuum mechanics and the parameters are not calibrated.

Jin et al.[25] studied the transverse vibrations of MT using finite elemental methods and modeling the cell's cytoplasm environment as a thermoset polymer. The dimensional effects on the flexural modulus of MT are studied. Also, experimental formulas based on cell mechanics, along with finite element methods, are exerted and their results are compared. In comparison with previous similar continuum-based studies, this study appears to be more exact. Because in addition to the finite element method, the empirical relations of cell mechanics have also been applied, and the results of these two methods obtained are close to each other. Heireche et al.[26] have applied the nonlocal theory to extract lateral vibration equations of MT. Then, they solved them using exact Navier solution for simple–simple support. Ghavanloo et al.[27] extracted the relations with nonlocal model and applied Navier–Stokes equations for modeling the cytoplasmic fluid around the MT; then, by considering the boundary conditions with torsional springs accompanied with simple–simple support, they obtained both analytically and numerically solved equations. In this study, the effects of changes in Young modulus on the damped frequency of MT are scrutinized too.[28]

4.1.3 CURRENT STUDY

The main target of this study is determining the mechanical and vibrational properties of MTs using molecular dynamics simulation. In general, experimental studies in the nanoscale have many limitations and problems that make empirical research on the mechanical behavior of the MTs so difficult. These restrictions include the following:

1. Limitations due to the nanometer dimensions of MT
2. Expenditure should be incurred for required laboratory equipment
3. The difficulty of applying precisely the boundary conditions and various support in nanometer dimensions
4. Lack of exact tools and suitable laboratory equipment and methods for loading MT protein and measuring the force and displacement
5. The restrictions for providing the same environmental conditions between laboratory studies and intracellular conditions

As a result, due to the restrictions and problems mentioned above, in experimental studies, computer modeling and simulation are important in studying the mechanical behavior of MTs. In the other words, simulation, without enduring of many expenditures allows modeling of the mechanical behavior of MT. Therefore, today, the study of molecular and atomic phenomena has attracted the attention of many researchers by simulating it. In order to use the beam model for MT, it is necessary to achieve dimensions corresponding to the beam model; with the expansion of the length of the studied system, the use of the all-atom model requires very long time and advanced hardware devices. As the need to reduce the cost, computational time, increase in speed, and also expand the length of the studied system, coarse-grained molecular dynamics method is applied in this research.

The molecular dynamics method is used to simulate systems with multiple angstroms up to 100 nm. In principle, in smaller scales, more precise methods such as quantum mechanics should be considered and in higher scales, semicontinuum methods are used. In the dimensions mentioned, there is an optimum rate of precision and speed for molecular dynamics simulations. Due to the fact that biological samples are often soluble in their equilibrium state, simulating various processes in an aqueous environment is very important. Therefore, methods of solvating system will be described as follows:

4.1.4 SOLVATED ENVIRONMENT

Consider the soluble environment by default. In general, the addition of water has two major effects on the system: first, the van der Waals forces between the water and our molecule make it more difficult to move, and in the second, the electrostatic forces are changed and reduced. In two ways, the solvent environment can be modeled as follows.

4.1.4.1 EXPLICIT SOLUTION

The first method that comes to mind is the most accurate way of exploring water explicitly in the environment. For this purpose, it is sufficient to use a simple code or VMD software to correctly determine the volumetric water molecules so that the energy of the system is minimal. The molecular density is approximately equal to 100–120 molecules of water in a box of 1 nm in 1 nm per 1 nm. Some energy minimization can be done to ensure the initial direction of the layout, then put it in a random manner. A very negative point of this method is the time-consuming nature of that. Simulation speed is usually slowed up even 10 times less than the vacuum environment. But, as compared with other methods of considering a fluid such as an implicit solution and other methods that introduce water into equations by applying variations in the equations, it gives priority.

4.1.4.2 COARSE-GRAINED SOLUTION

In general, the coarse-grained method acts in such a way that several atoms from a molecule are rigid in a bead and form relations on the basis. This will increase the speed of simulation in two directions. Firstly, because the number of atoms decreases, and secondly, it can increase the simulation time step. Therefore, this method can be used to model large systems or systems that require a lot of time. Of course, the term "coarse grains" is very general and refers to any method that considers the set of atoms as a group.

Figure 4.4 shows how to categorize atoms in shape-based coarse graining (SBCG) method and Figure 4.5 shows how to do residue-based coarse graining (RBCG) which is suitable for most of biomolecules because of their polymeric structures.

4.1.5 MOLECULAR DYNAMIC MODELING AND SIMULATION OF MT

The purpose of this part is to study the process of modeling and simulation of MT using molecular dynamics method and describing different parts of it. Initially, the process of constructing the model and extending it to the dimensions of the system will be dealt with. In order to simulate the system and the environment like the real condition of MT inside the cell, it is necessary to solvate the system and create the corresponding conditions related to the temperature and intracellular pressure for the simulated system. The energy minimization processes will be expressed in terms of temperature equilibrium and pressure setting stage for the system, and finally, the method of loading and extracting vibration data and mechanical properties will be described.

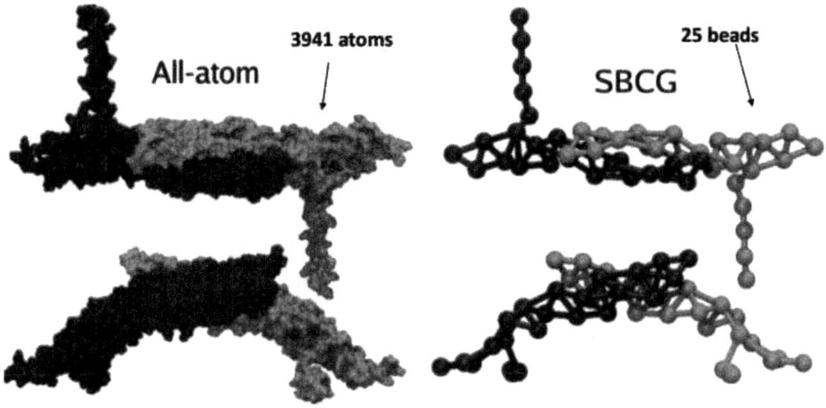

FIGURE 4.4 Atomic categorization in the SBCG method.

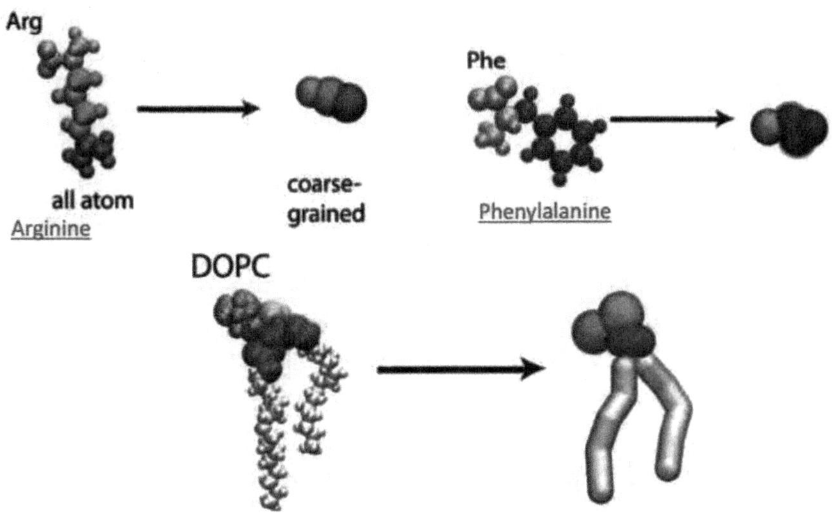

FIGURE 4.5 **(See color insert.)** Atomic categorization based on the RBCG method.

4.1.6 PRIMARY REQUIREMENTS FOR MODELING

In order to investigate the vibrations and mechanical properties of MT and extracting its natural frequencies, it is necessary to construct a precise model that is consistent with the reality of the nanostructure. For this purpose, the all-atom model of one loop of MT made by Wells et al.[3] was chosen which was carefully designed and validated using molecular dynamics studies. In order to construct the initial model, molecular dynamics software is required that would be compatible with biological systems and can be properly modeled on the systems. As the result, at first, the nanoscale molecular dynamics (NAMD) software (or not another molecular dynamic program) was considered, which is specifically designed to study and simulate biological systems. Hence, this software was considered as the initial choice. But, with the advent of research and study of various force fields and the decision to build a larger system, it was found that large-grain models of high dimension require the use of adaptive force fields that can be used with a large coarse grained system with high number of atoms.

The NAMD software is not a suitable choice for this research. Because all-atomic force fields, such as CHARMM, require a highly advanced computing and long run time for dimensions such as the system in question. As the next choice of Martini's force field, which is a suitable choice for coarse-grained models and adapted to the RBCG was chosen. The main idea of the Martini method is to divide the entities by polar or nonpolar. In this method, discontinuous interactions based on free energy between atoms are determined using laboratory studies. Also, interactions are modeled by matching with all-atom simulations.

The mapping used in this method is a four-to-one mapping. In a way, on average, all four heavy atoms, along with the hydrogen atoms attached to it, are mapped to a coarse aggregate. In this method, for simplification and modeling, only four possible states are defined: (1) polar, (2) non-polar with the ability to form a hydrogen bond, (3) non-polar without the ability to form hydrogen bond, and (4) with charge. These four main groups are also subdivided into other subgroups, based on the ability to bind hydrogen. Four subgroups: giver, receptors, both, and the fourth, neither giver and nor receptor, have been proposed. The polarity of each seed can also be graded in these subgroups.

Force fields to study constraints and maintain links, until the system saves its shape and state from stable and consistent with their reality, use algorithms that perform this must-do. One of the most important of these algorithms is linear constraint solver (LINCS). Another algorithm named SHAKE (it is a widely used algorithm to impose general holonomic constraints during molecular simulations) is less accurate than the LINCS and mainly focuses on angles whereas LINCS consider all angles and their links and every change in them. LINCS algorithm report any collapse or instability in the structure during the simulation process. Martini's force field also uses this algorithm. These algorithms have been encoded in another molecular dynamics software called the Groningen Machine for chemical simulations (GROMACS). Therefore, this software was chosen to develop this research. In addition, GROMACS molecular dynamics software is essentially a biologically applied one and has great application in the simulation of biological processes. It can be used to obtain precise information and details at the molecular level that could not be achieved by experimental methods. Specifically, GROMACS software is highly capable of simulating protein-based systems.

4.1.7 MODELING AND COARSE GRAINING OF MT

In order to construct the system, the MT proteins were needed. Therefore, by searching in the literature the all-atom model of one loop made by Wells et al.[3] was found (Fig. 4.6c).

As previously mentioned, in order to create a larger system of longer lengths, it is necessary to put a number of loops together for satisfying the conditions for the use of beam models and the benefit from beam equations. Expanding the system makes a high number of atoms and causes restrictions for simulation. In order to reduce the number of beads, it is necessary to convert the model of all-atom to a coarse-grained one. It decreases the number of beads and brings the system to possible simulation conditions.

In order to make coarse-grained model, first of all, it is necessary to recognize the second structure of proteins and the shape and exact coordinates of them. It could be done using the definition of secondary structure of proteins (DSSP) software. The software provides accurate information about structure of proteins. In the following sections, using the martinize.py software, the coarse-grained system is produced. This software makes coarse-grained model from the all-atom system using Martini method.[29] Following that, the number of beads per loop of the system decreased from about 174,000 to 24,635. By this stage, only one coarse-grained loop has been made (Fig. 4.7); then a larger system should be developed.

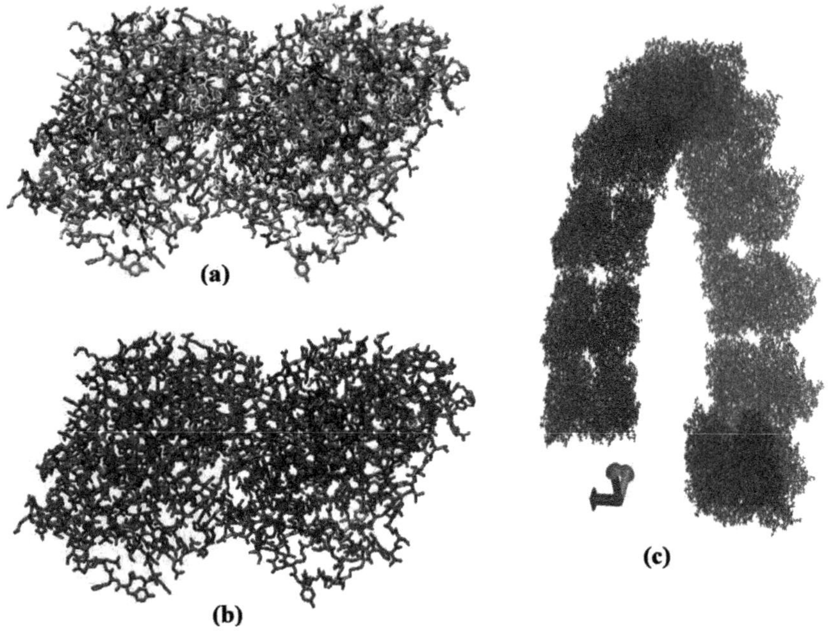

FIGURE 4.6 (See color insert.) (a) Aminoacids in α- and β-tubulin, (b) α- and β-tubulin as a substructure of one loop, and (c) all-atom model of one loop.

To do this, firstly, the distance between the loops should optimize and put the system in a state that energy minimization process could be done. To achieve the equilibrium and proper spacing between the loops, various values were tested from 10 Å to about 1 nm. In the following study, after the construction of a 12 loop systems of different sizes, the energy minimization process was performed using a parametric molecular dynamical file of the Martini model. In this stage, if the distance between the loops have been properly selected, energy of system would have been minimized correctly by eliminating undesirable disturbances due to the unbalanced position of the atoms and the additional forces. As the distance was not obvious, in this study, we had to use the try and error method to determine the desired distance. Different distances were tested and two events were observed:

1. The loops of the MT at a distance of less than 2 or 3 Å. It leads to additional forces stronger than van der Waals bond between the loops because beads are overlapped and it leads to having additional forces. Admittedly, when this happens, system is collapsed.
2. For distances more than 1.1 nm, the van der Waals bonds have not been properly established and the system is not uniformly formed.

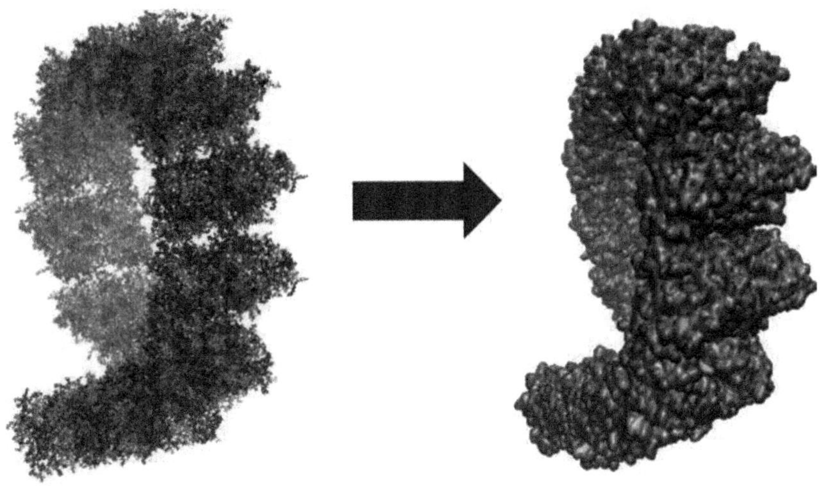

FIGURE 4.7 (See color insert.) All-atom loop (left) and coarse-grained loop (right).

At the end of this step, using various values, the 6.8 Å equilibrium distance was finally determined as the proper value for the distance between the loops.

Given that, the cut-off radius in Martini force field is 11 Å and van der Waals radius for the MT proteins is about 1.2 Å. The distance between the loops seems to be approximately the mean of these two numbers and it can be the logical result.

It should be noted that verification of the distance between the loops is integrating of the whole system and properly minimization of its energy, which was minimized with 6.8 Å distance and a stable and uniform system was achieved. The loops are connected to each other and form the MT. Figure 4.8 shows the isometric and front view of the MT.

To make the model stable and loadable, it needs to be solvated. Continuing the study, this will be attended.

FIGURE 4.8 (See color insert.) Front view of MT (left) and isometric view of MT (right).

4.1.8 SOLVATING THE SYSTEM

MT is naturally solvated in water in normal conditions. In this research, the model has been attempted to make the model conform to the actual and normal conditions of the system within the cell. As it is obvious, we need to use periodic boundary conditions for water modeling and simulation of the whole environment, which requires careful attention to following two points:

1. The distance between the water atoms from the periodic boundaries is very important because if this distance is low, the system is too volatile; because of the empty space, holes are created in the system. Therefore, with great effort and care, this distance should be carefully adjusted.
2. The water is such that the MT does not discharge water during the processes of balancing the temperature and pressure due to expansion and contraction. Therefore, after examining different values and their results, 0.23 nm distance, in the same amount as the van der Waals gradient radius in the system, was properly solved and minimized.

So far, several methods have been proposed for coarse-graining of water, in which usually three to five molecules of water are considered as an independent unit with a specific load in polar or nonloaded models, for example, in the Martini model. Following previous stages, to solvate the system, the water GRO (structure and trajectory) file was obtained from the Martini site[30]; using the file, the desired coarse-grained solvate model was made. Figure 4.9 shows the solvated system. For more clarity, the MT is shown by yellow color in Figure 4.9 and water molecules are shown by blue color.

FIGURE 4.9 (See color insert.) Solvated MT with water.

Since solubilizing the dry system requires stable and a minimum energy state, we implemented solvating process using a minimized dry system. After solubilizing, the number of beads reached to 1,335,145 of which 1,039,525 are water beads and the rest are protein beads. If the use of the all-atom model was done, the number of beads would reach to more than 8 million. This makes it impossible for any simulation process due to hardware constraints with simulated time required. In the following study, to bring the system to the desired temperature and pressure, it is necessary to carry out separate processes for the NVT and NPT, as will be described further.

4.1.9 NVT

System have been made and solvated in previous parts. Following the research, at this stage, by creating a temperature bath around the system, we bring it to the desired temperature inside the cell, equal to 310 K. To accomplish this, a molecular dynamics parameters (MDP) file was first created to enter the data for the simulation process into software. In this file, constraints on main polymer chains of proteins were applied because of the high volatility of the atoms at this stage due to the change of the temperature from 0 to 310 K; it is possible to disrupt and destroy the structure during the work. The number of steps was 50,000 and the Verlet algorithm was used for the cut-off radius. Brandson thermostat was applied for this process and periodic boundary conditions were considered. At the end of this stage, a GRO file was obtained from the software, which included the speed of the every bead in three directions at 310 K.

4.1.10 NPT

After extracting the output file of the NVT (the constant-temperature, constant-volume ensemble) stage, which will bring the system to the desired temperature and the speeds corresponding to that temperature for the beads, enter the next step to adjust the pressure and bring the system to the corresponding pressure in its natural and realistic condition. A MDP file for the NPT (the constant-temperature, constant-pressure ensemble) process was made. It is worth noting that the pressure inside the cell is fluctuating around 1 atm. Target pressure in the NPT process is set to 1 bar to allow the molecular dynamics software to overcome the atomic and molecular interactions of the system at the desired pressure. In this step, Parinlo–Rahman barostat and periodic boundary conditions were considered.

According to the previous steps, the output file of NPT process contains the correction speeds, coordinates, and dimensions of the new simulation box, which is the same as the previous step by applying a pressure bath around the system. In the following, by extracting the NPT phase output file, which is the system's specific temperature and pressure, the system is prepared to extract the desired mechanical parameters by performing simulations for determination of mechanical properties of MT. Practically, the transverse vibration of MT is similar to the vibrating behavior of a cylindrical nanobeam. Therefore, we can consider the model as a beam to investigate its beam-like behavior.

4.1.11 THE POSSIBILITY OF LOADING AND EXTRACTING MECHANICAL PROPERTIES USING A DRY SYSTEM

The steps for minimizing energy and the temperature equilibrium (NVT) described in the previous sections were initially performed for a dry and insoluble model, and then a dry system of 12 balanced rings at a temperature of 310 K. It is worth noting that in the absence of water, there was no need for the NPT stage because due to the presence of the system in a vacuum and absence of solvent, it is not possible to determine the pressure bath for the system, so instead of performing the NPT phase, repeat the NVT stage for the second time with NPT (MDP file). Of course, values and constants related to pressure and condensation are eliminated. The significance of this step is to measure the temperature desired for the system and deliver the model made at equilibrium and temperature conditions.

By performing this step, the desired dry system was obtained for loading. Using various parameters and constants, it was figured out that MT, as it was predicted, in the insoluble conditions loses its stability and suffered severe contraction and rupture. Because in the absence of a fluid environment, it is natural that internal molecules, due to intermolecular forces, tend to accumulate and fill the interior of the system, and the intra-system van der Waals bonds in the absence of water cannot maintain the shape of the system. It is noteworthy that the system's shape and structure can be kept stable by applying constraints to the system. However, since in this research aim is to extract mechanical properties and to investigate the vibrational behavior of MT, the additional constraints in addition to considered boundary conditions disorder the intrinsic structure of the system. Therefore, without considering any excessive constraints, loading and simulating of the system will be described.

4.1.12 BENDING VIBRATION OF MT

As previously mentioned, among the three major parts of the cellular cell structure, MT plays an important role in bearing mechanical loads and thus provides cellular strength. The mechanical and vibrational behaviors of the cytoskeleton and the cell itself are largely related to this nanostructure. Inside the cell, MTs are considered pathways for the transportation of protein engines such as kinesin and dynein, and in some way play the same role as railroads for public transport. Intracellular transport, such as entering foods, viruses, nanoparticles, drugs, etc., occurs with the transition upon MTs inside the cells.

It is obvious that transporting loads on the MT lead to the bending and transverse vibrations of this nanostructure. On the other hand, in recent years, a wide range of treatments for illnesses, such as cancer and AIDS, with the knowledge of drug-to-cell infiltration and control of cell proliferation has been considered, which could lead to vibration of MT and bilaterally interact with that. Therefore, this study emphasizes the transverse vibrations of this nanostructure and further elaborates on how to load and extract flexural rigidity and effective mass, and finally, the extraction of the natural frequencies of this nanostructure.

4.1.12.1 EXTRACTION OF FLEXURAL RIGIDITY AND NATURAL FREQUENCY FOR CLAMPED–CLAMPED STATE

We know from the theory of vibrations that the frequency and vibrational relations are necessary for a structure dependent on the dimensional and boundary conditions considered for the system. Therefore, in order to investigate the vibrational behavior of MT, it was necessary to consider various boundary conditions for this nanostructure. In this research, it was attempted to model the number of further loops and thus larger dimensions to reach a higher aspect ratio or acceptable length-to-thickness ratio for satisfying requirements to use Timoshenko beam model. It should be noted that adaptation of a theoretical model with biological reality is not quite exact, but with approximations, the actual behavior of the biologic structures can be approached nearly to theoretical models. MT could be modeled close to clamped–clamped boundary condition according to its natural situation inside the cells with different equivalent lengths. As a result, this considered boundary condition can be an appropriate approximation. Thus, by applying constraints on the proteins of first and last loops by hard springs in three directions with stiffness coefficient of 100 KJ/mol.nm^2 (in all directions), the clamped–clamped model was made.

In order to extract the first natural frequency, as it follows $\omega = \sqrt{k/m}$ relation, it needs to derive the flexural stiffness of the MT and its effective mass (for its first bending mode shape). In order to obtain the system-equivalent flexural stiffness, the relation $k = f/y$ was used. A distributed uniform load was exerted on all unconstrained proteins to reach the nanobeam to its first bending mode with an amount of deflection that does not cause to bring the nanobeam to its nonlinear state. In order to load MT using the GROMAX software and a solvent system in the desired and simulated intracellular temperature and pressure, we constructed a parameter file related to load loading on the system. Force was applied by connecting the spring to selected atoms and then pulling them. To implement this step, we first need to divide the atoms into three categories:

1. We want to load on them
2. Atoms that we do not want to load on them
3. Atoms that we want to stay fixed.

For the first case, all atoms of proteins from the second to the tenth loop were selected (because the system has 12 loops). The second group is all water atoms that are placed nine times on them and not bound. The third group was selected by applying the proteins of the loop to the beginning and the end (loop of 1 and 12). Following that, we have a clamped–clamped nanobeam with length of 95 nm. Then, it is necessary to select different speeds and stiffnesses for the pulling springs. For this reason, a stiffness of 1000 KJ/mol.nm^2 was selected and also the speed of 0.01 nm/ps was chosen for the pulling springs. The loading was carried out for 10 ns. Finally, the output was plotted, and force–time and velocity–time diagrams were drawn. It was determined that due to the high speed and spring coefficient of the striker, the noise level in the graphs was high and no satisfactory results were obtained. So, proceeding with different speeds and timing and different simulation time, a stiffness of 700 KJ/mol.nm^2 and speed of 1/1000 nm/ps results, submitted the least noisy diagrams. System after loading is shown in Figure 4.10.

FIGURE 4.10 MT after loading with solvent.

Using the linear approximation of the obtained graph, we can determine the flexural stiffness of the modeling system. As considering the area of the graph where the MT bending is closer to the linear behavior, it could be available to determine the stiffness carefully (Fig. 4.11).

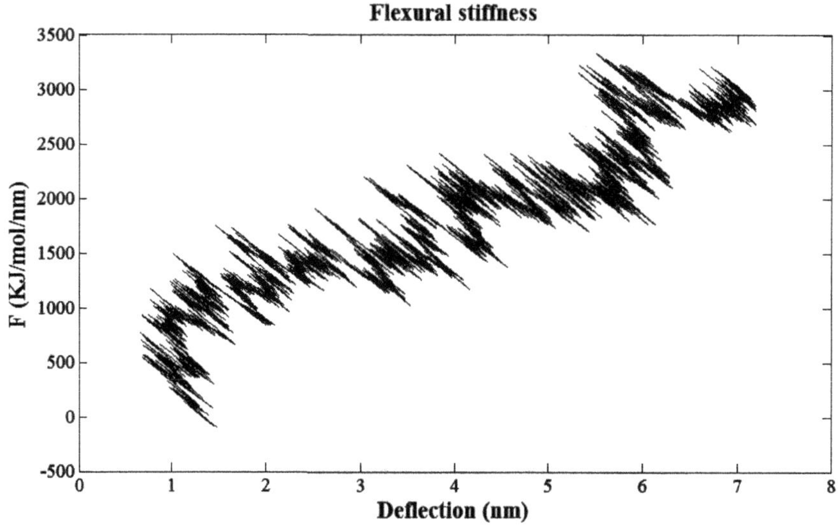

FIGURE 4.11 MT flexural stiffness.

Further, to calculate the natural frequency, it was necessary to obtain effective mass for the system. We consider the theory of vibrations $M_{\text{eff}} = \int_0^L \rho A \phi(x)^2 \, dx$, in which ρ is the volume mass and ϕ function of the first mode and A is the equivalent area. It should be noted that this approximation is related to the first bending mode of Euler beam for effective mass, but due to the proximity of the first frequency of Euler and Timoshenko beams, to prevent the complexity of calculating effective mass, it can be used with good precision only for the effective mass. Also for the bending vibrations, following equation is used:

$$X\phi = (\cos\text{h}(ax) - \cos(ax) - b(\sinh(ax)) - \sin(ax)) \tag{4.1}$$

Using eq (4.1), the effective mass was extracted for a length of 95 nm and the effective stiffness was obtained by simulating. In the following equation, natural frequency of clamped–clamped beam (f_{cc}) has been calculated:

$$f_{cc} = 1.474062051751557e+09 \text{ Hz} \tag{4.2}$$

By using the deflection and slope relations for clamped–clamped beam:

$$E = 577.4546706636134 \text{ MPa}, y = \frac{WX^2}{24El}(L-x)^2 \tag{4.3}$$

It should be noted that due to the orthotropic nature of MT, Young modulus obtained from this method is different from Young modulus obtained from the stress–strain test. And in fact, this is a linear combination of the modulus of elasticity in different directions.

After the clamped–clamped boundary condition, the second state close to the natural and realistic conditions of MT is clamped-free condition. On the other hand, simulating a simple joint using molecular dynamics simulation due to its weak bearing (and its possible high movement) gives a great probablity of error. As a result, using molecular dynamics simulation clamped–clamped and clamped-free conditions can be well modeled.

For modeling and simulating MT as a clamped-free beam, firstly, one loop should be constrained and other loops would be loaded. Due to the weakening of this state and for preventing large deformation and increasing the nonlinearization of results, a velocity equal to 0.001 nm/ps and a spring with a coefficient of KJ/mol.nm^2 were considered. Then, by using these data, the corresponding MDP file was constructed.

Using simulation for 10 ns, the flexural stiffness was calculated. Then, in order to calculate the effective mass involved in this state, the relationship between the effective mass for the cantilever beam in the first mode, $M_{eff} = \int_0^L \rho A \phi(x)^2 \, dx$ and the first mode shape function of clamped-free beam are used:

$$\phi(x) = \sinh \beta x - \sin \beta x - a(\cosh \beta x - \cos \beta x) \tag{4.3}$$

In the following equation, by using the relation $\omega = \sqrt{k/m}$, the first frequency of MT as a clamped-free nanobeam (f_{cf}) for 103 nm length is obtained:

$$f_{cf} = 4.948688515848953e+08 \text{ Hz} \tag{4.4}$$

The frequency of the clamped–clamped boundary condition, obtained in the previous section is, according to the prediction, significantly higher than the frequency obtained for this state because the clamped–clamped condition is more restrictive than the clamped-free one and has a higher flexural stiffness. It should be noted that due to the differences of applied boundary conditions on the loops for simulation and loading of each state, the characteristic length of the model is different for the two conditions: 95 nm for the clamped–clamped state and 103 nm for the clamped-free state.

4.1.13 DETERMINING MECHANICAL PROPERTIES

In following the study, we should have determined the mechanical properties of MT such as Young modulus and the shear modulus. Therefore, by applying constraints springs on the beads of the first loop and fixing and loading it along the longitudinal direction on the last loop, MT's stress–strain curve is extracted (Fig. 4.12). As shown in the stress–strain diagram, the behavior of the MT is initially elastic with a high stiffness, and with the continuation of loading over time, the slope decreases and enters the plastic range. Early deviation: the initial part of the graph is related to very small disturbances due to numerical error and the initial response of MT to the loading.

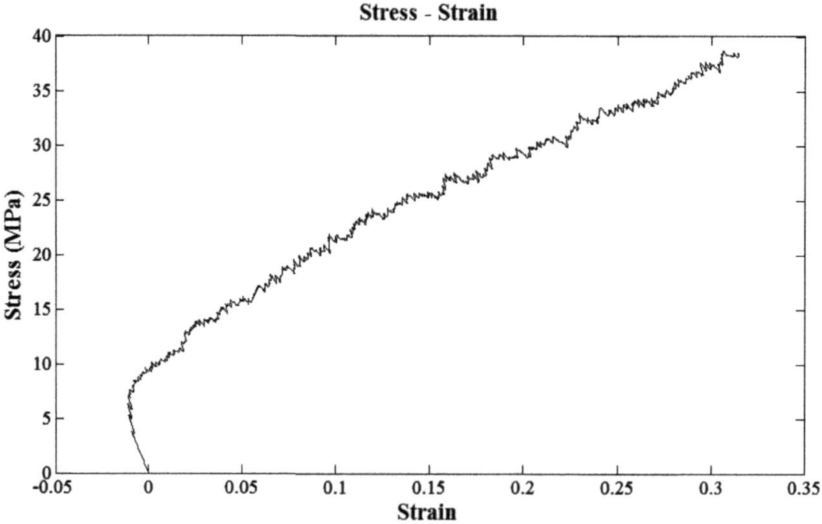

FIGURE 4.12 MT's stress–strain diagram.

By continuing the axial loading on the MT, Young's modulus decreases due to the nonlinear nature and complex behavior of the nanostructure. In the early stages of Young's modulus, this structure was about 1.51 GPa, followed by a linear approximation of about 1 GPa, and finally reached a plastic region of less than 1 GPa. This reduction process is fully justifiable due to the structure of MT since by increasing the force, the van der Waals bonds between the proteins are broken down and the stiffness decreases over time, but in the early stages of loading, there are still van der Waals connections and provide a high degree of stiffness. The numerical results obtained from this simulation are highly consistent with previous studies.

After extraction of the Young modulus, it is also necessary to extract the corresponding shear modulus. Using the following important relation and considering 0.3 for the Poisson coefficient[31] and 1 GPa approximation for Young modulus (according to the stress–strain graph for elastic region), the shear modulus was obtained:

$$G = \frac{E}{2(1+v)} = \frac{1}{2(1+0.3)} = 0.384 \tag{4.5}$$

4.1.13.1 LONGITUDINAL NATURAL FREQUENCY OF MT

From the simulation, the Young modulus was extracted and, given that loading is axial, it is possible to extract the longitudinal stiffness as follows:

$$k = \frac{AE}{l} = \frac{1.068141502220530e-16 \times 1e+9}{103e-9} = 1.03703058680126\frac{N}{m} \tag{4.6}$$

On the other hand, for extracting the effective mass, the relation, $M_{\text{eff}} = \int_0^L \rho A \phi(x)^2 \, dx$ and the first mode shape function of bar are used as follows:

$$\phi(x) = \sin\frac{\pi x}{2l} \tag{4.7}$$

And by the use of the relation $\omega = \sqrt{k/m}$, the longitudinal frequency (f_1) of the 103 nm bar is:

$$f_1 = 1.802350442056579e+09 \tag{4.8}$$

4.2 CONCLUSION

In this study, the mechanical properties and vibrational behavior of MT were investigated using molecular dynamics simulation. The results of the mechanical properties are in good agreement with the precise work done, and the results of the vibrational behavior have given proper accuracy in the molecular dynamics-based simulation method which can be a good basis for future researches in the field of cellular mechanics, biotechnology, nanobiomechanical medicine, and pharmaceutical studies. Identifying the properties and behavior of MT and the cell framework helps to understand the high spectrum of interstitial behaviors and intercellular interactions, as well as cell proliferation.

Eventually, successes were obtained to extract a range of mechanical properties of MT, and in addition, the complex and nonlinear nature of MT and its high sensitivity to the internal thermodynamic conditions were found. As MT plays an important role in cell profilation and deformation, MT's stability, birth and collapse, mechanical vibration, and its response to mechanical loading all do a significant task in the cellular processes. The importance of studies in the field of cellular mechanics help us not only to know more about life mechanisms but also to design and fabricate a wide range of important nanoapplications such as nanosensors, nanorobots, nanomachines, and many other important medical scientific tools. As MT plays a role similar to the railroad in the internal cells environment, the present study, by extracting the mechanical properties and analyzing the vibrational behavior in the range of about 100 nm, can be used by scientists involved in the field of injecting and transporting medicine inside the cells.

KEYWORDS

- **microtubule**
- **molecular dynamics simulation**
- **nano application**
- **cellular mechanics**

REFERENCES

1. Solomon, E.; Martin,, C.; Martin, D. W.; Berg, L. R. *Biology*, 10th ed; CA Brooks: Belmont, 2011.
2. Lodish, H.; Berk, A.; Kaiser, C. A.; Krieger, A.; Bretscher, A.; Ploegh, H.; Amon, A.; Martin, K. C. *Molecular Cell Biology, 8th ed.;* 2016.
3. Wells, D. B.; Aksimentiev, A. Mechanical Properties of a Complete Microtubule Revealed Through Molecular Dynamics Simulation. *Biophys. J.* **2010,** *99*, 629–637.
4. Zheng, W.; Chakraborty, S. Decrypting the Structural, Dynamic and Energetic Basis of Kinesin Interacting with Tubulin Dimer in Three ATPase States by All-atom Molecular Dynamics Simulation. *Biophys. J.* **2015,** *108*, 134a.
5. .Gurtin, M. E.; Murdoch, A. I. A Continuum Theory of Elastic Material Surfaces. *Arch. Ration. Mech. Anal.* **1975,** *57*(4), 291–323.
6. Cuenot, S.; Frétigny, C.; Demoustier-Champagne, S.; Nysten, B. Surface Tension Effect on the Mechanical Properties of Nanomaterials Measured by Atomic Force Microscopy. *Phys. Rev. B*, **2004,** *69*, 165410.1–165410.5.
7. Dingreville, R.; Qu, J.; Cherkaoui, M. Surface Free Energy and Its Effect on the Elastic Behavior of Nano-sized Particles, Wires and Films. *J. Mech. Phys. Solids* **2005,** *53* (8), 1827–1854.
8. McFarland, A. W.; Poggi, M. A.; Doyle, M. J.; Bottomley, L. A.; Colton, J. S. Influence of Surface Stress on the Resonance Behavior of Microcantilevers. *Appl. Phys. Lett.* **2005,** *87*, 053505.
9. Wang, G. F.; Feng, X. Q.; Yu, S. W. Surface Buckling of a Bending Microbeam Due to Surface Elasticity. *Europhys. Lett.* **2007,** *77* (4), 44002.
10. He, L., Lim, C.; Wu, B. A Continuum Model for Size-dependent Deformation of Elastic Films of Nano-scale Thickness. *Int. J. Solids Struct.* **2004,** *41* (3–4), 847–857.
11. Civalek, O.; Demir, Ç.; Akgöz, B. Freevibration and Bending Analyses of Cantilever Microtubules Based on Non Local Continuum Model. *Mathematical Comput. Appl.* **2010,** *15* (2), 289–298.
12. Wang, C. Y.; Ru, C. Q.; Mioduchowski, A. Vibration of Microtubules as Orthotropic Elastic Shells. *Physica E* **2006,** *35*, 48–56.
13. Ghorbanpour, A.; Abdollahian, A.; Ghorbanpour-Arani, A. H. Modified Couple Stress Theory for Vibration of Embedded Bioliquid-filled Microtubules Under Walking a Motor Protein Including Surface Effects. *J. Solid Mechanics* **2015,** *7* (4), 458–476.
14. Shen, H. S. Nonlinear Vibration of Microtubules in Living Cells. *Curr. Appl. Phys.* **2011,** *11*, 812–821.
15. Shen, H. S. Nonlocal Shear Deformable Shell Model for Postbuckling of Axially Compressed Microtubules Embedded in an Elastic Medium. *Biomech. Model Mechanobiol.* **2010,** *9*, 345–357.
16. Xiang, P. L.; Zhang, W.; Liew, K. M. A Mesh-free Computational Framework for Predicting Vibration Behaviors of Microtubules in an Elastic Medium. *Composite Struct.* **2016,** *149*, 41–53.
17. Motamedi, M.; Mosavi, M. M. Nanobiomechanical Properties of Microtubules. *Int. J. Nanosci. Nanotechnol.* **2015,** *11*, 179–184.

18. Gao, Y.; Lei, F. M. Small Scale Effects on the Mechanical Behaviors of Protein Microtubules Based on the Nonlocal Elasticity Theory. *Biochem. Biophys. Res. Commun.* **2009,** *387*, 467–471.
19. Ghorbanpour Arani, A.; Abdollahian, M.; Jalaei, M. H. Vibration of Bioliquid-filled Microtubules Embedded in Cytoplasm Including Surface Effects Using Modified Couple Stress Theory. *J. Theoretical Biol.* **2015,** *367*, 29–38.
20. Baninajjaryan, A.; Tadi Beni, Y. Theoretical Study of the Effect of Shear Deformable Shell Model, Elastic Foundation and Size Dependency on the Vibration of Protein Microtubule. *J. Theoretical Biol.* **2013,** *381*, 142–157.
21. Mokhtari, F.; Tadi Beni, Y. Free Vibration Analysis of Microtubules as Orthotropic Elastic Shells Using Stress and Strain Gradient Elasticity Theory. *J. Solid Mechanics* **2016,** *8*, 511–529.
22. Zeverdejani, M. K.; Tadi Beni, Y. The Nano Scale Vibration of Protein Microtubules Based on Modified Strain Gradient Theory. *Curr. Appl. Phys.* **2013,** 13 (8), 1566–1576.
23. Zeverdejani, M. K.; Tadi Beni, Y. The Nano Scale Vibration of Protein Microtubules Based on Modified Strain Gradient Theory. *Curr. Appl. Phys.* **2015,** *381*, 47–62.
24. Farajpour, A.; Rastgoo, M.; Mohammadi, M. Vibration, Buckling and Smart Control of Microtubules Using Piezoelectric Nanoshells Under Electric Voltage in Thermal Environment. *Physica B: Phys. Condensed Matter.* **2017,** *509*, 100–114.
25. Jin, M. Z.; Ru, C. Q. Localized Vibration of a Microtubule Surrounded by Randomly Distributed Cross Linkers. **2013,** *136*, 71–102.
26. Heireche, H.; Tounsi, A.; Benhassaini, H.; Benzair, A.; Bendahmane, M.; Missouri, M.; Mokadem, S. Nonlocal Elasticity Effect on Vibration Characteristics of Protein Microtubules. *Physica E* **2010,** *42*, 2375–2379.
27. Ghavanloo, E.; Daneshmand, D.; Amabili, M. Vibration Analysis of a Single Microtubule Surrounded by Cytoplasm. *Physica E* **2010,** *43*, 192–198.
28. http://alevelnotes.com/content_images/i3_peptide_bond.png (accessed Jan, 2017)
29. Monticelli, L. The MARTINI Coarse-grained Force Field: Extension to Proteins. *J. Chem. Theory Comput.* **2008,** *4* (5), 819–834.
30. www.cgmartini.nl/images/applications/water/water.gro. 2016. (accessed Jan, 2017)
31. De Pablo, P. J.; Shaap, I. A. T.; Mackintosh, F. C.; Shmit, C. F. Deformation and Collapse of Microtubules on the Nanometer Scale, *Phys. Rev. Lett.* **2003,** *91*, 098101.

CHAPTER 5

STATICALLY DISSIPATING AND LUMINESCENT ZINC OXIDE/LINEAR LOW-DENSITY POLYETHYLENE COMPOSITES

FILIZ ÖZMIHÇI ÖMÜRLÜ[1] and DEVRIM BALKÖSE[2,*]

[1]*Natural Anne Ekolojik Ürünler Çiğli İzmir, Turkey*

[2]*Department of Chemical Engineering, İzmir Institute of Technology, Gulbahce, Urla, Izmir, Turkey*

[*]*Corresponding author. E-mail: devrimbalkose@gmail.com*

ABSTRACT

Polyethylene is an electrically insulating material which accumulates static electricity. The electrostatic charging of polyethylene can be prevented by adding conductive additives. One of these additives is zinc oxide (ZnO) which could only effectively reduce the resistivity of linear low-density polyethylene (LLDPE) when added up to at least 20% in volume as the present study indicated. Since the mechanical properties were lowered at this high filler level, ZnO was modified for better dispersion in LLDPE by treating with aminopropyltriethoxysilane or polyethylene glycol 4000. However, mechanical and electrical properties were not sufficiently improved for modified ZnO/LLDPE composites than that of pristine ZnO/LLDPE composites. The composites were statically dissipating, white in color, and they emit light at 411, 425, 488–523 nm when irradiated at 375 nm. This property makes them as a new material which is white under visible light and blue–green under ultraviolet light.

5.1 INTRODUCTION

Polymeric materials are generally insulating materials in nature and likely to accumulate the electrostatic charge.[1] The application of conductive particles/nanoparticles to a polymer matrix is expected to induce electrical conductivity. The increase in conductivity of polymers depends on the volume fraction of conductive filler used. Filler particles in polymer start to contact with each other with the increasing of volume fraction of filler and create a continuous path for electron to pass electric current. This is called as percolation theory.[2] Introduction of conducting fillers into polymers may not give good results in terms of electrical properties due to mixing problems. Modifications are applied to filler surface to improve mixing with the polymer. There are many studies showing that modified fillers give better results in terms of conductivity compared to fillers without modification. To improve or change the distribution of the nanoparticles and the compatibility between the nanoparticles and other materials, physical or chemical methods are needed. Surface chemical modification of nanoparticles plays a very important role to reduce the agglomeration. Due to modifier adsorption or bonding on the particle surface, which reduces the surface force of hydroxyl groups, the hydrogen bonds between particles are eliminated preventing the agglomeration.[3]

Polyethylene (PE) can be low-density polyethylene (LDPE), high density polyethylene (HDPE), linear low-density polyethylene (LLDPE) and ultrahigh molecular weight polyethylene (UHMWPE). LLDPE is a substantially linear polymer, with significant numbers of short branches, commonly made by copolymerization of ethylene with short-chain alpha-olefins with density range of 0.915–0.925 g/cm^3.[3] It is an electrically insulating material with 1.02×10^{16} Ω cm resistivity.[4]

Zinc oxide (ZnO) is a most common semiconductive material that has a large surface area, high ultraviolet (UV) absorption, and long life span, and widely used as gas sensor, active filler for rubber and plastic, UV absorber in cosmetics, and antivirus agent in coatings. ZnO have dielectric, piezoelectric, pyroelectric, semiconducting, acousto-optic, optical, electrooptical, nonlinear optical, and photoelectrochemical properties.[5] The resistivity of ZnO was reported to be in the range of 2.9×10^{-2} to 1.55×10^{3} Ω cm depending on particle geometry as reported by Sundhi et al.[6]

ZnO is added to LLDPE as an antistatic filler.[4,7,8] ZnO nanoparticles are treated with some modifiers such as styrene,[9] aminopropyltriethoxysilane (AMPTES),[10] polyethylene glycol (PEG),[10] 3-methacryloxypropyl trimethoxysilane,[11] polymethacrylic acid,[12] oleic acid,[13] polymethyl methacrylate,[7] and amimosilane coupling agent (KH570)[14] to reduce the aggregation among nanoparticles and to improve the congruity between ZnO filler and the polymer matrix. In the case of AMPTES, pH $\approx$ 10.5, condensation into siloxane (Si-O-Si) polymer and the hydrolysis of silane into silanol are observed. However at this pH, the ZnO surface is negatively charged and a high condensation is observed, and after heating, the aminosilanoltriols forms aggregates which are later adsorbed onto ZnO surface.[5] The surface modification of ZnO by previous investigators[5,14] indicated that Zn–O–Si bonds were formed on the surface and maximum two of the four Si–O bonds of AMPTES could react with ZnO due to steric hindrances. Regarding to PEG and ZnO relation, PEG macromolecules bond with the solid surface of the ZnO by the aid of –OH end group and may interact with the PEG through hydrogen bonding. However, the pH of the solution affects the mechanism of the bonding since in low pH values the concentrations of $-ZnOH_2^+$ group and –ZnOH group increase and –ZnO group decrease. At low pH, more polymers could be adsorbed on the surface of the ZnO particles due to the presence of –OH groups in the solution.[15,16] Posthumus et al. modified various oxide nanoparticles using 3-methacryloxypropyl trimethoxysilane.[11] Tang et al. used polymethacrylic acid for nano-sized ZnO (nZnO) modification and dispersion of nZnO in aqueous system was improved.[12] Hong et al. modified nZnO by oleic acid and better compatibility was gained between inorganic nanoparticle and organic matrix.[13] Hong et al. used polymethyl methacrylate for grafting nZnO and applied it in nanocomposite preparation with styrene and better thermal behavior was obtained.[7] Ma et al. modified nZnO surface by KH570 silane coupling agent and the dispersion was improved.[14]

The surface modification of filler particles with AMPTES could be made using different techniques as shown in Table 5.1. Ethanol, toluene, and water are used as solvents.[17–21] The solutions with the AMPTES were added in different amounts, in amounts just sufficient to wet the particles or in excess amounts allowing adsorption of AMPTES. In the first case, the particles and the solution were first kneaded and then dried, whereas in the second case, the particles were separated by filtration. The surface-modified particles were dried at 60–110°C as summarized in Table 5.1.

TABLE 5.1 AMPTES Application Methods.

Solvent	Solvent/ water ratio	Silane % in solution	Filler/ solution ratio, g/cm^3	Filtration or solvent evaporation	Drying temperature, °C	References
Ethanol	1:1	10	1/0.3	Solvent evaporated	110	17
Ethanol	1:1	0.5, 1, 1.5, 2	1/1	Solvent evaporated	110	18
Ethanol	95:5	2.5	–	Filtered	70	19
Toluene	No water	1.25	2/80	Solvent evaporated	80	20
Water	No solvent	2	1.5/50	Filtered	60	21
Toluene	No water	0.5	2/200	Filtered	110	21

Hong et al. investigated the addition of nano- and micron-sized ZnO to low-molecular-weight PE.[7] The percolation threshold value for ZnO–PE composites were found approximately at 30 vol % and addition of 40 vol % ZnO decreased the resistivity to 10^{11} Ω cm. In another investigation, the particle size effect was studied by using nano- and micron-sized ZnO particles, and the electrical resistivity was found at the order of 0.3×10^{13} Ω cm after the addition of 60 vol % ZnO.[8] In contrast, Hong et al. reported the lowest resistivity as 10^9 Ω cm upon addition of 30 vol % ZnO.[7] In fact, Hong et al. also studied ZnO/LLDPE) composites to investigate the particle size effect. Micron-sized (300 nm) and nano-sized (49 and 24 nm) powders were used as fillers in their study. The percolation limit decreased as the particle size of ZnO was decreased. When the interparticle distance was decreased to below 40 nm, tunneling began to occur. The percolation onset occurred at a lower volume fraction as the particle size was decreased, due to decrease in interparticle spacing. The neat polymer volume resistivity was 10^{19} Ω cm and the lowest volume resistivity found was 10^9 Ω cm for 24 nm powder at above 30 vol %.[8] The room temperature conductivity evaluated from the straight-line fit in measured I-V plot and was found to be 1.3×10^{-5} S cm^{-1} and 4.7×10^{-3} S cm^{-1} for pure and 18% ZnO-filled film, respectively.[22] Ozmihci and Balköse[4] investigated the effects of particle size and electrical resistivity of ZnO on mechanical properties and electrical and thermal conductivities of composites made with LLDPE. Micron- (mZnO), submicron- (sZnO), and nZnO powders having resistivities of 1.5×10^6, 1.5×10^9, and 1.73×10^8 Ω cm were used to prepare composites with 5–20 vol % filler. Rather than the particle size of the ZnO, its initial resistivity and aspect ratio affected the resistivity of composites. The resistivity of the LLDPE was lowered from 2.3×10^{16} Ω cm down to 1.4×10^{10} Ω cm with mZnO addition. The composites can be used for electrostatically dissipating applications due to their decreased electrical resistivity.[4]

There are also studies showing decrease of electrical conductivity of LDPE by addition of nano-ZnO.[23,24] Experiments showed that, the trap density was increased by three to five times in the nanocomposite with 0.5–7% ZnO as compared to LDPE. The conduction current of the nanocomposite is decreased to 0.5–0.25 of the value of LDPE both at low and high electric fields.[24] This was explained by fixation of polymer molecules at the polymer–ZnO interphase.

Chang et al. studied the silane-modified ZnO/UHMWPE composites. The elongation at break and tensile strength of modified and unmodified ZnO-filled UHMWPE was decreased with increasing of ZnO filler loading compared to pure UHMWPE.[25] Li et al. investigated the mechanical properties of silane-modified nano-ZnO–LLDPE composites. They found that the elongation at break and the tensile strength of the composite films was increased firstly and then was decreased with increasing ZnO content. The optimal ZnO content was found to be 0.3 wt %, and the maximum tensile strength and the elongation at break were found to be 43.2% and 39.4% higher than pure LLDPE's, respectively.[26] Although 3, 6, 12 wt % concentrations were more successful in terms of particle distribution, at concentrations over 12 wt % particle content, many particles were not embedded in the matrix.[22] At these high concentrations of ZnO, the particles could not make an interaction with LDPE and could not be wetted by the matrix. Consequently, the extrusion process, which was carried out without using any compatibilizer, allowed good dispersions of ZnO nanoparticles in the LDPE matrix at concentrations up to 12 wt %. Over this amounts of ZnO nanoparticles, the use of a compound as compatibilizer can be needed in order to obtain LDPE/ZnO nanocomposites having fine dispersions. As the particle content increased, Young modulus increased and elongation at break decreased due to the hardness of the particles. Moreover, the change in mechanical properties of the LDPE films depends on particle dispersion of nano-ZnO particles. The presence of the well dispersed particles in the matrix provides good interfacial adhesion, and this increased the ability of the interfacial structure to show higher elastic modulus and tensile strength. On the other hand, although the amount of particle content was increased from 3 to 36 wt % within a systematic order, mechanical properties did not changed obviously. This is due to the agglomeration effect of the ZnO particles as verified in the SEM scanning electron microscope (SEM) images of the films. Inorganic nanoparticles tend to agglomerate more than microparticles. The agglomerated particles induce stress concentration and they act as microcracks initiator, and in this regard, high amount of particle concentration could not improve the mechanical properties. Furthermore, the high amount of particle concentration inhibits the flexibility of the molecular chain movement resulting in a more brittle structure with sharp decrease in elongation. As a result, up to 12 wt% of nano-ZnO addition could be accepted as a threshold in improving the tensile strengths of the LDPE/ZnO nanocomposites.[22] Ozmihçı and Balkose[4] indicated that composites up to 20 vol % ZnO had sufficient tensile strength to be used as an engineering material.

Luminescence behavior of ZnO changes with the preparation method and presence of impurities. For example, solvo thermal and amine-aided sol–gel synthesis methods in ethanol, ethylene glycol, 1,4-butanediol and PEG 600 were used by Rezapour et al.[27] The obtained ZnO had green and yellow emission peaks appeared at about 530–535 nm and 640–645 nm, respectively. The origins of the green were recently attributed to oxygen-related defects.[27] In one other study, uniform rod-like and multi-pod-like ZnO whiskers were synthesized by a simple hydrothermal process. The precursor used in the hydrothermal process is a mixture of $Zn(OH)_2$ and PEG. The as-synthesized ZnO whiskers show UV (around 401 nm), blue (around 443 nm) and blue–green (around 484 nm) emission bands when they were excited at 325 nm.[28] Egbuchunam et al.[29] obtained ZnO from zinc chloride and sodium hydroxide. The UV–Visible (VIS) absorption spectrum of ZnO showed an absorption band at 375 nm. The photoluminescence (PL) spectrum of ZnO excited at 380 nm exhibited three emission peaks: one at 424 nm and 490 nm corresponding to band gap excitonic emission and another located at 520 nm due to the presence of singly ionized oxygen vacancies. PL spectra of ZnO nanorods various excitation power have similar features including a sharp UV emission band, two broadened visible emission bands and a red emission band with peaks centered at 375.8, 486.8, 647.8, and 754.6 nm, respectively. The sharp UV emission band is related to the near band edge emission from the direct band gap of ZnO. The two broad visible emission bands can be attributed to the deep level emissions related to the intrinsic defects, such as zinc and oxygen vacancies, zinc interstitials and oxygen interstitials, etc. The origin of the red emission band is still in controversial. Some researchers attributed it to the defects in ZnO as the two broad visible emission band. Some others believed that it is a second order feature of the UV emission band. With the excitation power increasing, the intensity of all emission peaks increased gradually and no peak shifts were observed.[30]

The polymer composites having ZnO had the UV shielding and the luminescence properties of ZnO. The HDPE nanocomposite containing 2.5 wt % ZnO nanoparticles with an average particle size of 25.22 nm had the most optimal optical properties, namely high-visible light transparency and high-UV light shielding efficiency.[31] ZnO nanoparticles of 2 nm in size could be dispersed in PE–co-vinyl acetate evenly. For 265 nm UV irradiation, the transmittances of the films containing 0, 0.2, 0.50, 1.0, and 2.0 wt% of the castor-oil-stabilized ZnO nanocrystals were 71%, 60%, 48%, 38%, and 4%, respectively.[32] Nano-ZnO particles not only absorbed UV light but also scattered visible light. Therefore, the UV absorbance increased with increasing nano-ZnO content of the HDPE films containing up to 2% nano-ZnO, whereas the transmission of the visible light lowered.[33]

In this chapter, composites with statically dissipating and florescent composites were aimed to be prepared from LLDPE and ZnO. The mechanical properties were targeted to be improved by modifying ZnO with AMPTES and PEG 4000. The ZnO–polymer composite sheets were prepared by mixing ZnO and PE in a rheometer and pressing the mixture in a hot press. Morphology, density, water contact angle, electrical conductivity, modulus of elasticity, elongation at break, and the tensile strength and fluorescence spectra of composites were measured.

5.2 EXPERIMENTAL

ZnO powder (Ege Kimya AŞ), characterized in our previous study,[4] was used in the experiments. It has 3.86 μm particle size and 1.6×10^6 Ω cm volumetric resistivity. The particle size distribution of ZnO dispersed in 1:1 w/w water–glycerol solution was determined by using Sedigraph (Micromeritics-5100).

The surface modification of ZnO powders was made with 3-AMPTES (Fluka) and PEG 4000 (Aldrich). 1.08 g PEG was dissolved in 9 cm^3 1:1 v/v ethanol and water solution by mixing with magnetic stirrer with a constant stirring rate by 20 min at 25°C. This solution was kneaded with 30 g of ZnO in a porcelain mortar with a porcelain pestle for 15 min. This sample was dried in fume hood at 25°C for 24 h and then it was kept in an air-circulating oven at 40°C for 24 h. The same procedure was repeated with 2.16 g PEG with 9 cm^3 solution. Modified ZnO particles having AMTES in the same proportions were prepared similarly. The modified ZnO were called as 3% and 6% PEG 4000 and AMPTES samples.

LLDPE with melting point of 111.8°C from Sigma Aldrich was used for preparation of the composites. ZnO particles and PE were mixed using a Haake Rheomixer at 50 rpm for 10 min at 160°C. Pristine ZnO and polymer composites were prepared in 5, 10, 15, and 20 vol %. Composites with LLDPE matrix containing 20 vol % modified ZnO were also prepared by using surface-modified ZnO powders. The mixtures were then uniaxially pressed and well shaped with dimensions of 15 cm × 15 cm × 0.1 cm by using Carver hot press at 6800 kg force and 160°C with a 10 min hold.

ZnO powders were characterized by SEM (Philips XL-30S FE) after coating with gold, Fourier Transform Infrared (FTIR) spectroscopy by KBr disc method (Shimadzu FTIR-8201) and thermal gravimetric (TG) (Perkin Elmer Diamond TG/DTA) analysis by heating up to 1000°C at 10°C/min rate under nitrogen gas flow. The morphology of fracture surface of composites obtained by breaking them after immersing in liquid nitrogen at −190°C were characterized by using SEM (Philips XL-30S FE). Density of the composites was measured by using density kit apparatus (Sartorious). Water contact angles of the composites were measured using Attension theta optical tensiometer with attached camera of KSU CAM 101.

Electrical resistivity values of the composites were measured according to American Society for Testing and Materials (ASTM) Standard Test Methods for DC Resistance or Conductance of Insulating Materials (ASTM D257) using Keithley 6517A electrometer/high-resistance meter connected to 8009 Resistivity Text Fixture sample holder.

The tensile tests of the composites were made according to ASTM standard 638 by Shimadzu AG-I 250kN using 5 mm min^{-1} stretching rate at room temperature. Dog bone samples were cut from composite plates by using CEAST automatic hollow die punch for the determination of mechanical properties of the composites.

The absorption spectrum of ZnO dispersed in water was determined using UV–Vis spectrometer (Perkin Elmer Lambda 45). The fluorescence spectra of ZnO pellet that was obtained by pressing and heat treating at 1100°C, and composite sheets were determined by using the fluorescence spectrometer Varian Cary Eclipse by exciting the samples at 380 nm and the data were recorded in the 390 and 600 nm intervals.

5.3 RESULTS AND DISCUSSION

5.3.1 CHARACTERIZATION OF ZnO

ZnO is a white and fluffy powder which is insoluble in water. The particle size distribution of ZnO used in the present study is shown in Figure 5.1. The mean particle size of ZnO powders was found to be 3.86 μm with 98.6% under 10 μm, 82.2% under 5 μm, 28.1% under 2 μm, and 16% under 1.5 μm.

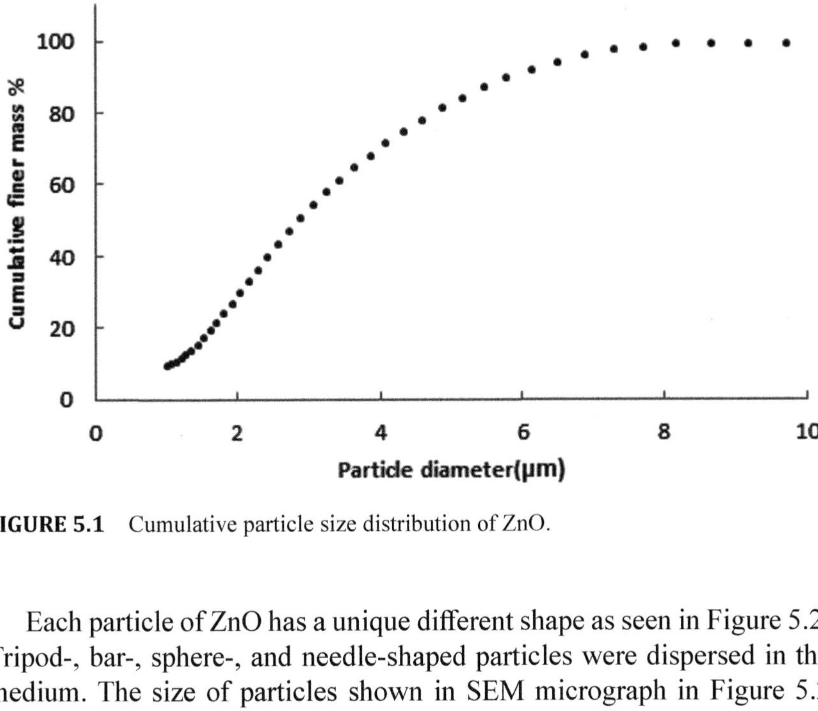

FIGURE 5.1 Cumulative particle size distribution of ZnO.

Each particle of ZnO has a unique different shape as seen in Figure 5.2. Tripod-, bar-, sphere-, and needle-shaped particles were dispersed in the medium. The size of particles shown in SEM micrograph in Figure 5.2 was much smaller than the average particle size (3.8 μm) determined by Sedigraph. This shows that the particles were in an agglomerated state in 1:1 water glycerol solution. The average aspect ratio of these ZnO crystals was 4.35 as reported by Ozmihci and Balkose.[4] They increased the conductivity of LLDPE at higher extent than ZnO crystals with smaller aspects ratio values, such as 2.29 and 1.96.[4]

FIGURE 5.2 SEM micrograph of ZnO.

The X-ray diffraction (XRD) diagram of ZnO is seen in Figure 5.3. The peaks at 2θ values of 31.7°, 34.4°, 36.2°, 47.5°, 56.5°, 62.8°, 67.9°, and 69.3° are identical with the peaks at the XRD powder pattern of ZnO in Joint Commitee on Powder Diffraction Standards (JCPDS) Card No: 79-0206. The relative intensities of the peak at 2θ 34.4° of (002) planes to 2θ 36.2° peak of (101) planes for ZnO in the present study and in JCPDS Card No: 79-0206 are 0.36 and 0.42, respectively. Since 0.36 is smaller than 0.42, the ZnO crystals were oriented in perpendicular to c direction in the present study.

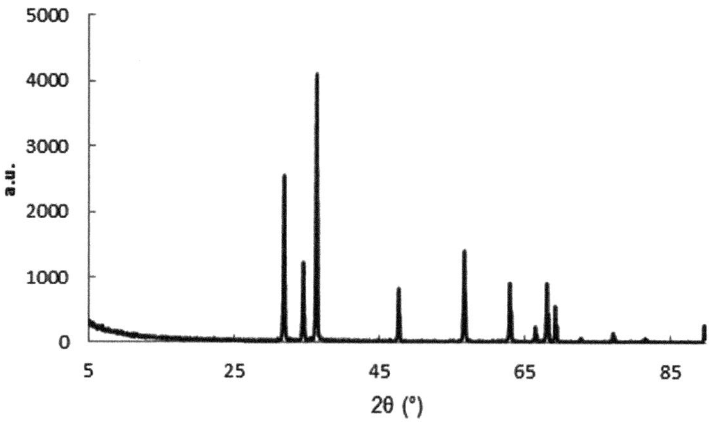

FIGURE 5.3 X-ray diffraction diagram of ZnO.

Characteristic peaks of (ZnO) at 473 and 532 cm^{-1} are seen in the FTIR spectrum of ZnO in Figure 5.4. There were hydrogen-bonded OH groups in ZnO as indicated by the peak at 3400 cm^{-1}. ZnO was entirely dry since no peaks related with H_2O bending vibration at 1600 cm^{-1} were seen.

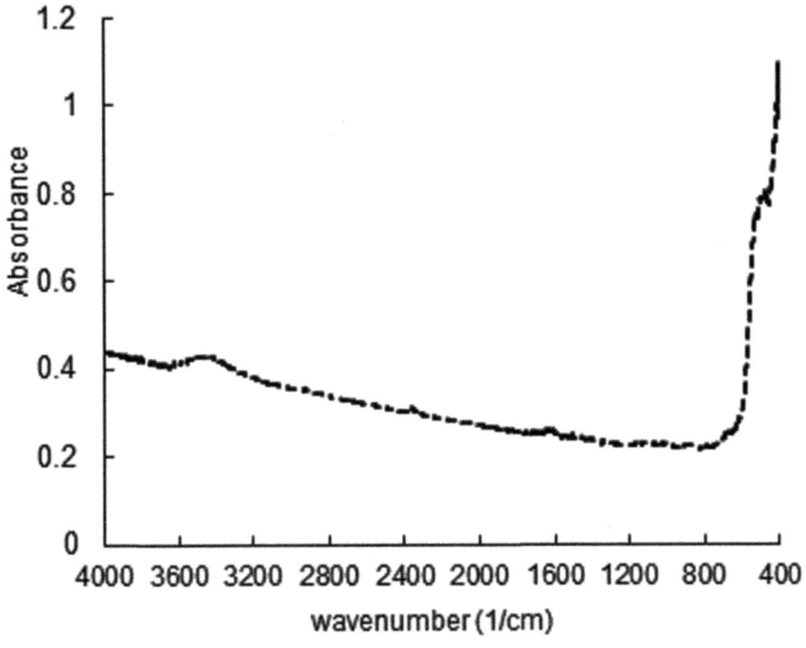

FIGURE 5.4 FTIR spectrum of ZnO.

5.3.2 CHARACTERIZATION OF MODIFIED ZnO

The SEM micrographs of modified ZnO particles are seen in Figure 5.5. As seen in Figure 5.5, the particles were agglomerated during surface modification with PEG and AMTES. It was not possible to disperse the agglomerates by simple grinding with mortar and pestle, since they were very strongly bound to each other. The mixing of ZnO with PEG and AMPTES in 1:1 ethanol water solution and then drying may be the reason of agglomeration. Using higher ethanol concentration or another solvent with low surface tension would prevent the agglomeration of ZnO particles due to capillary action of water in solid–water interface.

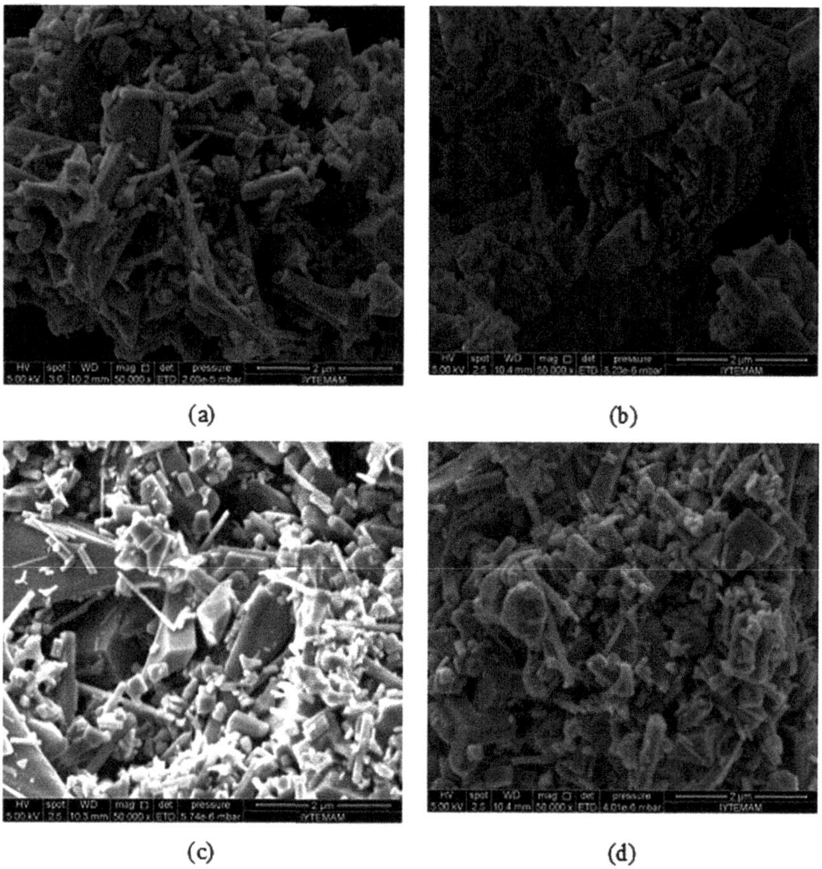

FIGURE 5.5 SEM micrographs of (a) 3% AMTES, (b) 6% AMPTES, (c) 3% PEG, and (d) 6% PEG-modified ZnO at 50,000× magnification.

FTIR spectra of AMPTES and PEG-modified ZnO are given in Figure 5.6. Characteristic peaks of (ZnO) at 473 and 532 cm^{-1} are seen in the spectra of modified ZnO. The hydrogen-bonded OH group vibrations of pristine ZnO shifted from 3400 to 3375 cm^{-1} in surface-modified ZnO indicating that it had hydrogen bonds with the modifiers. The absorption peaks at 2875 and 2808 cm^{-1} for asymmetric and symmetric stretching vibrations of CH_2, at 1591 cm^{-1} vibration of amine groups, at 1004 and 1055 cm^{-1} for Si–O stretching, and at 811 cm^{-1} Si–O–Zn vibrations[20] are seen in FTIR spectra of AMPTES-modified ZnO in Figure 5.6a. This indicated that AMPTES was coupled to the surface of the ZnO. The 2800 cm^{-1}

C–H stretching, 1100 cm^{-1} C–O–C stretching, 1300 cm^{-1} CH_2 bending, and 1200 cm^{-1} CH_2 twisting vibration[18] are present in the FTIR spectra of PEG 4000-modified ZnO in Figure 5.6b.

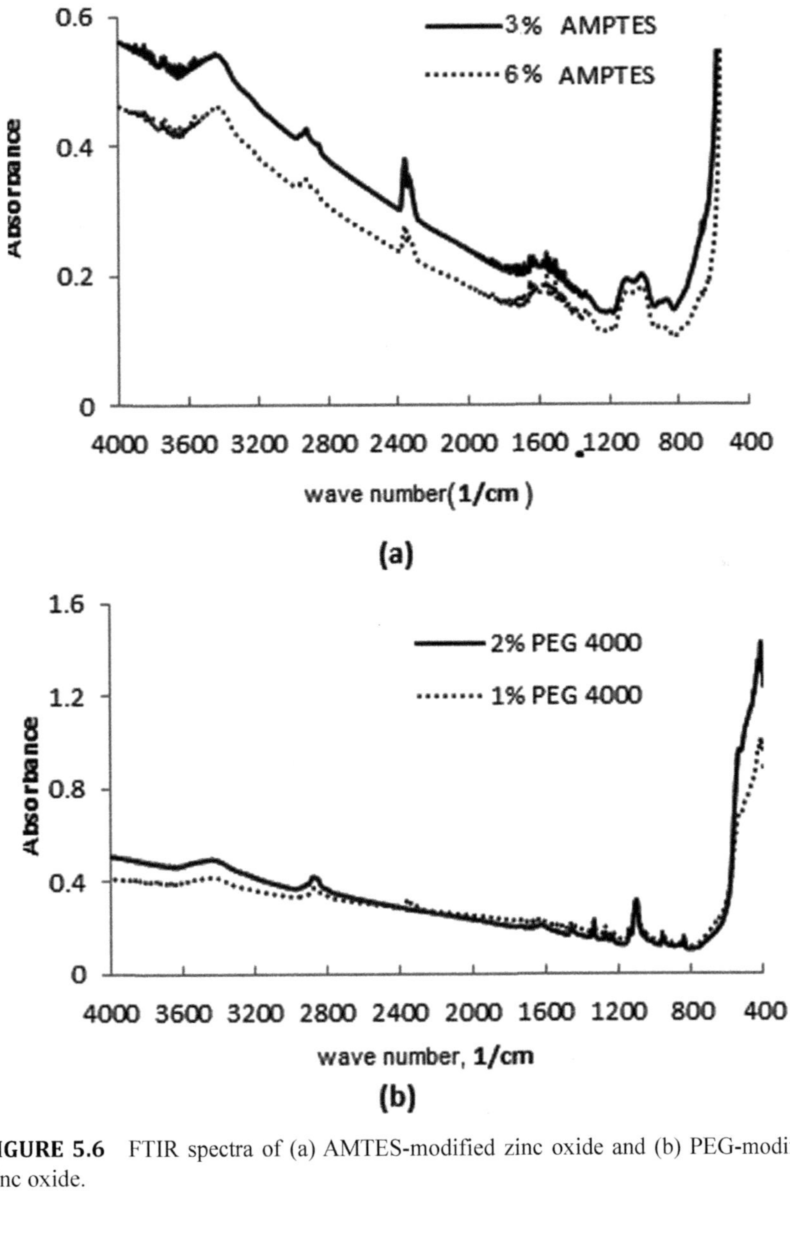

FIGURE 5.6 FTIR spectra of (a) AMTES-modified zinc oxide and (b) PEG-modified zinc oxide.

TG analysis indicated AMPTES and PEG content of the modified ZnO. TG curves for PEG- and AMPTES-modified ZnO are seen in Figure 5.7a,b, respectively. While PEG-modified ZnO started to a fast mass loss at around 200°C, AMPTES-modified ZnO continuously lost its mass starting from room temperature. PEG entirely decomposes to volatile products up to 415°C.[34] Considering the mass loss of PEG-modified ZnO samples at 1000°C, their actual PEG content were 4.2% and 6% for 3% and 6% PEG samples, respectively. Assuming AMTES decomposed to volatile products, [35,36] and solid SiO_2 by heating up to 1000°C, AMPTES content were found as 3.47% and 4.86% for 3% and 6% AMPTES containing ZnO.

5.3.3 CHARACTERIZATION OF COMPOSITES

5.3.3.1 MORPHOLOGY OF FRACTURE SURFACES

Micrographs of fracture surfaces of LLDPE and composites having 20% pristine and modified ZnO are seen in Figure 5.8. The composites with PEG- and AMPTES-modified ZnO had granular particles at their fracture surfaces. These granules could be agglomerated ZnO particles formed during surface modification process. However, their morphologies were different than the modified ZnO granules as shown in Figure 5.5. The granules formed spherical larger particles and were coated with LLDPE during mixing of the polymer melt in rheometer. The fracture surface of LLDPE and composite with 20% pristine ZnO had no agglomerated particles. While composite made with unmodified ZnO showed brittle fracture at the liquid nitrogen temperature (−190°C), the composites made form AMPTES- and PEG-modified ZnO showed ductile fracture with a fibrillated structure indicating deformation of polymer phase during fracture as seen in Figure 5.8c,f,l, respectively.

5.3.3.2 XRD OF COMPOSITES

XRD diagrams for LLDPE and ZnO/LLDPE composites having different vol % of pristine ZnO are shown in Figure 5.9. In XRD diagram of LLDPE in Figure 5.9, curve 1 diffraction peaks due to C planes are seen. In Figure 5.9, curves 2–4 show the XRD diagrams of the ZnO/LLDPE composites. The mass absorption coefficient (μ/ρ) of Zn and oxygen are much higher

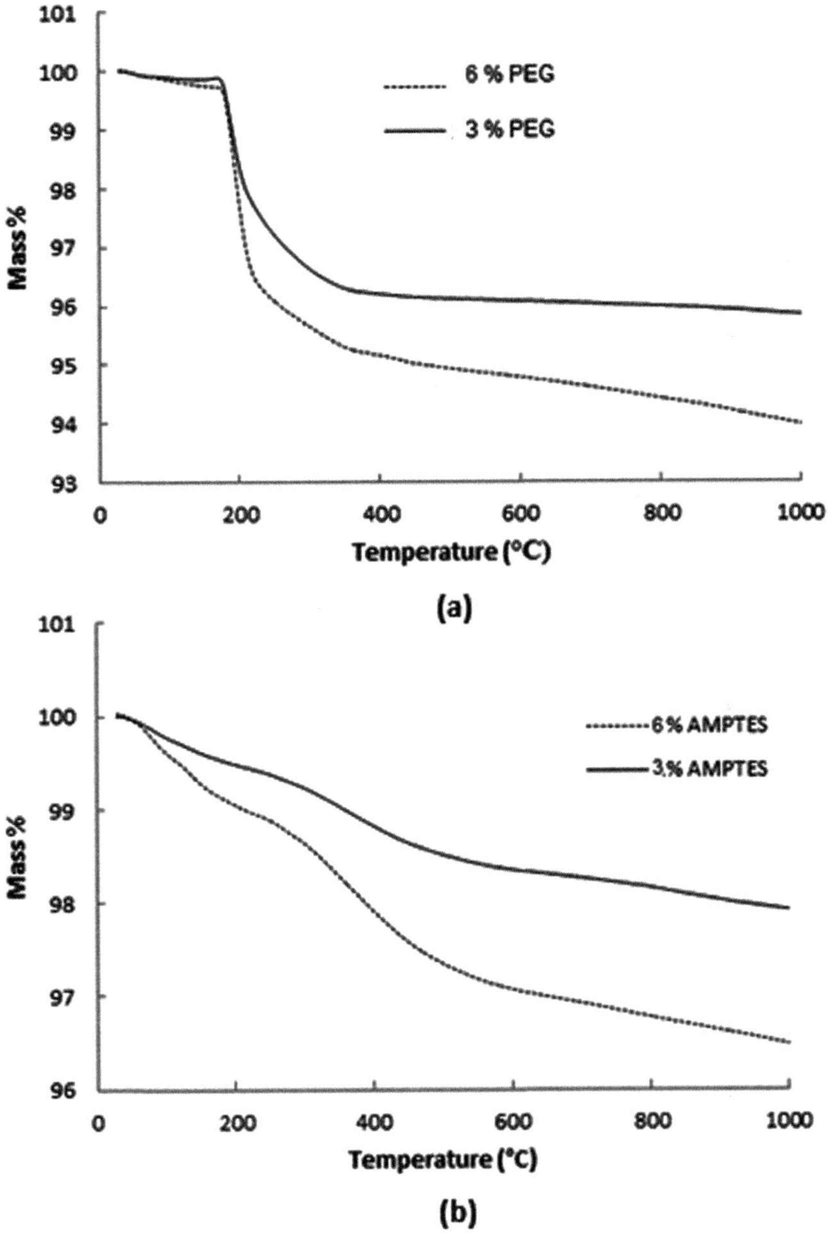

FIGURE 5.7 TG curves of (a) PEG-modified ZnO and (b) AMPTES-modified ZnO.

than the mass absorption coefficient of *C*. For example, they are 28.9, 0.8, and 0.4 $cm^2 g^{-1}$ for Zn, O, and C, respectively, for Ag K_α radiation.[37] Thus in ZnO and PE composites, the intensity of diffraction peaks due to PE were reduced due to absorption of X-rays by ZnO in a very thin layer of sheet preventing exposure of X-rays to LLDPE layer throughout the thickness of the sheet.

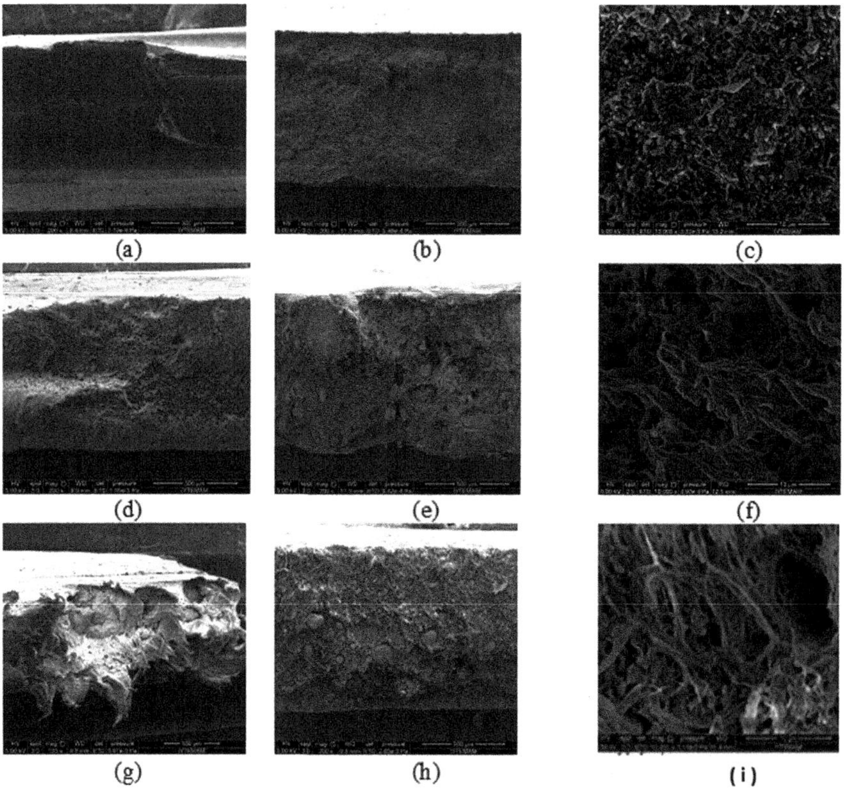

FIGURE 5.8 (a) Pure LLDPE at 200×, 20 vol % composites with (b) pristine ZnO at 200×, (c) pristine ZnO at 10,000×, (d) 3% PEG-modified ZnO at 200×, (e) 6% PEG-modified ZnO at 200×, (f) 3% PEG-modified ZnO at 10,000×, (g) 3% AMTES-modified ZnO at 200×, (h) 6% AMTES-modified ZnO at 200×, and (i) 3% AMPTES-modified at 10,000× magnification.

The XRD pattern of LLDPE had one dominant fairly sharp peak at 21.4°, a weak broad peak at 23.6°, and a weak broad third peak centered on 19.5°. The dominant peaks (21.4° and 23.6°) and a small broad peak (19.5°) arose from crystalline and amorphous regions in LLDPE.[38]

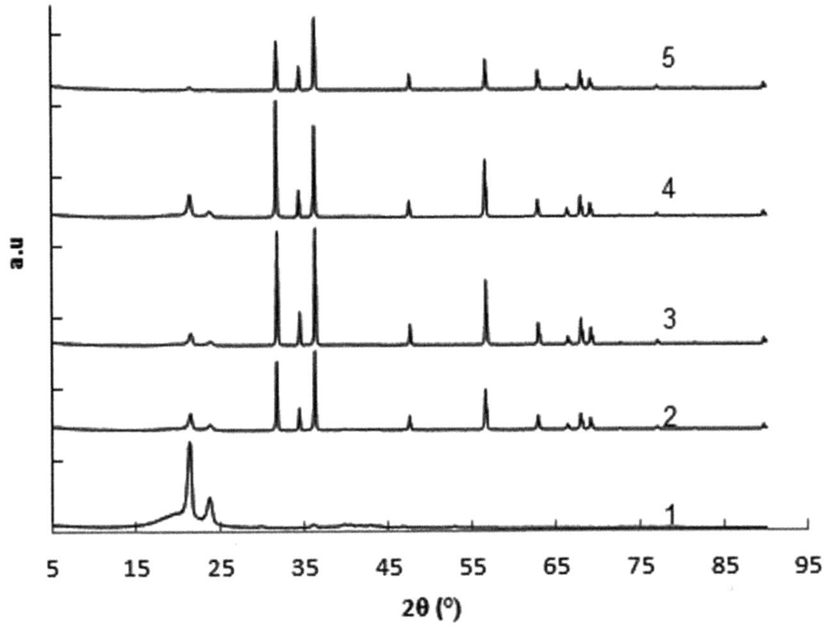

FIGURE 5.9 X-ray diffraction diagram of (1) LLDPE composites with (2) 5 vol %, (3) 10 vol %, (4) 15 vol %, and (5) 20 vol % ZnO.

XRD diagrams of composites having 20 vol % ZnO with 3% PEG- and 3% AMPTES-modified ZnO are seen in Figure 5.10. The ratio of the intensities of the peak at 2θ 34.2° of (002) planes to 2θ 36.1° peak of (101) planes indicates the orientation of ZnO crystals. This ratio is reported for pristine and modified ZnO composites in Table 5.2. It increases from 0.29 to 0.33 as the volume fraction of ZnO is increased. It is 0.33, 0.36, and 0.51 for pristine, 3% PEG, and 3% AMPTES-modified ZnO composites having 20 vol % of ZnO, respectively. This ratio is reported as 0.42 in JCPDS card number 79-0206 for randomly oriented bulk ZnO. The orientation of the ZnO crystals are very close to that of pristine ZnO in composites with 3% and 6% PEG and 6% AMPTE-modified ZnO. However, since this ratio (0.51) for 3% AMPTES-modified ZnO is greater than 0.42, 3% AMPTES-modified ZnO crystals were oriented in c direction in composites.

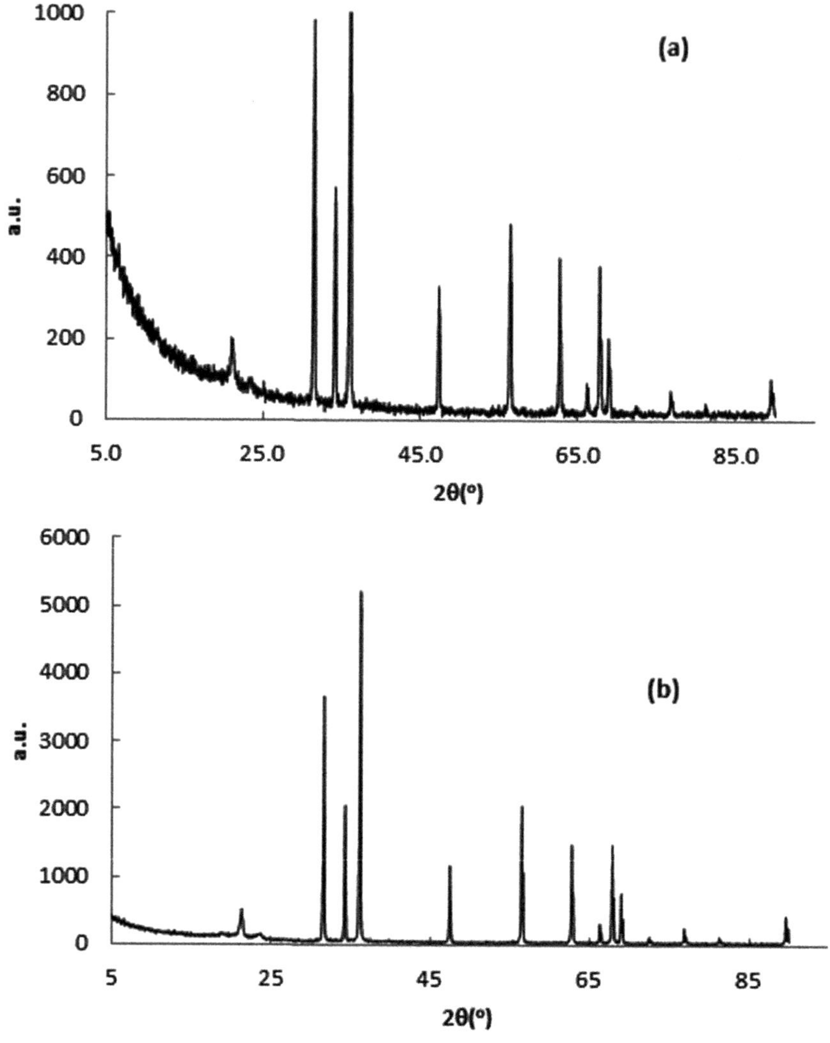

FIGURE 5.10 X-ray diffraction diagram of 20% (a) 3% AMPTES-modified composites and (b) 3% PEG-modified composites.

5.3.3.3 TG ANALYSIS OF COMPOSITES

TG curves of composites are seen in Figure 5.11. The onset, end set temperature of mass loss and the mass loss % at 1000°C for 20 vol % pristine and modified ZnO/LLDPE composites are reported in Table 5.3.

The onset temperature of the mass loss of pristine ZnO-added PE (423°C) is higher than that of composites with surface-modified ZnO (399–420°C) which could be due to lower onset temperature of surface modifiers. The remaining mass at 1000°C (40.4–38.6%) is lower for PEG-modified composites than that of pristine ZnO composite (41.5%). Since transformation of PEG to volatile components is expected to occur up to 415°C,[34] higher mass loss values were expected to be observed. On the other hand AMPTES-modified ZnO composites had higher mass loss values at 1000°C than pristine ZnO composites, which can be attributed to different degradation behavior of AMPTES-modified ZnO in LLDPE.

TABLE 5.2 Orientation of ZnO Crystals in Composites.

Modification	ZnO vol %	$I_{34.2}/I_{36.2}$
Non	5	0.29
Non	10	0.29
Non	15	0.30
Non	20	0.33
3% PEG	20	0.36
6% PEG	20	0.37
3% AMPTES	20	0.51
6% AMPTES	20	0.32

5.3.3.4 DENSITY OF COMPOSITES

Theoretical density of the composites was calculated by using density (ρ) for ZnO powders (5.6 g/cm^3) and LLDPE (0.92 g/cm^3) matrix materials. The theoretical density calculation of composites was made using eq 5.1.

$$\rho_{Theoretical} = \rho_{PE}(1-\Phi) + \rho_{ZnO}\Phi \tag{5.1}$$

where Φ is volume fraction of ZnO. The calculated density values and the measured density values were not identical but they were close to each other as seen in the Table 5.4. However for composites having modified

ZnO, lower density values were observed accept for 6% PEG-modified composite as seen in Table 5.4. Lower experimental density indicated that the composite sample had empty spaces (2–3%), whereas higher density indicated higher crystallinity in polymer phase.

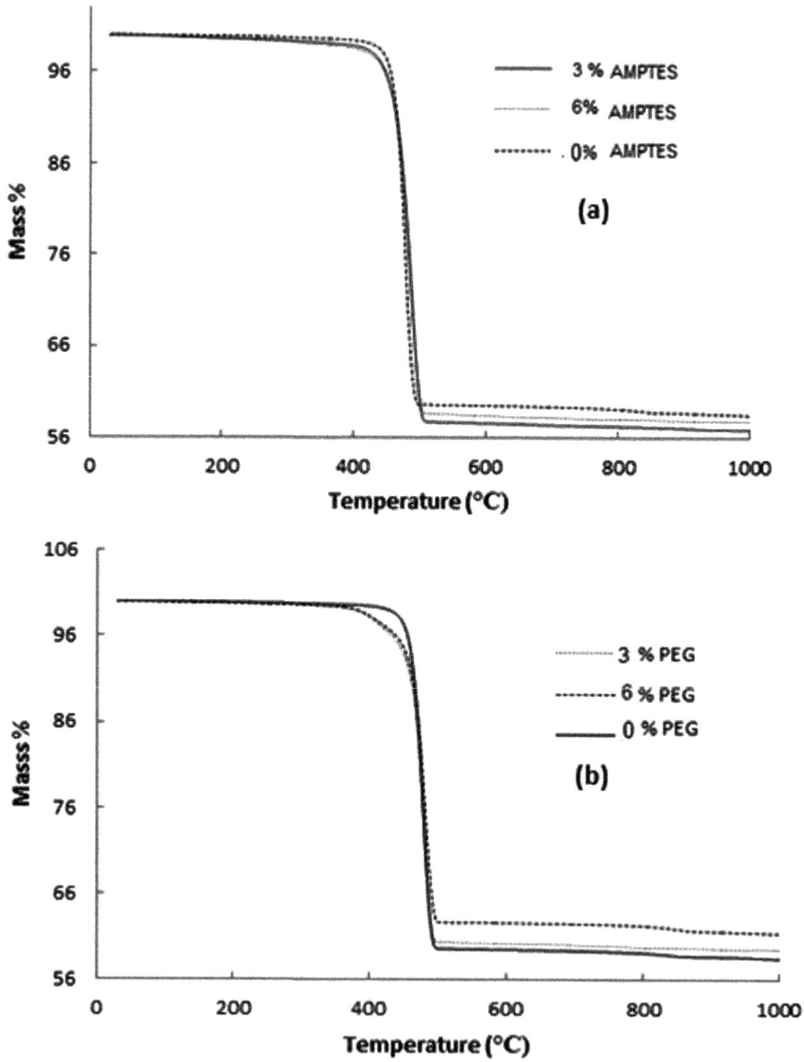

FIGURE 5.11 TG curves of LLDPE and LLDPE/20 vol % (a) AMPTES-modified ZnO composites and (b) PEG-modified ZnO composites.

TABLE 5.3 Onset of Mass Loss, End of Mass Loss, Mass Loss at 1000°C for Composites of 20% Pristine ZnO/LLDPE and 20% Modified ZnO/LLDPE.

Modification	Onset of mass loss temperature, °C	End of mass loss temperature, °C	Mass loss at 1000°C
Non	423	491	41.5
3% PEG	416	490	40.4
6% PEG	399	492	38.6
3% AMPTES	416	501	43.1
6% AMPTES	420	499	42.2

5.3.3.5 *CONTACT ANGLE OF WATER*

Contact angle of water droplet on the surface of the composites were measured to investigate the effect of the surface modifier on wettability of the surface. The hydrophilic ZnO surface was modified with an organic compound (AMPTES and PEG) to make the ZnO surface hydrophobic to make the polymer and ZnO surface compatible. The contact angle of on the surface of LLDPE was found to be 87° as seen in Table 5.4. The contact angles of the composites with different volume fraction of unmodified ZnO and 20 vol % modified ZnO are also shown in Table 5.4. They were in the range of 85.2–92.0° showing that ZnO or modified ZnO had no significant effect on the wettability of LLDPE with water.

5.3.3.6 *RESISTIVITY OF COMPOSITES*

The volume resistivity of LLDPE was 4.57×10^{16} Ω cm as seen in Table 5.4. For the addition of 5 vol % ZnO to the polymer, the volume resistivity value decreased to 1.02×10^{16} Ω cm. The volume resistivity of PE decreased to 3.16×10^{14}, 1.26×10^{12}, and 2.06×10^{10} Ω cm for 10%, 15%, and 20% ZnO, respectively, as shown in Table 5.4. When the concentration of filler increased and reached the percolation threshold value, the resistivity of the composites decreased dramatically. After reaching threshold value, electron passing through the composite started hopping, and resistivity of the composite decreased. The resistivity of the LLDPE was lowered from

2.3 × 10^{16} Ω cm down to 1.4 × 10^{10} Ω cm with ZnO addition as reported in our previous study[4] which are very close to the present study showing the reproducability of our work. The composites should have an electrical conductivity in the range of 10^{-12} and 10^{-8} S/cm for electrostatic dissipation applications, 10^{-8} and 10^{-2} S cm $^{-1}$ for moderately conductive applications 10^{-2} S cm^{-1} and higher for shielding applications.[4] Thus, the composites prepared in the present study are electrostically dissipating materials at 15 vol % ZnO and 20% ZnO loading levels as seen in Figure 5.12. ZnO filler with 1.6 × 10^6 Ω cm resistivity decreased the resistivity of electrically insulating LLDPE to the level of statically dissipating material.

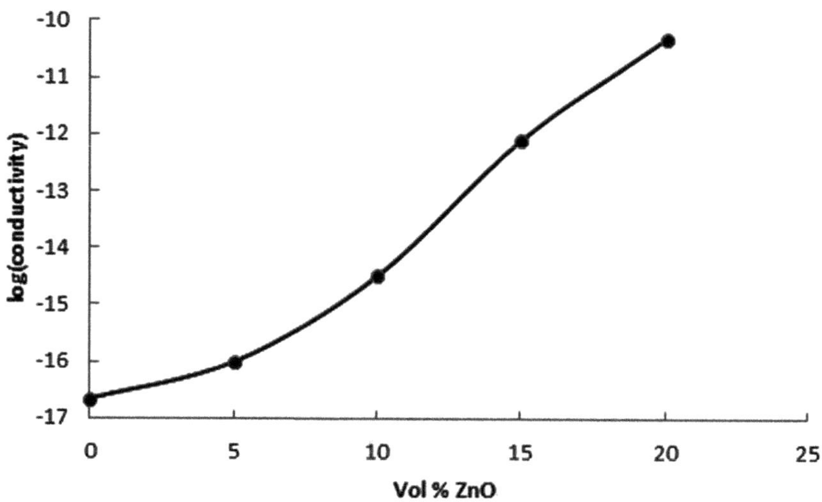

FIGURE 5.12 Change of Log [conductivity (S cm^{-1})] with vol % of pristine ZnO.

The resistivity values of LLDPE composites prepared from surface-modified ZnO are given in Table 5.4 and Figure 5.13. The resistivity of the composites from PEG- and AMPTES-modified ZnO composites was higher than that of the unmodified composite but still in the range of statistically dissipating materials. The resistivity increase could be attributed to the coating of ZnO particles with an insulating layer of surface modifier.

TABLE 5.4 Theoretical Density, Archimedes Density, Contact Angle, and Electrical Resistivity of Composites with Pristine and modified ZnO.

ZnO %	Modification	Theoretical density, g cm^{-3}	Archimedes density, g cm^{-3}	Contact angle, °	Electrical resistivity, Ω cm
0.0	Non	0.92	1.01	86.8	4.57×10^{16}
5.0	Non	1.15	1.14	85.2	1.02×10^{16}
10.0	Non	1.39	1.38	87.6	3.16×10^{14}
15.0	Non	1.62	1.61	92.2	1.26×10^{12}
20.0	Non	1.86	1.85	93.8	2.06×10^{10}
20.0	3% PEG	1.86	1.82	87.6	5.43×10^{10}
20.0	6% PEG	1.86	1.88	86.6	1.97×10^{11}
20.0	3% AMPTES	1.86	1.82	89.0	1.78×10^{11}
20.0	6% AMPTES	1.86	1.79	86.4	1.30×10^{13}

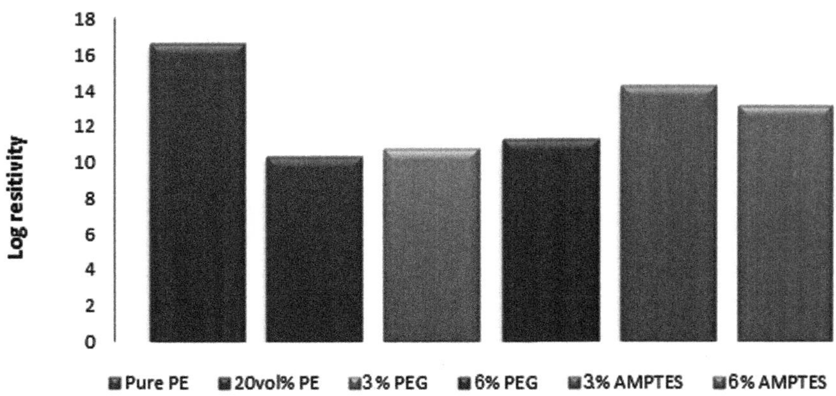

FIGURE 5.13 **(See color insert.)** Log resistivity comparison of LLDPE, 20 vol % ZnO/LLDPE composites.

5.3.3.7 MECHANICAL PROPERTIES OF COMPOSITES

Representative force stroke curves of LLDPE and 20 vol % of pristine and modified ZnO are shown in Figure 5.14. The mechanical properties of the composites with different vol % of ZnO are reported in Table 5.5. Tensile strength of LLDPE decreased from 18.3 kg/cm^2 down to 9.9 kg/cm^2 with ZnO addition up to 20% in volume. The elongation at break of the composites with up to 15% of ZnO was above 1100%, but it decreased to 12.3% when 20% ZnO was added. The elastic modulus of the composites generally increased with ZnO volume fraction. The mechanical properties of composites prepared from 20% pristine and modified ZnO are also reported in Table 5.5. When the surface of ZnO was modified with PEG or AMPTES, both tensile strength and elongation at break was not sufficiently improved.

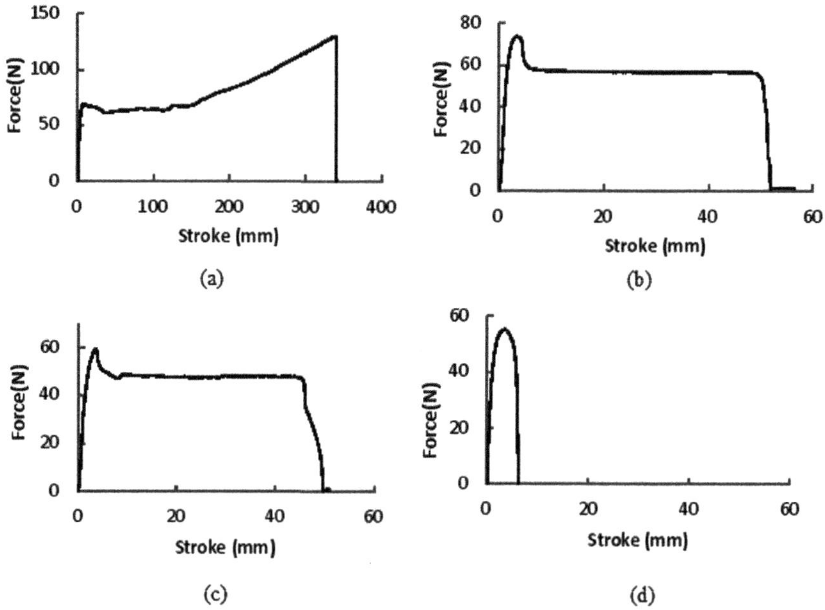

FIGURE 5.14 Force stroke curves of (a) LLDPE, (b) 20 vol % ZnO/LLDPE, (c) 20 vol % 3% PEG-modified ZnO/LLDPE, and (d) 3% AMPTES-modified ZnO/LLDPE.

In the present study it was aimed to get statically dissipating material with sufficient mechanical properties. Composite with 20% ZnO is statistically dissipating but does not have sufficient tensile strength. Thus ZnO was surface modified with AMPTES and PEG 4000 to make a better dispersion of ZnO particles at this high loading. However, the improvement obtained was not sufficient to make the material to be used in applications where high tensile strength is required. Other modification methods such as treating ZnO with AMPTES in toluene[20] or in water in acidic and basic conditions[21] or other modifiers should be used in further studies.

TABLE 5.5 Mechanical Properties of Composites Filled with Pristine and Modified ZnO.

% ZnO	Modification	Tensile stress, N mm^{-2}	Elongation at break, %	Elastic modulus, N mm^{-2}
0.0	Non	18.3 ± 3.37	1190.9 ± 197.0	125.6 ± 4.5
5.0	Non	17.0 ± 2.49	1140.1 ± 157.0	118.9 ± 8.9
10.0	Non	17.7 ± 0.02	1165.5 ± 5.74	122.3 ± 5.48
15.0	Non	14.4 ± 1.53	1105.9 ± 126.0	171.5 ± 43.3
20.0	Non	9.9 ± 0.2	12.3 ± 0.4	216.0 ± 23.1
20.0	3% PEG	10.2 ± 0.5	15.0 ± 0.3	159.3 ± 11.5
20.0	3% AMTES	11.0 ± 0.4	11.9 ± 1.2	291.9 ± 37.6
20.0	6% PEG	8.7 ± 0.2	12.0 ± 0.6	154.3 ± 3.1
20.0	6% AMTES	10.0 ± 0.3	12.3 ± 0.4	216.0 ± 23.1

5.3.3.8 OPTICAL BEHAVIOR OF ZnO AND ZnO–LLDPE COMPOSITES

ZnO is a luminescent material that can emit light in different regions of electromagnetic spectrum when optically or electrically excited. Also it is a chemically stable, inexpensive, and environmental friendly material. Its luminescent property can be useful to construct solid-state lamps for illumination or as UV emitter.

5.3.3.8.1 The Absorption Spectrum of ZnO

The absorbtion spectrum of an aqueous suspension of ZnO is seen in Figure 5.15a. ZnO absorbs light at 378 nm which is close to the values (374–376 nm) of nanoparticless of ZnO reported by Misra et al.[39]

5.3.3.8.2 The Luminessence Spectra of Zn, LLDPE, and ZnO/LLDPE Composites

The luminesence spectrum of ZnO pellet excited at 380 nm has a weak narrow UV band around 380 nm due to radiative annihilation of exciton and a broader band in the green part of the spectrum with a maximum at 500 nm due to the defects in the sample caused by doubly ionized zinc or oxygen vacancy as seen in Figure 5.15b.

UV-irradiation is a frequently encountered factor that can affect photodegradation of polymers. ZnO powders absorbed nearly 100% of in the UV region. LLDPE has luminescence spectrum shown in Figure 5.15c. It has a peak at 425 nm due to the luminessence of unsaturated ketone or benzene ring of the antioxidant added by the producer.[40]

The optical properties of composites were investigated if composites posses the optical properties of ZnO powders. The visible light emission from the composites was investigated to understand effect of ZnO addition in LLDPE. Fluorescence spectra of ZnO-filled LLDPE composites are shown in Figure 5.15d. In the optical perspective of ZnO–polymer composites system, ZnO absorbed light in UV region and it screened the polymer matrix from the affect of UV light and the light could not penetrate into the polymer matrix. Thus, ZnO fluorescence spectrum rather than the fluorescence spectra of polymer became important. As seen in Figure 5.14d, ZnO–LLDPE composites and those of their ZnO (filler) spectra are very similar with each other.

Composites that were excited at 380 nm showed two sharp peaks at 411 and 425 nm and a broader band in the green part of the visible spectrum with a maximum between 488 and 523 nm. The LLDPE matrix influences the optical properties of composites since the emission peak of ZnO at 376 nm shifts to 411 and 425 nm under the influence of LLDPE. When the electron in the valence band was excited to the conduction band it could be trapped in a shallow level and then recombined with the hole left in the valence band, this process gave a rise of the band centered at 420 nm.[41] The emission peak at 425 nm can be attributed to electron hole plasma recombination emission of ZnO particles. The broad peak around 488–523 nm corresponds to green and blue shifts of composite.

The intensities of the peaks in PL spectra did not increase systematically with concentration of ZnO in the composites which might be due to strong absorption of UV light at 376 nm by ZnO at its surface shielded the composite from penetration of the light to its entire thickness. The absorption UV light by ZnO will prevent the UV degradation of LLDPE.[40]

ZnO and LLDPE composites showed the luminescence behavior of ZnO. When irradiated with UV light at 480 nm the composites emitted blue–green light. This property makes them as a new material which is white under visible light and blue–green under UV light. It can be used in stamps which can be seen under UV light only.

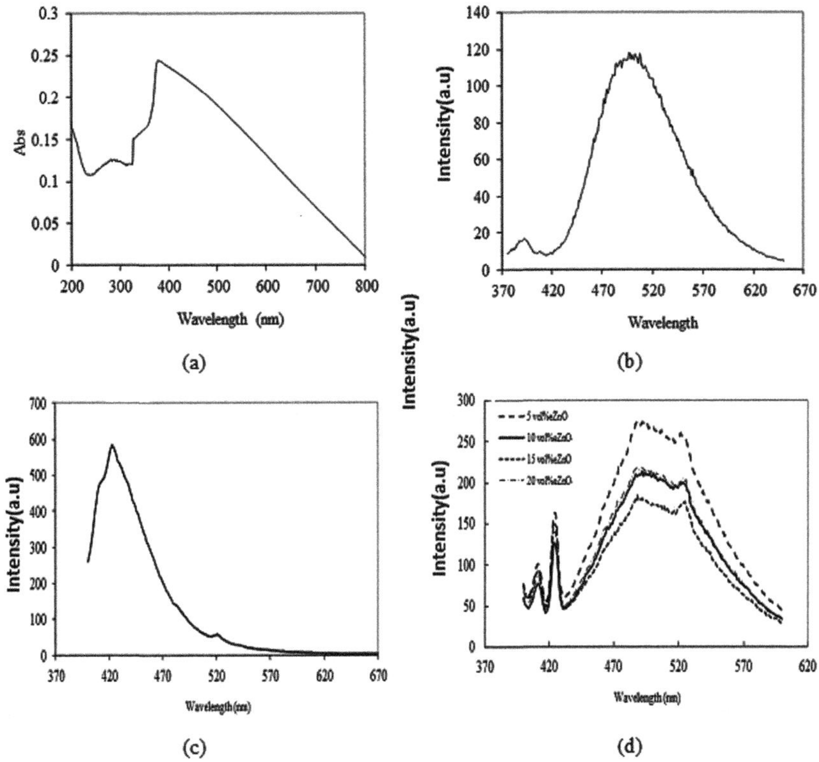

FIGURE 5.15 (a) Absorption spectrum of ZnO suspension, (b) luminescence spectrum of ZnO pellet, (c) luminescence spectrum of LLDPE sheet, and (d) luminescence spectrum of ZnO/LLDPE composites.

5.4 CONCLUSION

ZnO powder was used to improve the electrical conductivity and mechanical property of the PE matrix composites. To increase the homogeneity of ZnO particle in PE, two different modifier AMPTES and PEG 4000 were used with two different percentages, 3% and 6%. Rheometer was used to have dispersed polymer composites. Composites with 5%, 10%, 15%, and 20% in volume of ZnO in LLDPE were prepared. Also, 20 vol % modified ZnO/LLDPE composites were prepared.

When amount of ZnO was increased, the density values of the composites were also increased. However, the surface modification caused agglomeration of ZnO particles giving hard to grind large particles. This resulted in composites with 2–3% empty volume. Unmodified ZnO and modified ZnO did not have a significant effect on water contact angle on the surface of composites. Addition of AMPTES and PEG 4000 modifiers also did not significantly improved the mechanical properties contrary to expectations. This was caused by the agglomeration formation during surface modification of ZnO.

The electrical conductivity of the composites increased from insulating to statistically dissipating material by addition of conducting ZnO. When the surface of ZnO was modified with PEG or AMPTES both tensile strength and elongation at break was not sufficiently improved. Further studies on ZnO surface modification should be made to improve electrical and mechanical properties.

ZnO and LLDPE composites showed the luminescence behavior of ZnO. When irradiated with UV light at 380 nm the composites emitted blue–green light.

ACKNOWLEDGMENT

The authors acknowledge the contribution of K. Cagatay Dirik and Ismail Gureler to experimental work and Dr. Senem Yetgin for SEM microphotographs.

KEYWORDS

- **optical properties**
- **morphology**
- **surfaces and interfaces**
- **composites**
- **thermoplastics**

REFERENCES

1. Sethi, R. S.; Goosey, M. T. *Special Polymers for Electronics and Optoelectronics*; Chilton, J. A., Goosey, M. T., Eds.; Chapman & Hall: London, 1995; pp 14–46.

2. Özmihci, F. Enhancement of Thermal, Electrical and Optical Properties of Zinc Oxide Filled Polymer Matrix Nano Composites. PhD Dissertation, Izmir Institute of Technology, 2009.
3. Wang, L. S.; Hong, R. Y. *Synthesis, Surface Modification and Characterization of Nanoparticles in Advances in Nanocomposites—Synthesis, Characterization and Industrial Applications*; Reddy B., Ed., InTech. DOI: 10.5772/10540, 2011; pp 295–297.
4. Ozmihci, F. O.; Balkose, D. Effects of Particle Size and Electrical Resistivity of Filler on Mechanical, Electrical, and Thermal Properties of Linear Low Density Polyethylene-zinc Oxide Composites. *J. Appl. Polymer Sci.* **2013,** *130* (4), 2734–2743.
5. Chu, S. Y.; Yan, T. M. Characteristics of Sol-gel Synthesis of ZnO-based Powders. *J. Mater. Sci. Lett.* **2000,** *19*, 349–152.
6. Sindhu, H. S.; Joishy, S.; Rejendra, B. V.; Rao, A.; Goankar, M.; Kulkani, S. D., Babu, P. D. Tuning Optical, Electrical and Magnetic Properties of Fiber Structured ZnO Film by Deposition Temperature and Precursor Concentration. *Mater. Sci. Semiconductor Process.* **2017,** *68*, 97–107.
7. Hong, R. Y.; Qian, J. Z.; Cao, J. X. Synthesis and Characterization of PMMA Grafted ZnO Nanoparticles. *Powder Technol.* **2006,** *163*, 160–168.
8. Hong, J. I.; Schadler, L. S.; Siegel, R. W; Martensson, E. Rescaled Electrical Properties of ZnO/Low Density Polyethylene Nanocomposites. *Appl. Phys. Lett.* **2003,** *82*, 1956–1958.
9. Hong, R. Y.; Li, J. H.; Chen, L. L.; Liu, D. Q.; Li, H. Z.; Zheng, Y.; Ding, J. Synthesis, Surface Modification and Photocatalytic Property of ZnO Nanoparticles. *Powder Technol.* **2009,** *189*, 426–432.
10. Grasset, F.; Saito, N.; Li, D.; Park, D.; Sakaguchi, I.; Ohashi, N.; Haneda, H.; Mornet, S. Surface Modification of Zinc Oxide Nanoparticles by Aminopropyltriethoxysilane. *J. Alloys Compound.* **2003,** *360*, 298–311.
11. Poshumus, W.; Magusin, P.; Broken-Zijp, J. C.; Tinnemans, A. H. A.; Linde, R. Surface Modification of Oxidic Nanoparticles Using 3-methacryloxypropyltrimethoxysilane. *Colloid Interface Sci.* **2004,** *269*, 109–119.
12. Tang, E.; Cheng, G.; Ma, X.; Pang, X.; Zhao, Q. Surface Modification of Zinc Oxide Nanoparticle by PMAA and Its Dispersion in Aqueous System. *Appl. Surf. Sci.* **2006,** *252*, 5227–5232.
13. Hong, R.; Pan, T.; Qion, J; Li, H. Synthesis and Surface Modification of ZnO Nanoparticles. *Chem. Eng. J.* **2006,** *119*, 71–80.
14. Ma, S.; Shi, L.; Feng, X.; Yu, W.; Lu, B. J. Graft Modification of ZnO Nanoparticles with Silane Coupling Agent KH570 in Mixed Solvent. *J. Shanghai Univ.* **2008,** *12* (3), 278–282.
15. Liufu, S.; Xiao, H.; Li, Y. Investigation of PEG Adsorption on the Surface of Zinc Oxide Nanoparticles. *Powder Technol.* **2004,** *145*, 20–24.
16. Taghaevi-Ganjali, S.; Malekzadeh, M.; Farahani, M.; Abbasian, A.; Khosravi, M. Effect of Surface-modified Zinc Oxide as Cure Activator on the Properties of a Rubber Compound Based on NR/SBR. *J. Appl. Polymer Sci.* **2011,** *122*, 249–256.
17. Ozmihci, F.; Balkose, D.; Ulku, S. Natural Zeolite Polypropylene Composite Film Preparation and Characterization. *J. Appl. Polymer Sci.* **2001,** *82* (12), 2913–2921.

18. Metin, D.; Tihminhoglu, F.; Balkose, D.; Ulku, S. The Effect of Interfacial Interactions on the Mechanical Properties of Polypropylene/Natural Zeolite Composites. *Composites Part A-Appl. Sci. Manufact.* **2004,** *35* (1), 23–32.
19. Demir, H.; Atikler, U.; Balkose, D.; Tihminlioglu, F. The Effect of iFber Surface Treatments on the Tensile and Water Sorption Properties of Polypropylene-luffa Fiber Composites. *Composites Part A-Appl. Sci. Manufactu.* **2006,** *37* (3), 447–456.
20. Jaramillo, A. F.; Baez-Cruz, R.; Montoya, L. F.; Medimam, C.; Perez-Tijerina, E.; Salazar, F.; Rojas, D.; Melendrez, M. F. Estimation of the Surface Interaction Mechanism of ZnO Nanoparticles Modified with Organosilane Groups by Raman Spectroscopy. *Ceramics Int.* **2017,** *43* (15), 11838–11847.
21. Grasset, F., Saito, N.; Li, D.; Park, D.; Sakaguchi, I.; Haneda, H.; Roisnel, T.; Mornet, S., Deguet, E. Surface Modification of Zinc Oxide Nanoparticles by Aminopropyltriethoxysilane. *J. Alloys Compound.* **2003,** *360* (1–2), 298–311.
22. Yamak, H. B.; Altan, M.; Altındag, A. Electrical, Morphological, Thermal and Mechanical Properties of Low Density Polyethylene/Zinc Oxide Nanocomposites Prepared by Melt Mixing Method. *Optoelectro. Advan. Mater.-Rapid Commun.* **2016,** 10 (11–12), 925–934.
23. Fleming, R. J.; Ammala, A.; Casey, P. S.; Lang, S. B. Conductivity and Space Charge in LDPE Containing Nano- and Micro-sized ZnO Particles. *IEEE Trans. Dielectrics Electric. Insulation* **2008,** *15* (1), 118–126.
24. Tian, F.; Lei, Q.; Wang, X.; Wang, Y. Investigation of Electrical Properties of LDPE/ZnO Nanocomposite Dielectrics. *IEEE Trans. Dielectrics Electric. Insulation* **2012,** *19* (3), 763–769.
25. Chang, B. P.; Akil, H. M.; Peng, B.; Muhammad Ghaddafy Affendy, M. G.; Abbas Khan, A.; Nasir, R. B. M. Comparative Study of Wear Performance of Particulate and Fiber-reinforced Nano-ZnO/Ultra-high Molecular Weight Polyethylene Hybrid Composites Using Response Surface Methodology. *Mater. Design.* **2014,** *63*, 805–819.
26. Li, S. C.; Li, B.; Qin, Z.-J. The Effect of the Nano-ZnO Concentration on the Mechanical, Antibacterial and Melt Rheological Properties of LLDPE/Modified Nano-ZnO Composite Films. *Polym.-Plastics Technol. Eng.* **2010,** *49* (13), 1334–1338.
27. Rezapour, M.; Talebian, N. Comparison of Structural, Optical Properties and Photocatalytic Activity of ZnO with Different Morphologies: Effect of Synthesis Methods and Reaction Media. *Mater. Chemist. Phys.* **2011,** *129* (1–2), 249–255.
28. Wang, J. M.; Gao, L. Synthesis of Uniform Rod-like, Multi-pod-like ZnO Whiskers and Their Photo Luminescence Properties. *J. Crystal Growth* **2004,** *262* (1–4), 290–294.
29. Egbuchunam, T. O.; Yetgin, S.; Filiz Ozmihci Omurlu, F. Ö.; Balkose, D. Optical and Surface Properties of Zinc Oxide Nanoparticles Dried by Conventional and Super-critical Ethanol Drying Techniques. *Int. J. Appl. Eng. Res.* **2014,** *9* (21), 11631–11646.
30. Xian, F. L.; Zheng, G. G.; Xu, L.; Kuang, W.; Pei, S.; Cao, Z.; Li, J.; Lai, M. Temperature and Excitation Power Dependence of Photoluminescence of ZnO Nanorods Synthesized by Pattern Assisted Hydrothermal Method. *J. Alloys Compound.* **2017,** *710*, 695–701.
31. Mwafy, E. A.; Abd-Elmgeed, A. A.; Kandil, A. A.; Elsabbagh, I. A., Elfass, M. M.; Gaafar, M. S. High UV-shielding Performance of Zinc Oxide/High-density Polyethylene Nanocomposites. *Spectrosc. Lett.* **2015,** *48* (9), 646–652.

32. Zhou, L.; He, B. Z.; He, B.; Wu, F. Castor Oil-stabilized Magnetic Fe_3O_4 and Luminescent ZnO Nanocrystals: One-step Green Synthesis and Application for Polymer Composites. *Advan. Powder Technol.* **2016,** *27* (4), 1839–1844.
33. Li, S. C.; Li, Y. N. Mechanical and Antibacterial Properties of Modified Nano-ZnO/High-density Polyethylene Composite Films with a Low Doped Content of Nano-ZnO. *J. Appl. Polymer Sci.* **2010,** *116* (5), 2965–2969.
34. Kwon, S. K.; Kim, D. H. Effect of Process Parameters of UV-assisted Gas-phase Cleaning on the Removal of PEG (Polyethyleneglycol) from a Si Substrate. *J. Korean Phys. Soc.* **2006,** *49*, 1421–1427.
35. Vunain, E.; Opembe, N. N.; Jalama, K.; Ajay, K.; Mishra, A. K.; Meijboomet, R. Thermal Stability of Amine-functionalized MCM-41 in Different Atmospheres. *J. Thermal Anal. Calorimetry.* **2014,** *115* (2), 1487–1496.
36. Wang, X. G.; Lin, K. S. K.; Chan, J. C. C.; Cheng, S. Direct Synthesis and Catalytic Applications of Ordered Large Pore Aminopropyl-functionalized SBA-15 Mesoporous Materials. *J. Phys. Chem. B.* **2005,** *109* (5), 1763–1769.
37. https://www.webelements.com/periodicity/x_ray_mass_absorption_Ag_k/index.html (accessed Nov. 11, 2017)
38. Bilmayer, F. W. *Textbook of Polymer Science*; John Wiley and Sons: New york, 1984, p 288.
39. Mishra, S. K.; Srivastava, R. K.; Prakash, S. G. ZnO Nanoparticles: Structural, Optical and Photoconductivity Characteristics. *J. Alloys Compound.* **2012,** *539*, 1–6.
40. Ito, T.; Fuse, N.; Ohki, Y. Effects of Additives, Photodegradation, and Water-tree Degradation on the Photoluminescence in Polyethylene and Polypropylene. *Trans. Inst. Elect. Eng. Jpn. A*, **2004,** *124* (7), 624–630.
41. Lima, S. A. M.; Cremona, M.; Davolos, M. R.; Legnani, C.; Quirino, W. G. Electroluminescence of Zinc Oxide Thin-films Prepared via Polymeric Precursor and via Sol–Gel Methods. *Thin Solid Films*. **2007,** *516*, 165–169.

SECTION II
Electronic and Ionic Composites

CHAPTER 6

MATERIALS FOR ORGANIC TRANSISTOR APPLICATIONS

ANDREEA IRINA BARZIC[1,*] and RAZVAN FLORIN BARZIC[2]

[1]*"Petru Poni" Institute of Macromolecular Chemistry, 41A Grigore Ghica Voda Alley, 700487 Iasi, Romania*

[2]*Faculty of Mechanics, "Gheorghe Asachi" Technical University of Iasi, 43 Dimitrie Mangeron, 700050 Iasi, Romania*

[*]*Corresponding author. E-mail: irina_cosutchi@yahoo.com*

ABSTRACT

Organic electronics is found in a continuous development owing to discovery of new organic compounds and multiphase materials that exhibit excellent electrical, mechanical, and thermal performance. Among the devices that revolutionized the electronic products one can mention organic thin-film transistors (OTFTs). The chapter attempts to present the current state of art concerning the OTFT parameters, structures and models. A great attention is given to the types of materials utilized in construction of OTFTs, such as dielectrics, semiconductors, conductors, passivation, and substrates. Several aspects that affect the device performance are described, including dielectric properties, scaling of dielectric thickness, processing conditions, contact effects, and environmental stability. The main utility of OTFT devices in the electronic industry and biomedical field is reviewed. The chapter concludes with future research directions that could help to accomplish OTFTs with enhanced functionality.

6.1 CURRENT APPROACHES FOR ORGANIC THIN-FILM TRANSISTORS

6.1.1 GENERAL ASPECTS ON TRANSISTORS

Transistors can be viewed as the fundamental part of the modern circuitry, where they are able to amplify or switch electronic signals and electrical power.[1] Since its implementation in 1947,[2] the field of electronics has known an enormous progress since it led to smaller and cheaper devices, such as radios and computers. There are several types of transistors, which can be categorized by considering their structure, electrical polarity, power rating, amplification factor, or physical packaging. The most important factor is the transistor structure, which divides these building blocks as follows:[3]

- BJT: *bipolar junction transistor* conducts current by using both majority and minority carriers. It is mainly a sandwiched structure of *n–p–n* or *p–n–p* semiconductors;
- FET: *field-effect transistor* can be viewed as unipolar transistor, which uses either electrons (in *n*-channel FET) or holes (in *p*-channel FET) for conduction. The system presents four terminals and the body is connected to the source inside the package;
- IGFET (MOSFET): *metal-oxide-semiconductor field-effect transistor* is a type of field-effect transistor. It is composed of an insulated gate, whose voltage affects the conductivity of the device. This ability to modify conductivity with the quantity of applied voltage is suitable for amplifying or switching electronic signals;
- IGBT: *insulated-gate bipolar transistor* is a three-terminal power semiconductor device mainly having the role of an electronic switch, that is able to combine high efficiency and fast switching. It is found to control electric power in many applications: air-conditioners, variable-frequency drives, electric automobiles, speed refrigerators, and sometimes in stereo systems with switching amplifiers. The IGBT is a semiconductor device, composed of four alternating layers (*p–n–p–n*) that are directed by a metal-oxide-semiconductor gate structure without regenerative action.

Among these transistor structures, a great deal of attention was given to FET. The field effect is a phenomenon in which the conduction property of a semiconductor is modified as a result of an electric field, applied normally to its surface.[4] The system presents four terminals of the FET, which are named source, gate, drain, and body (substrate), as well as a semiconductor layer and an insulating sheet placed between the gate and semiconductor. The electric field is applied through metallic gate in the device. The latest developments in FET area involved the replacement of the inorganic semiconductor with a thin film of an organic material.

The main advantages are low thermal budget and facile manufacturing process.[5] Thus, utilization of *organic thin-film transistors* (OTFT) has led to a considerable advance of electronic circuits and corresponding devices. From this point, organic electronics is found in a continuous development owing to discovery of new organic compounds that exhibit excellent electrical, mechanical, and thermal performance.

The OTFT, one the most remarkable devices among the organic derived ones, is now proving to have advanced or comparable performance to hydrogenated amorphous silicon TFTs (a-Si:H TFTs)[6] commonly encountered in the pixel drivers in active matrix flat-panel displays. The supplementary degrees of freedom for the OTFTs highlight that it is important to carefully select materials for a suitable OTFT platform and to be able to evaluate its potential in organic analog and digital circuits.[7,8] The basic operating principle is: on applying a bias between the gate and source, a spatial layer of mobile charge carriers is drawn near the semiconductor–dielectric interface that permits the flow of current through the active layer for an adequate drain to source potential. From a fundamental point of view, an OTFT works in a similar manner as a capacitor that generates an electric field in the dielectric at negative/positive source-gate potential for *p*/*n* type soluble organic semiconductors.[6–8] It results in gathering of holes/electrons through the positioning of the metal's Fermi level close to the highest occupied molecular orbital (HOMO) or lowest unoccupied molecular orbital (LUMO) levels of the semiconductor.

The chapter presents the current state of art concerning the OTFT parameters, structures, and models. A great attention is given to the types of materials utilized in construction of OTFTs, such as dielectrics, semiconductors, conductors, passivation, and substrates. Several aspects that affect the device performance are described, including material features, processing conditions, contact effects, and environmental stability. The main utility of OTFT devices in electronic industry and biomedical field is reviewed. The chapter concludes with future research directions that could help to accomplish OTFTs with enhanced functionality.

6.1.2 BASIC PARAMETERS OF OTFTs

There are some significant parameters that influence the practical utilization of an organic transistor. In this section, certain of them are briefly presented. In order to achieve the best results from an organic transistor, several criteria must be accomplished, such as enhanced mobility, big on/off current ratio (I_{on}/I_{off}), low threshold voltage, and steeper subthreshold slope ($S_{3\text{-}th}$).[6,8–12] These parameters are mainly affected by certain factors, particularly by device geometry, properties of constituting materials, grain size of organic semiconductor, and structural dimensions. The performance of OTFTs must be quantified based on the parameters described in the following paragraphs.

6.1.2.1 CHARGE CARRIER MOBILITY AND CONTACT RESISTANCE

The charge carrier mobility (μ) represents the mean carrier drift velocity per unit electric field and it reflects how easy they are travelling through the conduction path. This is different from the intrinsic charge carrier mobility (μ_o), which is a feature of the material and thus it is a constant. The effective charge carrier mobility is related to the *contact resistance* (*Rc*), but at the same time, it the perquisite for achieving large on-current. The mobility can be evaluated from the measured linear or saturation transfer characteristics. One can control the effective charge carrier mobility through the channel length, gate capacitance per unit area, and the active layer thickness.[13,14] This parameter is gate-biased dependent, and it can be enhanced when higher gate bias is applied.[15] Another factor that influences the carrier mobility is represented by grain dimension of the active layer, which can be optimized through manner in which it is deposited.[16,17]

6.1.2.2 ON/OFF CURRENT RATIO

The I_{on}/I_{off} reveals the degree in which the current can be modulated by the gate-source voltage. This OTFT characteristic depicts the ratio of accumulation to depletion region currents. In other words, it can be viewed as the on-state drain current at maximum gate-source voltage (~3 V) referenced to the off-state drain current at zero gate-source voltage.[9] The parameter is determined by material properties (especially the thickness of the semiconductor and dielectric), charge density and channel length. Additionally, a bigger I_{on}/I_{off} can be achieved with enhancement of the permittivity of insulator and a high gate capacitance. The I_{on}/I_{off} can be obtained from the experimentally determined input transfer characteristics.[18]

6.1.2.3 TRANSCONDUCTANCE

The transconductance (g_m) indicates the response of the drain current to variations of the gate-source voltage at a fixed drain-source voltage. This parameter is proportional to the gate-overdrive voltage in the saturation regime, while remains unchanged in the linear regime.[8]

6.1.2.4 THRESHOLD VOLTAGE

The threshold voltage (V_{th}) represents the minimum gate voltage demanded for agglomerating the charge carriers at the interface between organic semiconductor and dielectric interface. In this way, a conducting path between the source and the drain is created, enabling the control of the switching behavior of the device. The V_{th} factor ranges with gate insulator capacitance or thickness of the organic layer. It is also affected by the permittivity of the insulator, doping concentration, and channel length. In metal-oxide-semiconductor FETs, the threshold voltage presents an exponential variation with gate bias and the drain source bias since under the V_{th} value the free carrier density is strongly influenced by the local bias.[11] This is determined from transfer characteristics of saturation zone. The devices with shorter channel length and thicker film conduct to reduced threshold voltages.[19] The aspect is optimal for lowering the device power consumption and for creating portable devices.

6.1.2.5 SUBTHRESHOLD SLOPE AND TURN-ON VOLTAGE

The $S_{s\text{-}th}$ describes the transition between the off- and on-states. The parameter is for gate-source voltages under V_{th} and above the turn-on voltage (V_{on}), which delimits the point where off-state current suffers a sudden increase. The $S_{s\text{-}th}$ can be connected with impurity concentration, interface state, and trap density. By varying the drain and gate biasing, there are considerable modifications of this slope owing to a change in the conductivity of the channel. The poor quality of an active layer obtained during the manufacturing step influences this slope significantly since the presence of defects increases the trap states and implicitly the $S_{s\text{-}th}$. A steep $S_{s\text{-}th}$ can be mainly realized by enhancing as much as possible the gate dielectric capacitance and/or by employing a gate dielectric with few trap states at the dielectric/semiconductor interface. The $S_{s\text{-}th}$ and V_{on} can be estimated from input transfer characteristics.[9]

6.1.2.6 CHANNEL CONDUCTANCE AND SHEET/CONTACT RESISTANCE

The channel conductance (g_{ch}) is also known as the output conductance. It depicts the response of the drain current to modifications of the drain-source voltage at a fixed gate-source voltage. In the saturation zone, the g_{ch} is almost zero as a result of the effect named channel length modulation.[6] Moreover, at sufficiently low drain-source voltage, the gate-generated accumulation channel is presumed to exhibit a nominally uniform thickness. So, the sheet resistance (R_{sheet}), which expresses the resistance of the channel, can be obtained if the g_{ch} is determined. It is desirable that the contact resistance to be ohmic and small to allow the whole voltage applied to the OTFT to contribute to the transport current. For top contact configurations, it is affected by gate bias and abruptly increases at low gate-source voltage, while is almost independent of the gate bias in bottom contact devices.

6.1.2.7 INSULATOR CAPACITANCE

To obtain the insulator capacitance per unit area, the metal–insulator–metal (MIM) structures are examined using an inductance–capacitance–resistance (LCR) meter.[9] Basically, the MIM is a two-terminal pendant of the OTFT imparting the identical layer structure excluding the semiconductor.

6.1.2.8 CURRENT-GAIN CUTOFF FREQUENCY

The dynamic performance of organic transistors is reflected through the current-gain cutoff frequency (f_T), also named relaxation frequency. This parameter denotes the frequency at which the current gain is equal to 1. In this case, the solution relies on using an organic semiconductor with a enhanced carrier mobility or in diminishing device lateral dimensions.[20]

6.1.2.9 GATE LEAKAGE

For an OTFT to work in ideal manner, the gate current must be null at all bias conditions. However, in practice, the gate dielectrics have imperfections and together with other factors, such as surface conduction or the lack of semiconductor patterning determines undesired gate leakage. The performance of OTFT is better when drain current is equal to the source one. The leakage can be estimated when the drain-source voltage is zero over several V_{GS}, while the gate current can be obtained for various bias conditions.

6.1.2.10 HYSTERESIS/BIAS-STRESS

A serious issue in drive characteristics is represented by hysteresis. Such responses are, by their nature, problematic to model because the device retains the history of drive and the response will be affected by that. Thus, it is desirable to reduce hysteresis as much as possible. Hysteretic device characteristics are connected to the transport of carriers, which depending on the applied bias, are released by dielectric relaxation, conducting to errors in parameter evaluation. Among the short-term hysteretic impacts one can mention[5]: dielectric charge storage, slow relaxation of the gate material, and presence of traps in the semiconductor. Moreover, irreversible storage and bias-stress can also be noticed in OFET systems. Deep characterization through the application of the stressor can reveal the outcome of these effects.

6.1.3 OTFT STRUCTURES AND MODELS

There are several types of OTFTs structures, which can be classified based on the ordering of the layers regardless of the materials and sizes. Even the structures are differentiated based on the relative position of source (S), drain (D), and gate (G) contacts with regard to the organic semiconductor. The first one was made in 1960s and from that point, scientists attempted to create other prototypes with better electrostatic control of gate over the channel. Specific merits and demerits are ascribed to each of them. The most important categories of OTFT structures are the following ones[6,21]:

- *Single-gate OTFT structure* is discerned on the basis of the place where the gate is positioned. It can be found either on the top or at the bottom, and thus named as top gate and bottom gate structures, respectively. In the case of OTFT in top gate structures the performance is not so high since the semiconductor layer is contaminated during the deposition of a metal gate at high temperature. For this reason, bottom gate structures are more advantageous. The position of a source and drain contacts in regard with the active layer further divides them into the top contact and bottom contact structures, while the gate element remains at the same position. The performance of the device is affected by its structure and less by constituting materials because more attention should be given to the path crossed by the charge carriers between the source and drain.[22] A top contact structure is characterized by a larger current owing to the big injection area for the charge carriers determining a smaller contact resistance. On the other hand, the bottom contact structure is not so reliable if considering the raised metal–organic semiconductor contact resistance. This is a result of contact barriers as well as a reduced uniformity of casted semiconductor around the prepatterned S/D contacts;
- *Dual gate OTFT structure*: through the new configuration the device allows to accomplish enhanced charge carrier modulation in the semiconductor layer. Literature reports several devices based on CdSe, a-Si:H, and pentacene.[23–25] The best improved feature of this OTFT in comparison to single gate consists in better control on the threshold voltage. Other advantages include higher on-current, steeper and $S_{s\text{-}th}$.[6] In this OTFT configuration, one gate is placed in the front (or bottom) along with its front dielectric, source/drain contacts, organic semiconductor, and the other back (or top) gate is near the back dielectric. The carriers are drawn to the front gate in the channel, while a bias on the back gate enhances the conductivity of the channel electrostatically;[6]
- *Vertical channel OTFT structure* appeared as an attempt to solve the issues of the other types of structures, namely morphological disorders, reduced carrier mobility and high channel length. The vertical structure of OTFT could represent a good alternative to produce smaller length devices.[26] Such OTFT is made of five distinct layers that include metallic sheets of the source, drain, and gate in conjunction with two semiconductor layers. The

performance of the vertical transistor is considerably enhanced by creation of a Schottky contact at the interface between the active layer and gate,[27] diminishing the contact resistance,[28,29] introduction of a semiconductor layer at the contact–organic semiconductor interface, and fine controllability of the channel by using a meshed structure for the source electrode;[6]

- *Cylindrical gate OTFT structure* is made of a metal core of yarn which functions as the gate electrode, thereafter it is covered with a thin dielectric material. Then, the organic semiconductor is deposited, and finally S/D contacts (metal or conductive polymer) are accomplished through either thermal evaporation or soft lithography procedures. This kind of structure can be achieved on a single fiber substrate or at the intersection of two isolated fibers. Literature shows that cylindrical gates can be made of poly(vinyl cinnamate) and poly(4-vinyl phenol).[30] Cylindrical gate OTFT have attested to be adequate to manufacture circuits for wearable electronics owing to their hysteresis-free operation and raised bending stability. Moreover, cylindrical structures are optimal for size diminution, thus aiming for high packing density.

Several analytical models are often incorporated in the simulators to evaluate the performance of electronic devices and the circuits.[6,8] These models must be accurate for ensuring the device simulation along with an adequate proportion of convergence in the circuit functioning. Essentially, the model must consider the material features and the physical bases of a device structure. In addition, the model must be easily put into practice, upgradable, and reducible. Some OTFT models are shortly reviewed in the following paragraphs, namely, compact direct current model, charge drift model, and charge drift model for subthreshold region. More details about them are found in Refs. [5–8].

6.2 MATERIALS USED IN FABRICATION OF OTFTs

The developments made in the past decades on organic transistor area were focused on only on optimization of transistor structures and models, but also on synthesis of new materials that would raise the performance of such devices. Figure 6.1 depicts the scheme of an organic transistor and the constituting elements.

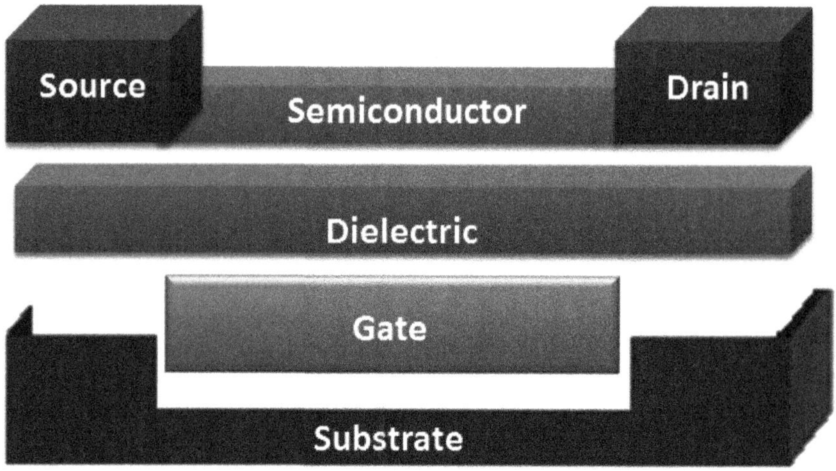

FIGURE 6.1 Schematic representation of the main materials from which an organic transistor is made: substrate, dielectric, semiconductor, and electrodes.

In this way, it was noticed that the applicability of the organic transistors has a known fast increase. Besides their properties, the constituting elements must be judged from the point of view of their cost effectiveness and processing. The reliability of an OTFT depends on the features of compounds used for the active layer, dielectric film, and electrodes. Moreover, the adequate selection of substrate makes possible the implementation of device in flexible circuitry at low cost. In the next paragraphs, a short description of the most used materials, including semiconductors, dielectrics, electrodes, and substrates, is discussed.

6.2.1 ELECTRODE MATERIALS

The proper choosing of the electrode materials is essential for good operation of the device. When selecting the contact metal for the source and drain electrodes, one must pursue low contact resistance. Thus, the contact should exhibit a reduced interface barrier with the active layer, ensuring a huge number of carrier injections. In *p*-type OTFTs, the electrodes are made of gold. The rare metal is preferred owing to high work function (5.1 eV), which is close to the HOMO level of most organic semiconductors (4.9 eV).[31] Therefore, a low interface barrier (~0.2 eV) is achieved. Utilization of other metals, such as titanium, nickel, palladium, and chromium together with gold enhances the adhesive properties.[6] Some metals, such as magnesium and aluminum are characterized by low work function and render relatively higher electron mobility in single crystals.[32]

The materials used for gate electrode must induce good adhesion and patterning capabilities with substrate and gate dielectric, respectively. The work function of compounds designed for gate should be similar to the semiconductor for achieving low threshold voltage. For such electrode fabrication, one should use materials that are doped with silicon, aluminum, and indium tin oxide.[7,29,33] It is proven that for cathode terminal, the compounds with small values of work function (magnesium, lithium, and calcium) are preferred.

Other trends reveal that graphite-based inks[34] and conducting polymers[35] are also good candidates for fabricating the electrodes, allowing accomplishment of flexible organic devices. Among the macromolecular compounds with good charge transport abilities, one can mention poly-3,4-ethylenedioxythiophene (PEDOT) and polyaniline. Derivate materials from these polymers are proved to be suitable for electrode fabrication, namely, poly(3,4-ethylenedioxythiophene):styrene sulfonic acid (PEDOT:PSS), polyaniline doped with camphorsulfonic acid (PANI-CSA), and poly(styrene sulfonate) (PSS).[35–38]

6.2.2 ORGANIC SEMICONDUCTORS

The utilization of organic semiconductors (conducting polymers and small molecules) has led to notable progress in transistors area, particularly through enhancement of the charge transport properties. In case of polymers, one may remark that mobility is smaller comparatively with their counterparts because of their high molecular weight. To overcome this issue, it is desirable to obtain large grains of semiconductors. The production of highly ordered thin films represents a good alternative for increasing the performance of the device. Low-molecular-weight compounds exhibit better mobility than the polymers (>0.9 cm^2/Vs), but in the past decade, this gap is diminishing because of developments in fabrication methodologies.

The organic semiconductors can be divided in two main categories:

- *n-type*: There are fewer efforts made to synthesize n-type semiconductors, but such materials are mandatory when designing a complementary inverter. Most of the *n*-type semiconductors have a certain degree of instability in the air which is determined by the free energy of activation related with the chemical reaction with either oxygen and sometimes with water. The properties of this category of semiconductors are influenced by operational conditions. Introduction of –Cl, –CN, and –F groups to the outermost molecular orbitals enhances the ability to withdraw electrons, thus improving the stability and mobility[39,40];

- *p-type*: They exhibit the hole as the charge carrier in their doped form. Latest research is focused on polymers and π-conjugated oligomers. Among the most used *p*-type conducting organic semiconductors, one can indicate pentacene,[41] thiophene,[42] and their derivatives.[43–47] Copper phthalocyanine and tetracene are other small molecules used in OTFTs.[6] Though low-molecular-weight organic semiconductors have higher mobility than polymers, the majority are almost insoluble in the organic solvents. Pentacene is shown to be an excellent material in terms of hole mobility as a result of good overlapping of the molecular orbitals in the crystal lattice. In addition, this compound has high chemical stability including in adverse environmental conditions, orderly formation in thin-film structure, and suitable adhesion at interface with most electrode metals. However, its reduced solubility limits its utilization in applications that demand low-cost printing methods. In comparison to pentacene, poly(3-hexylthiophene) has less carrier mobility, but owing to its remarkable solubility in several organic solvents, it is easier to process from solution phase. A recent report[48] showed that liquid crystals, such as thienylethynyl-terthiophene or phenyl-terthiophene derivatives can be used to construct single-domain crystalline OTFTs by a phase transition from a nematic phase on an alignment layer with a mobility of 10^{-2} cm^2/Vs. Other devices were based on di-thienyl-naphthalene, hexabenzocoronene, bis-(5′-hexylthiophen-2′-yl)-2,6-anthracene, and phenyl-benzothiazole derivatives.[48] The device mobilities are enhanced twice by thermally annealing of the liquid crystal phases. In low-molecular-weight liquid crystalline molecules, annealing at the temperature in low-ordered smectic phases is more pronounced comparatively with that at the temperature in highly ordered smectic phases.

The mobility of *p*- and *n*-type organic semiconductors has been enhanced in the past decade. A continuous growth is noticed for both types of materials, ascribed to the unremitting developments in the synthesis and manufacturing process. Pentacene is the best performing *p*-type material that showed outstanding increase in mobility. On the other hand, fullerene (*n*-type) has also proved notable semiconducting features. Many high-performance soluble fullerene-based materials have been analyzed for testing *n*-type OTFTs.

6.2.3 DIELECTRIC LAYER

The dielectric layer should be characterized by an elevated electrical resistivity to avoid the leakage between gate metal and semiconductor channel. In addition, the insulating film should have a big dielectric constant to exhibit enough capacitance for channel current flow. Dielectrics with high permittivity determine low switching voltage of the OTFTs. Utilization of tantalum oxide or zirconate titanate led to devices with low-voltage operating characteristics.[49,50]

Other studies[51,52] show that strontium titanate- and barium titanate-based dielectrics are also suitable for dielectric layers in organic transistors. Polymethyl methacrylate and cyanoethyl pullulan tend to adhere to semiconductor and boost carrier mobility. In comparison with classical silicon dioxide insulator, one can observe that the other dielectrics augment the mobility of almost 10 times.[53] The interactions at the interface of dielectric with active layer are essential for the transistor operation, so smooth surface with minimum amount of defects is pursued.[54] Recent studies were focused on new polymer materials as dielectrics for OTFT, such as *p*-sexiphenyl/vanadyl-phthalocyanine,[55] poly(vinylidene fluoride-trifluoroethylene-chlorofloroethylene),[56] and poly(vinylidene fluoride–trifluoroethylene).[57] Implementation of these materials in organic transistors led to improvements in their performance.

6.2.4 GATE DIELECTRIC MATERIALS

The gate dielectric material has the role to prevent current leakage between the gate and semiconducting layer. The accumulation of charge carriers at the semiconductor–insulator interface is mainly affected by the permittivity and the thickness of the insulating layer. Good interfacial adhesion of gate to substrate and semiconductor is desired for preventing creation of dipoles and trap states. Additional features that are also mandatory are high dielectric constant and high resistance. High permittivity materials, such as La_2O_3, HfSiOx, HfLaO, ZrO_2, $Pr6O_{11}$, HfO_2, and Al_2O_3, are suitable for a steep $S_{s\text{-th}}$ and small threshold voltage owing to their low band gaps.[6] For submicron devices, it is preferable to reduce the dielectric thickness for eliminating the short channel effects. At the same time, the gate material should be characterized by high breakdown voltage and long-term stability. These features are found in many polymers, poly(methyl methacrylate), polyvinyl pyrrolidone, poly(4-vinylpyridine), polyvinylidene fluoride, and propylene. Other macromolecular compounds exhibit in addition good adhesion properties, easy processing, and low manufacturing costs. Among them, one can mention polystyrene, polyvinyl alcohol, and polyimides.

The devices made of dielectrics with low permittivity need a high operating voltage. A solution would be to cast a very thin layer of dielectric, but the current techniques cannot avoid the formation of defects during processing and they lead to undesired leakage current.[58,59] In the past years, it was noticed that polymer nanocomposites represent an important option for gate dielectrics.[60] Introduction of high permittivity nanoparticles in polymer matrices leads to materials that meet the dielectric criteria and at the same time, are easy to process in order to ensure the adhesion with other transistor layers. In this context, gate dielectrics were made from barium strontium titanate and barium zirconate nanoparticles dispersed in poly(vinylidene fluoride-co-hexafluoropropylene),[61] benzocyclobutene/barium titanate polymer nanocomposites,[62] TiO_2/polystyrene core–shell nanoparticles,[63] TiO_2/polyimide,[64] poly(4-vinylpyridine)/aluminum nitride,[65] polystyrene/barium titanate,[66] and poly(4-vinylpyridine)/barium titanate.[67] However, such multiphase materials present some issues that arise from nanoparticle dispersion and poor wettability with polymer matrix that determined scientists to find solutions.[68-71] A first alternative was to chemically modify the matrix to enhance the contact between the two phases. For example, hydroxyl-containing polyimide/barium titanate were found to lead to bottom-gate top-contact OTFTs, which present low threshold voltages (−4.09–2.62 V), relatively high field-effect mobility rates (3.36×10^{-2}~2.32×10^{-1} cm^2 V^{-1} s^{-1}), and high I_{on}/I_{off} ratios (~10^5).[72]

Another aspect that should be addressed is connected to the percolation effects. For gate dielectric processed from solution phase, the rheological behavior is very sensitive to microstructural changes in material as a result of occurrence of nanoparticle networks in the continuous phase. These modifications are easily observed in the shear viscosity dependence on shear rate, which gains the form typical to thinning fluid at low shearing. The slope of thinning curve becomes pronounced as the amount of introduced filler increases. Temperature is an important factor that affects the nanocomposite microstructure since polymer chains are more mobile and causes disruption of the percolation network.[67] Moreover, the microstructure variations have implications on the material morphology.[73] Atomic force microscopy examination of a polystyrene/barium titanate nanocomposite reveals that addition of the nanofiller determines the shrinkage of the distance between ceramic particles which after a certain concentration begin to form pseudonetwork. At high percentages of reinforcement, the materials present heights of at least the ceramic particle size. The utilization of such composites in OTFTs is adequate since they have a smooth surface, that is, root mean square roughness ranging from 16 to 64 nm.[73]

A similar investigation pursued the effect of surface roughness of the nanocomposite gate dielectric on the correlation between grain size and mobility of pentacene-based device.[74] The high capacitance dielectric was prepared by inserting in a polyimide TiO_2 particles from 0 to 5 vol%. This led to a roughness of 0.8–20 nm for the nanocomposite, which reduced the grain dimension. At reinforcement of 0–2 vol%, the mobility is not much modified with decreasing grain size of semiconductor. This is similar to the behavior of the polymer gate dielectric. For 3–5 vol% TiO_2, it was revealed that mobility is significantly reduced with diminishing the grain dimension of pentacene, analogous to the behavior of the gate material.[74] Another solution to raise the efficiency of OTFTs, when composites are used as gate materials relies on compositing of over-coated polyimide on self-assembled layer of mixed HfO_2 and ZrO_2 nanoparticles.[75] This conducted to higher dielectric constant of the gate and lower leakage characteristics.

Recent progress on the preparation of organic-inorganic hybrid thin layers as gate insulating components proved that polymethylsilsesquioxane (PMSQ) are suitable for OTFT design.[76] This polymer can be prepared by sol–gel condensation of methyltrimethoxysilane. In order to achieve highly cross-linked PMSQ films with minimum concentration of residual silanol groups that lower electric resistivity, PMSQ was obtained in distinct reaction conditions.[77] The solvent polarity and the amount of water or methanol molecules are affecting the condensation reaction rate, and implicitly the concentration and stability silanol. It was revealed that the electrical resistivity was changed in the range from 10^{12} to 10^{14} Ω cm as a result of different nature of the organic solvent. The PMSQ prepared in methanol presented a high resistivity of about 10^{12} Ω cm, which is of two orders of magnitude lower than that of material formed in toluene. Electrical characterization indicated that inside the PMSQ, dense siloxane networks appeared which were cured at 150°C. The dielectric revealed a high breakdown field (over 3.0 MV cm^{-1}) together with a very low leakage current; the latter being much lower than that of polyvinyl pyrrolidone and polyimide.[78] Further improvements can be accomplished through the control of surface hydrophobicity of PMSQ. This is done by employing a complex chemical procedure involving octadecyltrichlorosilane functionalization and thermal treatment. The increase in gate hydrophobicity was reflected in the OTFT performance, namely, four times higher mobility and larger on/off ratio compared to that with the nontreated PMSQ. Another route toward hydrophobic surface consists in cocondensation of methyltrimethoxysilane and a small amount of alkyltrimethoxysilanes, resulting copolymethylsilsesquoxanes (co-PMSQs).[79] It was noticed that the thin surface of co-PMSQ became more hydrophobic comparatively with that of pristine PMSQ. As the surface free energy of the gate material is reduced, the mobility increased. Moreover, the permittivity of the PMSQ was improved by reaction with alkyltrialkoxysilanes and methyltrimethoxysilane. In this way, the material contains more polarizable groups, such as cyano or epoxy, increasing the dielectric constant. However, the device performance was limited by the hydroxyl groups produced by ring opening of epoxy groups that affected the charge trapping.[76] Conversely, co-PMSQ containing cyano group (PMSQ-CN), prepared from cocondensation with cyanoethyltrimethoxysilane, exhibited both improved permittivity and increased drain current. Furthermore, PMSQ gains photocurability by introducing photofunctional acrylic groups.[80] The sol–gel cocondensation of methyltrimethoxysilane and acryloxypropyltrimethoxysilane led to a copolysilsesquioxane. In combination with Darocur 1173 photoradical initiator, one may observe after UV exposure the development of a negative pattern. The amount of acrylic groups in PMSQ affected the electric resistivity, but as required for gate-insulating materials. Also, given the lack of dielectric constant changes revealed no existence of ionic impurities. The features of OTFT with PMSQ-acryl as a gate-insulating material were similar to that with pristine PMSQ.[76]

6.2.5 SUBSTRATE AND PASSIVATION MATERIALS

In order to choose the adequate substrate material for any device, one must pay attention to on the kind of application the device is meant for. For substrates, it can be stated that silicon is suitable for electronics not only because of its intrinsic characteristics but also for its capacity to form an oxide layer during thermal oxidation. On the other hand, organic substrates with considerable degree of pliability are necessary for flexible electronics. In 1990, the first OTFT was fabricated on the glass substrate.[81] Then, in the same year, an organic transistor was manufactured on a flexible imidic polymer substrate that proved a performance similar to that of OTFTs deposited on silicon or glass.[82] After 7 years, the scientists made a fully printed P3HT-based device which had a substrate constituted of ITO-coated polyester.[83] Meanwhile, several fully printed organic transistors were realized, and for the construction of their flexible substrates, the following materials were reported: polyethylene naphthalate (PEN), polyethylene terephthalate (PET), polyimide, polyethylene, plastic, paper, and fiber. All these have opened a novel era of flexible low-cost printed electronics.[6]

Passivation materials (encapsulation) are needed to secure the devices from the effects created by the environmental factors, such as mechanical damaging (scratches) and degradation as a result of water, oxygen, or electromagnetic radiations.[9] Such encapsulation materials should have the following characteristics: smooth surface, barrier capacity, temperature resistance, and dimensional and mechanical stability during deformation. Among the materials that meet these demands, one can mention: polyvinyl acetate/photoacryl,[84] polyvinyl alcohol and photoinitialized acryl,[85] polyvinyl alcohol/layered silicate nanocomposite,[86] and SnO_2 thin films achieved by ion-beam-assisted deposition.[87]

The effect of different passivation layers on properties of pentacene OTFTs was investigated.[88] Four types of organic materials were examined: photopatternable acryl (PA), polyvinyl alcohol (PVA)/dichromated polyvinyl alcohol (D-PVA), PVA/D-PVA/PA, and parylene-C. In the first stage, the mobility was found to be 1 cm^2Vs and the I_{on}/I_{off} of 10^7. These parameters were affected for device with PA as passivation layer because of the damage to the pentacene film by the organic solvent in PA solution. At the same time, almost no change was remarked for PVA/D-PVA material. This was performed to diminish the amount of metal ionic impurity at the interface of the pentacene and PVA. The OTFT covered with PVA has the advantage of reducing damage to the pentacene and avoid undesired effects of the organic solvent in PA solution. Utilization of PVA/D-PVA/PA material best retains the initial performance of the device in terms of field-effect mobility and I_{on}/I_{off}.[88]

Table 6.1 summarizes the main material types used in manufacturing OTFTs for a clearer perspective of the progresses made in OTFT area.

TABLE 6.1 The Most Used Materials for Construction of OTFTs Elements.

Elements of OTFTs	**Material characteristic**		**References**
Electrode materials	*Work function (eV)*		
Copper	4.70		6
Platinum	5.65		6
Gold	5.10		31
Nickel	4.10		6
Doped silicone	3.90		6
Indium tin oxide	5.30		6
Organic semiconductors	Mobility (cm^2/VS)	I_{on}/I_{off}	
Pentacene	1.00	1×10^5	89
Poly(3-hexylthiophene)	0.08	7×10^3	83
Poly(2-methoxy-5-(2′-ethyl-hexyloxy)-1,4-phenylene vinylene)	0.00016	1×10^4	6
C10-{Di-n-decyldinaptho[2,3-b:2 0, 3 o-f] thieno[3,2-b] thiophene}	2.400	1×10^7	6
Poly(3-hexylthiophene)/carbon nanoparticles (1–43 wt%)	0.00036–0.0123	2.26×10^5–2.10	90
N,N′-dioctyl-3,4,9,10-perylene tetracarboxylic diimide	0.60	$>10^5$	40
Fullerene	5.1	10^6	6
Perfluoropentacene	0.22	10^5	6
Dielectric layer and gate materials	*Permittivity*		
$BaTiO_3$	~500		91
TiO_2	80		6
La_2O_3	30		6
ZrO_2	25		92
HfO_2	22		6

TABLE 6.1 *(Continued)*

Elements of OTFTs	Material characteristic	References
Y_2O_3	15	92
$HfSiO_4$	11	92
Poly(4-vinyl phenol)	5.30	6
Polymethyl methacrylate	3.60	6
Polyimide	3.33	93
Poly(arylene ether)	2.90	92
Poly(arylene ether oxazole)	2.80	92
Poly(phenyl quinoxaline)	2.80	92
Polystyrene	2.40	94
Substrate and passivation materials	*Glass transition temperature (°C)*	
Polyvinyl acetate	30	95
Poly(vinyl alcohol)	85	95
Poly(ethylene isophthalate)	51	95
Poly(ethylene terephthalate)	76	95

6.3 FACTORS AFFECTING THE PERFORMANCE OF OTFTs

6.3.1 DIMENSIONAL PARAMETERS AND PROCESSING CONDITIONS

The dimensional aspect of OTFTs influences their reliability. For this reason, scientists focused their attention on some aspects such as channel length, device width, and the thicknesses of active and insulating layers in order to check the manner in which they can help to improve the performance of instrument. A high capacitance will determine a high drain current at low switching voltage, which can be accomplished by reducing the dielectric film thickness. This will also induce the augmentation of the carrier mobility, concomitantly with reduction the off-current that, in turn, lowers the leakage power and noise margins.[96,97] Other attempts were focused on tuning the performance of OTFT, through the channel length. It was revealed that scaling down the length from 125 to 10 nm has determined an enhancement of current by 13 times.[6,9]

The processing conditions are also important in creating reliable OTFTs since they influence charge injection efficiency and contact resistance. For instance, during metal deposition on the active layer one may notice variations of contact resistance. The sputtering of metals on organic layer can produce in-diffusion of the metal, modifications in the morphology of the organic film, and may cause disruption of chemical bonds in the macromolecules[98] These aspects can generate significant interfacial resistances in OTFTs, which can augment the contact resistance and damage device performance. Many efforts should be directed toward optimizing the processing to prevent chemical, physical, and morphological modifications generated by metal casting on the organic semiconductor layer. So, a good solution is to use solution-processable organic conductors, and thus, process-induced contact effects could be minimized.

The processing medium can influence the OTFT parameters due to the induced variable ambient-air stability.[99] The substrate surface/interface modification using self-assembled monolayers have significant impact on the carrier transport. Processing the materials in chloroform and hexamethyldisilazane could provide only 10 min of stability for the off current, while for 1,2,4-trichlorobenzene and *n*-octadecyltrichlorosilane the exposure to air revealed no change during 120 min. To achieve insight into the enhanced reliability, it would be preferable to use more hydrophobic, longer-chain self-assembled monolayer and solvents with high boiling-point. When manufacturing OTFTs of soluble organic semiconductors by spray-deposition, one should monitor the effect of processing parameters on film morphology and device mobility.[100] There is an effect of the type of solvent, the pressure of the carrier gas used in deposition, and the distance at which spraying occurs. Surface morphology analysis indicated that the molecules pack along the π-stacking direction, which is the favorite one for charge transport. The mobility of spray-deposited devices can be controlled through fabrication parameters and thus tuned up to two orders of magnitude. Regarding the spraying distance, it was shown that at short distances, the air brush is very close to the substrate and it has an analogous impact such as high gas pressure. Thus, the solution in the contact with the substrate is partially blown away by the incoming flow. This effect is no longer prevalent as the distance is higher, determining a better mobility. Further, increase of the spraying distance produces mobility drops to lower values. This could be explained by considering generation of lower film crystallinity as a result of the fact that the solvent evaporates prior to the tiny aerosolized solution droplets contact with the substrate. It is expected that higher distances will be optimal at higher pressures, to reduce as much as possible the solution elimination by the incoming flow.

6.3.2 ENVIRONMENTAL STABILITY

Environmental factors can deteriorate the friability of OTFTs since the majority of organic semiconductors are not stable during exposure to oxygen and water. Therefore, the main purpose was to eliminate these elements from organic semiconductor layers and encapsulate against their penetration. A facile solution would be to laminate a thin plastic coating on top of the device with a suitable adhesive.[101] Thus, the OFETs can be covered with a sheet of teon, parylene, or another polymer with suitable wettability properties, and perhaps, the device should be sealed through a metal–vapor barrier material.[102] A distinct option is to subject OFET components to high temperatures for dehydration in an inert ambient, but the procedure is limited by the material's temperature stability and reactivity with water.[103] For rigid structures, one can use an epoxy seal with a rigid lid and inside the cavity can be inserted a desiccant to remove water after passing the epoxy layer. Other sophisticated barrier layers are represented by transparent flexible substrates.

6.3.3 CONTACT EFFECTS AND MATERIALS FEATURES

The interface phenomena are influencing in a significant manner the working features of the OTFTs. Two main aspects must be closely examined when establishing the construction schemes and the materials for the device, namely, contact–semiconductor and semiconductor–dielectric interactions at interface. In ideal conditions, source and drain contacts of transistor are ohmic, thus the value of the contact resistance is negligible in comparison with the electrical resistance of the active layer. The resistances ascribed to carrier injection and collection steps can be combined into the contact resistance, whereas the resistance attributed to crossing the channel length in the semiconductor denotes the channel resistance. Maintaining the contact resistance value below to the channel resistance is essential for accomplishment of "ohmic contacts" in organic transistors.[54] Bao and Locklin[54] stated: "Importantly, nominally high-resistance contacts can still be ohmic as long as they are able to source or sink the current driven through an even more resistive channel." Moreover, contact resistance is affected by band lineup at the limit between the metal–organic layer. The effect of metal–active interface is less obvious in *p*-type semiconductors owing to their smaller barrier height in regard with n-type compounds.[6] The presence of dipoles at electrode–semiconductor interface seems to enhance the contact resistance. When morphological disorders are found in active layer, the mobility is diminished. Among the solutions for this issue, one can include casting of a supplementary thin organic active layer at the interface between contacts and semiconductor.[6]

On the other hand, the interface between the semiconductor and gate dielectric should be adapted to obtain proper ordering of the active layer and optimal charge carrier accumulation. The key factor here is the resistivity of the dielectric gate which is preferable to present high values in order to allow reducing interface trap density. The latter is further responsible for bigger mobility and current. In addition, by exposing the surface of dielectric to a specific treatment (with octyltrichlorosilane or NH_2 groups), the trap states can be diminished and thus more carriers are accumulated at the semiconductor–dielectric interface, improving device performance.[6]

Materials features are impacting the devices performance. The impact of the dielectric thickness on mobility in top-contact organic thin-film transistor was investigated by Singh and Mazhari.[104] They used poly(methylmethacrylate) (PMMA) and cross-linkable poly(4-vinylphenol) (PVA) as dielectric materials and attempted to make a correlation between polymer thickness and field-effect mobility. The band mobility is superior for pentacene/PMMA comparatively with pentacene/PVP device. They showed that at a fixed gate voltage, the mobility was significantly higher as thickness of both dielectrics is diminished. Careful analysis at a constant gate electric field or analogous generated accumulation charge proved that only a part of the enhancement takes place due to simple scaling of dielectric thickness, and the interface quality also plays an important role. Nanoscale morphological examination of the dielectric surface indicated that the roughness is smaller for thicker films. Variation of the capacitance of the dielectric layer led to a distinct behavior for top-contact and bottom-contact architectures.[105] Top-contact transistors exhibit almost constant contact voltage in the linear regime determining to an apparent mobility reduction and the opposite is noticed for bottom-contact structures. The source and drain contact thickness is also reported to influence on the performance of organic thin-film transistors.[106]

6.4 PRACTICAL IMPORTANCE OF OTFTs

6.4.1 ELECTRONIC INDUSTRY

The organic transistors are encountered in electronic and optoelectronic devices mostly because they are less expensive, can be manufactured at low temperature processing, and represent the molecular engineering approach to electronic industry. The main applications in this domain are:

- *Backplanes for displays*: printed OTFTs have deep implications in the production of displays.[5,6,107] In the first stages, a backplane for flexible displays consisted in array of 16 × 16 (256) organic transistors, which are able to drive an electrophoretic display. Such devices are mainly field (voltage)-driven and do not imply much current flow to work. Therefore, they are bistable, resulting that they can retain their state (image) without power. The latter is only demanded to switch the state of the device. The display material is constructed from little spheres which are filled with white-charged substance and a colored (black) liquid. In the presence of electric field, the white spherical particles are travelling either toward the top or the bottom of the liquid. When this component is positioned toward the observer, the display looks (white), whereas when it is placed at the other side (bottom) of the display, the color of the liquid (black) is noticed. The spheres and liquid can render any color. The contrast is not affected by viewing angle;
- *Radio-frequency identification* (RFID): an antenna connected to a rectifier is utilized to draw a dc power supply from RF radiation that falls on the tag.[107,108] This dc voltage is useful in proper functioning of a logic circuit on the tag, which is able to read out the tag memory and submits the data stored in the tag toward to a remote reader through load modulator. Operating domain of majority low-cost RFID instruments is somehow limited by power delivery from the reader to the passive tag;
- *Logic*: organic transistors can be combined in order to realize integrated circuits, which can be utilized to execute complex logic functions. The ability to combine transistors together into circuits is dependent on the ability to achieve voltage amplification;[107,108]
- *Gas sensors*: OTFT can be used to detect gases. Organic materials are sensitive to room temperature and selectivity is monitored

through an array of sensors (i.e., OTFTs) which pursues a single parameter (voltage, frequency) providing a pattern of data which is used to recognize the chemical species[109];

- *Mechanical sensors*: the mechanical flexibility of OTFTs makes them adequate to switch and amplify mechanical actuations. The initial prototype of OFETs-based mechanical sensor was obtained by laminating an OFET backplane with an elastomer conductor whose resistance ranged as a result of the applied pressure. By switching through the transistor matrix and monitoring the current low from a basic power source using a variable resistor, one may construct a force map that can be implemented as flexible skin sensor.[5]

6.4.2 BIOMEDICAL FIELD

The main biomedical applications where OTFTs have found utility are the following:

- *Organic bioelectronic tools*: active organic bioelectronic surfaces play an essential role in controlling the cell attachment and release, but also in cell signaling through electrical, chemical, or mechanical actuation.[110] The special properties of conductive polymers make them excellent materials for labeled or label-free biosensors. OTFTs are connected to the current techniques used for electronically controlled ion transport in organic bioelectronic devices;[110]
- *Biological sensors:* Bhaskar studied the implications of a pentacene-based transistor as a biological sensor to Fragile X Syndrome.[111] This is a genetic disorder ascribed to congenital mental retardation. The humans affected by this are not able to produce fragile X mental retardation protein (FMRP). However, the precise function of the FMRP is not elucidated. The transistor was brought in contact to biological and control environments in order to describe function of the FMRP. These environments were based on of certain components: water, bovine serum albumin, transfer ribonucleic acid (RNA), a salt solution, RNase inhibitor, FMRP, and FMR1 RNA. The device was exposed to each of these media, with the exception of the RNAse inhibitor, the FMR1 RNA, and the FMRP. In addition, the OTFT made contact with combined solution of all the components (an RNA binding assay),

subsequently the current–voltage characteristics from observing the transistor's response were registered before and after the exposure. The experiments were aiding to detect changes in fabrication that could be essential for optimal sensing, as well as changes that should be done to the environment to maintain the transistor's response in the proper range for the pursued component of the solution. In this context, OTFT can contribute to significant developments for the FMRP patients by clarifying mechanism that leads to fragile X syndrome;[111]

– *Biomedical sterilization*: a major challenge in this area is to overcome the issue of thermal instability. This is determined by low melting temperatures and high thermal expansion coefficients of organic components, which can produce thermal degradation.[112] Kuribara et al.[112] made a thermally stable OTFT based on a high-mobility organic semiconductor, dinaphtho[2,3-b:2′,3′-f]thieno[3,2-b]thiophene and 2 nm-thick gate dielectric. The device is suitable for medical sterilization purposes, which are occurring under saturated steam or hot water. To work properly under these conditions, the instrument was covered with a triple-layer stack of 300-nm-thick parylene, 200-nm-thick gold (Au), and 1-μm-thick parylene.

6.5 GENERAL REMARKS AND FUTURE PERSPECTIVES

The developments in the area of organic transistors are in continuous progress not only from the point of view of the construction materials but also from the direction of the design and processing technologies. Many efforts are still devoted to optimization of the OTFT working parameters to improve their performance and functionality.

The implementation of nanostructured materials in such devices will allow higher control on the pursued properties. There is also grown interest in using self-assembly of nanostructured dielectrics that behave as templates for the definition of growth areas for nanowires and nanostructures. However, this will involve a better comprehension of how to achieve a wanted size, orientation, and electronic features. Chemists can obtain novel materials, but metrology is limited to analysis of the structure of a very small number of these nanostructured materials. Thus, the relation between chemistry and nanostructure properties is not fully elucidated and remains as an aspect for future research direction.

The next significant step will be the further development of performant devices that will exploit unique features of new organic semiconductors. In this way, multifunctional systems that in the present are easy to be fabricated from inorganic semiconductors will be prepared. All technologies demand improvements in charge mobility to reduce the drive voltage. On the other hand, ample development is required to improve the quality and patterning methods for contacts based on soluble organic conductors. It seems that OTFTs will impact the advances for the next few years bringing technology to performance levels that will allow these devices to be incorporated in a wide range of modern products with outstanding performance, including light-weight, unbreakable, flexible displays on plastic substrates, biological sensor, and other unexplored applications.

ACKNOWLEDGMENT

This work is dedicated to the anniversary of "Petru Poni" Institute of Macromolecular Chemistry of Romanian Academy of Sciences.

KEYWORDS

- **transistors**
- **basic parameters**
- **materials**
- **electronics**
- **biomedicine**

REFERENCES

1. Sun, S. S.; Dalton, L. R. *Introduction to Organic Electronic and Optoelectronic Materials and Devices*; CRC Press: Boca Raton, 2017.
2. Milestones: Invention of the First Transistor at Bell Telephone Laboratories, Inc., 1947. IEEE Global History Network. http://ethw.org/Milestones:Invention_of_the_First_Transistor_at_Bell_Telephone_Laboratories,_Inc.,_1947. (accessed Jan 4, 2017)
3. https://en.wikipedia.org/wiki/Transistor#Types.
4. Pierret, R. F. *Semiconductor Device Fundamentals*; Addison Wesley Longman: Reading, MA, 1996, pp 525–732.

5. Kymissis, I. *Organic Field Effect Transistors: Theory, Fabrication and Characterization*; Springer: New York, 2009.
6. Kaushik, B. K.; Prajapati, B. K. S.; Mittal, P. *Organic Thin-film Transistor Applications. Materials to Circuits;* CRC Press: Boca Raton, 2017.
7. Chandar Shekar, B.; Lee, J.; Rhee, S. W. Organic Thin Film Transistors: Materials, Processes and Devices. *Korean J. Chem. Eng.* **2004,** *21*, 267–285.
8. Kumar, B.; Kumar Kaushik, B.; Negi, Y. S. Organic Thin Film Transistors: Structures, Models, Materials, Fabrication, and Applications: A Review. *Polymer Rev.* **2014,** *54*, 33–111.
9. Zaki, T. *Short-channel Organic Thin-film Transistors Fabrication, Characterization, Modeling and Circuit Demonstration;* Springer: New York, 2015.
10. Kumar, B.; Kaushik, B. K.; Negi, Y. S.; Mittal, P.; Mandal, A. In *Organic Thin Film Transistors Characteristics Parameters, Structures and Their Applications,* Recent Advances in Intelligent Computational Systems (RAICS), 2011 IEEE, doi: 10.1109/RAICS.2011.6069402, 2011.
11. Mittal, P.; Kumar, B.; Negi, Y. S.; Kaushik, B. K.; Singh, R. K. In *Organic Thin Film Transistor Architecture, Parameters and Their Applications*, 2011. International Conference on Communication Systems and Network Technologies. DOI: 10.1109/CSNT.2011.96.
12. Sirringhaus, H. 25th Anniversary Article: Organic Field-Effect Transistors: The Path Beyond Amorphous Silicon. *Adv. Mater.* **2014,** *26*, 1319–1335.
13. Ante, F.; Kälblein, D.; Zaki, T.; Zschieschang, U.; Takimiya, K.; Ikeda, M.; Sekitani, T.; Someya, T.; Burghartz, J. N.; Kern, K.; Klauk, H. Contact Resistance and Megahertz Operation of Aggressively Scaled Organic Transistors. *Small* **2012,** *8*, 73–79.
14. Sun, Y.; Liu, Y.; Zhu, D. Advances in Organic Field-effect Transistors. *J. Mater. Chem.* **2005,** *15*, 53–65.
15. Dimitrakopoulos, C. D.; Malenfant, P. R. L. Organic Thin Film Transistors for Large Area Electronics. *Adv. Mater.* **2002,** *14*, 99–117.
16. Horowitz, G.; Hajlaoui, M. E. Grain Size Dependent Mobility in Polycrystalline Organic Field-effect Transistors. *Synth. Metal* **2001,** *122*, 185–189.
17. Knipp, D.; Street, R. A.; Volkel, A.; Ho, J. Pentacene Thin Film Transistors on Inorganic Dielectrics: Morphology, Structural Properties, and Electronic Transport. J. *Appl. Phys.* **2003,** *93*, 347–355.
18. Gupta, D.; Katiyar, M.; Gupta, D. An Analysis of the Difference in Behavior of Top and Bottom Contact Organic Thin Film Transistors Using Device Simulation. *Organ. Electron.* **2009,** *10*, 775–784.
19. Subramanian, V. In *Toward Printed Low Cost RFID Tags—Device, Materials and Circuit Technologies,* 2nd Advanced Technology Workshop on Printing an Intelligent Future: Printed Organic and Molecular Electronic Technologies, Boston, MA, 2003.
20. Ante, F. Contact Effects in Organic Transistors. Ph.D. dissertation, Swiss Federal Institute of Technology in Lausanne, Lausanne, Switzerland, 2011.
21. Mittal, P.; Negi, Y. S.; Singh, R. K. A Depth Analysis for Different Structures of Organic Thin Film Transistors: Modeling of Performance Limiting Issues. *Microelectron. Eng.* **2016,** *150*, 17–18.
22. Pal, A.; Kumar, B.; Tripathi, G. S. In *Single Gate Based Different Structures of OTFTs: Prospective and Challenges*, IEEE Emerging Trends in Communication Technologies (ETCT), International Conference. DOI: 10.1109/ETCT.2016.7883001, 2017.

23. Luo, M. F. C.; Chen, I.; Genovese, F. C. A Thin Film Transistor for Flat Panel Displays. *IEEE Trans. Electron Devices* **1981,** *28*, 740–743.
24. Tuan, H. C.; Thompson, M. J.; Johnson, N. M.; Lujan, R. A. Dual Gate a-Si:H Thin Film Transistors. *IEEE Electronic Device Lett.* **1982,** *3*, 357–359.
25. Kaneko, Y.; Tsutsui, K.; Tsukada, T. Back Bias Effect on the Current Voltage Characteristics of Amorphous Silicon Thin Film Transistors. *J. Non-Crystalline Solids* **1992,** *149*, 264–268.
26. Nishizawa, J.; Terasaki, T.; Shibata, J. Field Effect Transistor Versus Analog Transistor (Static Induction Transistor). *IEEE Trans. Electron Devices* **1975,** *22*, 185–197.
27. Kudo, K.; Wang, D. X.; Lizuka, M.; Kuniyoshi, S.; Tanaka, K. Organic Static Induction Transistor for Display Devices. *Thin Solid Film* **2000,** *111–112*, 11–14.
28. Chen, Y.; Shih, I. Fabrication of Vertical Channel Top Contact Organic Thin Film Transistors. *Org. Electron.* **2007,** *8*, 655–661.
29. Watanabe, Y.; Kudo, K. Vertical Type Organic Transistor for Flexible Sheet Display. *Proc. SPIE* **2009,** *7415*, 741515-1–741515-10.
30. Jang, J.; Nam, S.; Park, J. J.; Im, J.; Park, C. E.; Kim, J. M. Photocurable Polymer Gate Dielectrics for Cylindrical Organic Field Effect Transistors with High Bending Stability. *J. Mat. Chem.* **2012,** *22*, 1054–1060.
31. Li, C.; Pan, F.; Wang, X.; Wang, L.; Wang, H.; Wang, H.; Yan, D. Effect of the Work Function of Gate Electrode on Hysteresis Characteristics of Organic Thin-film Transistors with Ta_2O_5/Polymer as Gate Insulator. *Org. Electron.* **2009,** *10*, 948–953.
32. Schön, J. H.; Kloc, Ch.; Batlogg, B. On the Intrinsic Limits of Pentacene Field-Effect Transistors. *Org. Electron.* **2000,** *1*, 57.
33. Klauk, H.; Zschieschang, U.; Halik, M. Low Voltage Organic Thin Film Transistors with Large Transconductance. *J. Appl. Phys.* **2007,** *102*, 074514-1–074514-7.
34. Yang, W.; Wang, C. Graphene and the Related Conductive Inks for Flexible Electronics. *J. Mater. Chem. C* **2016,** *4*, 7193–7207.
35. Garnier, F. Thin-Film Transistors Based on Organic Conjugated Semiconductors. *Chem. Phys.* **1998,** *227*, 253.
36. Tiwari, S. P.; Namdas, B.; Rao, V. R.; Fichou, D.; Mhaisalkar, S. G. Solution Processed n-type Organic Field Effect Transistors with High ON/OFF Current Ratios Based on Fullerene Derivatives. *IEEE Electron Device Lett.* **2007,** *28*, 880–883.
37. Cosseddu, P.; Bonfiglio, A. A Comparison Between Bottom Contact and Top Contact All Organic Field Effect Transistors Assembled by Soft Lithography. *Thin Solid Films* **2007,** *515*, 7551–7555.
38. Maccioni, M.; Orgiu, E.; Cosseddu, P.; Locci, S.; Bonfiglio, A. Towards the Textile Transistor: Assembly and Characterization of an Organic Field Effect Transistor with a Cylindrical Geometry. *Appl. Phys. Lett.* **2006,** *89*, 143515-1–143515-3.
39. Bao, Z.; Lovinger, A. J.; Brown, J. New Air Stable n-channel Organic Thin Film Transistors. *J. Am. Chem. Soc.* **1998,** *120*, 207–208.
40. Malenfant, P. R. L.; Dimitrakopoulos, C. D.; Gelorme, J. D.; Kosbar, L. L.; Graham, T. O.; Curioni, A.; Andreoni, W. n-Type Organic Thin Film Transistor with High Field Effect Mobility Based on a N,N'-dialkyl-3,4,9,10-perylene Tetracarboxylic Diimide Derivative. *Appl. Phys. Lett.* **2002,** *80*, 2517–2519.
41. Sarma, R.; Saikia, D.; Saikia, P.; Saikia, P. K.; Baishya, B. Pentacene Based Thin Film Transistors with High-k Dielectric Nd_2O_3 as a Gate Insulator. *Braz. J. Phys.* **2010,** *40*, 357–360.

42. Deng, Y.; Sun, B.; Quinn, J.; He, Y.; Ellard, J.; Guo, C.; Li, Y. Thiophene-S,S-dioxidized Indophenines as High Performance n-type Organic Semiconductors for Thin Film Transistors. *RSC Adv.* **2016,** *6,* 45410–45418.
43. Horowitz, G. Organic Field-effect Transistors. *Adv. Mater.* **1998,** *10* (5), 365–377.
44. Schön, J. H.; Kloc, Ch.; Batlogg, B. Hole Transport in Pentacene Single Crystals. *Phys. Rev. B* **2001,** *63,* 245201.
45. Afzali, A.; Dimitrakopoulos, C. D.; Breen, T. L. High-performance, Solution-processed Organic Thin Film Transistors from a Novel Pentacene Precursor. *J. Am. Chem. Soc.* **2002,** *124,* 8812.
46. Gundlach, D. J.; Klauk, H.; Sheraw, C. D.; Kuo, C. C.; Huang, J. R.; Jackson, T. N. High-mobility, Low Voltage Organic Thin Film Transistors. *Int. Elect. Devices Meeting Technical Digest* **1999,** *111,* 521–524.
47. Nelson, S. F.; Lin, Y. Y.; Gundlach, D. J.; Jackson, T. N. Temperature-independent Transport in High-mobility Pentacene Transistors. *Appl. Phys. Lett.* **1998,** *72,* 1854.
48. Iino, H.; Hanna, J. Liquid Crystalline Organic Semiconductors for Organic Transistor Applications. *Polymer J.* **2016,** *49,* 23–30.
49. Bartic, C.; Jansen, H.; Campitelli, A.; Borghs, S. Ta_2O_5 as Gate Dielectric Material for Low-voltage Organic Thin-film Transistors. *Org. Electron.* **2002,** *3,* 65.
50. Dimitrakopoulos, C. D.; Purushothaman, S.; Kymissis, J.; Callegari, A.; Shaw, J. M. Low-voltage Organic Transistors on Plastic Comprising High-dielectric Constant Gate Insulators. *Science* **1999,** *283,* 822.
51. Yu, Y. Y.; Jiang, A. H.; Lee, W. Y. Organic/Inorganic Nano-hybrids with High Dielectric Constant for Organic Thin Film Transistor Applications. *Nanoscale Res Lett.* **2016,** *11,* 488.
52. Cai, Q. J.; Gan, Y.; Chan-Park, M. B.; Yang, H. B.; Lu, Z. S.; Li, C. M.; Guo, J.; Dong, Z. L. Solution-processable Barium Titanate and Strontium Titanate Nanoparticle Dielectrics for Low-voltage Organic Thin-film Transistors. *Chem. Mater.* **2009,** *21,* 3153–3161.
53. Okubo, S. Organic Transistors Expedite Flexible EL Displays, Asia Bitz Tech. 2001, http://neasia.nikkeibp.com/nea/200112/peri_161078.html (accessed Feb 9, 2017).
54. Bao, Z.; Locklin, J. *Organic Field-Effect Transistors;* CRC Press: Boca Raton, 2007.
55. Li, Y.; Wang, H.; Shi, Z.; Mei, J.; Wang, X.; Yan, D.; Cui, Z. Novel High-k Polymers as Dielectric Layers for Organic Thin-film Transistors. *Polym Chem.* **2015,** *6,* 6651–6658.
56. Li, J.; Liu, D.; Miao, Q.; Yan, F. The Application of a High-k Polymer in Flexible Low-voltage Organic Thin-film Transistors. *J. Mater. Chem.* **2012,** *22,* 15998–16004.
57. Müller, K.; Paloumpa, I.; Henkel, K.; Schmeißer, D. Organic Thin Film Transistors with Polymer High-k Dielectric Insulator. *Mater. Sci. Eng. C* **2006,** *26,* 1028–1031.
58. De Angelis, F.; Cipolloni, S.; Mariucci, L.; Fortunato, G. High-field-effect-mobility Pentacene Thin-film Transistors with Polymethylmetacrylate Buffer Layer. *Appl. Phys. Lett.* **2005,** *86,* 203505-1–203505-3.
59. Chou, W. Y.; Kuo, C. W.; Cheng, H. L.; Chen, Y. R.; Tang, F. C.; Yang, F. Y.; Shu, D. Y.; Liao, C. C. Effect of Surface Free Energy in Gate Dielectric in Pentacene Thin-film Transistors. *Appl. Phys. Lett.* **2006,** *89,* 112126-1–112126-3.
60. Dong, Y.; Umer, R.; Kin-Tak Lau, A. *Fillers and Reinforcements for Advanced Nanocomposites*; Elsevier: Amsterdam, 2015.

61. Faraji, S.; Hashimoto, T.; Turner, M. L.; Majewski, L. A. Solution-processed Nanocomposite Dielectrics for Low Voltage Operated OFETs. *Organ. Elect.* **2015,** *17*, 178–183.
62. Lu, J.; Moon, K. S.; Wong, C. P. In *High-k Polymer Nanocomposites as Gate Dielectrics for Organic Electronics Applications*, 57th Proceedings on Electronic Components and Technology Conference, 2007. ECTC '07. DOI: 10.1109/ECTC.2007.373836, 2007.
63. Maliakal, A.; Katz, H.; Cotts, P. M.; Subramoney, S.; Mirau, P. Inorganic Oxide Core, Polymer Shell Nanocomposite as a High K Gate Dielectric for Flexible Electronics Applications. *J. Am. Chem. Soc.* **2005,** *127*, 14655–14662.
64. Lee, W. H.; Wang, C. C. Effect of Nanocomposite Gate-dielectric Properties on Pentacene Microstructure and Field-effect Transistor Characteristics. *J. Nanosci. Nanotechnol.* **2010,** *10*, 762–769.
65. Barzic, R. F.; Barzic, A. I. Dumitrascu, Gh. Percolation Network Formation in Poly(4-vinylpyridine)/Aluminum Nitride Nanocomposites: Rheological, Dielectric, and Thermal Investigations. *Polym Compos.* **2014,** *35*, 1543–1552.
66. Barzic, R. F.; Barzic, A. I.; Dumitrascu, Gh. Percolation Effects on Dielectric Properties of Polystyrene/$BaTiO_3$ Nanocomposites. *U.P.B. Sci. Bull., Series A* **2014,** *76*, 225–234.
67. Barzic, A. I. Temperature Implications on the Rheological Percolation Threshold in Poly(4-vinylpyridine)/Barium Titanate Nanocomposites. *Rev. Roum. Chim.* **2014,** *59*, 515–519.
68. Thomas, S.; Zaikov, G. E.; Valsaraj, S. V. *Recent Advances in Polymer Nanocomposites*; Brill NV: Leiden, Netherlands, 2009.
69. Thomas, S.; Zaikov, G. E. *Progress in Polymers Nanocomposites Research*; Nova Publishers: USA, 2008.
70. Harrats, C.; Thomas, S.; Groeninckx, G. *BMicro- and Nanostructured Multiphase Polymer Blend Systems: Phase Morphology and Interfaces*, CRC Press: USA, 2005.
71. Thomas, S.; Yang, W. *Advances in Polymer Processing: From Macro- to Nanoscales*; Woodhead Publishing Ltd: UK, 2009.
72. Yu, Y. Y.; Liu, C. L.; Chen, Y. C.; Chiu, Y. C.; Chen, W. C. Tunable Dielectric Constant of Polyimide–Barium Titanate Nanocomposite Materials as the Gate Dielectrics for Organic Thin Film Transistor Applications. *RSC Adv.* **2014,** *4*, 62132–62139.
73. Barzic, A. I.; Stoica, I.; Barzic, R. F. Microstructure Implications on Surface Features and Dielectric Properties of Nanoceramics Embedded in Polystyrene. *Rev. Roum. Chim.* **2015,** *60*, 809–815.
74. Lee, W. H.; Wang, C. C. Effect of Nanocomposite Gate Dielectric Roughness on Pentacene Field-effect Transistor. *J. Vacuum Sci. Technol. B, Nanotechnol. Microelectron. Mater. Process. Measure. Phenomena* **2009,** *27*, 1116.
75. Kim, J. H.; Hwang, B. U.; Kim, D. I.; Kim, J. S.; Seol, Y. G.; Woong Kim, T.; Lee, N. E. Nanocomposites of Polyimide and Mixed Oxide Nanoparticles for High Performance Nanohybrid Gate Dielectrics in Flexible Thin Film Transistors. *Electron. Mater. Lett.* **2017,** DOI: 10.1007/s13391-017-6345-9.
76. Matsukawa, K.; Watanabe, M.; Hamada, T.; Nagase, T.; Naito, H. Polysilsesquioxanes for Gate-insulating Materials of Organic Thin-film Transistors. *Int. J. Polym. Sci.* **2012.** DOI:10.1155/2012/852063.
77. Tomatsu, K.; Hamada, T.; Nagase, T., et al. Fabrication and Characterization of Poly(3-hexylthiophene)-based Field effect Transistors with Silsesquioxane Gate Insulators. *Jpn. J. Appl. Phys.* **2008,** *47*, 3196–3199.

78. Yang, S. Y.; Kim, S. H.; Shin, K.; Jeon, H.; Park, C. E. Low-voltage Pentacene Field-effect Transistors with Ultrathin Polymer Gate Dielectrics. *Appl. Phys. Lett.* **2006,** *88*, 1–3.
79. Watanabe, M.; Muro, K.; Hamada, T., et al., Surface Modification of Organic-inorganic Hybrid Insulator for Printable Organic Field-effect Transistors. *Chem. Lett.* **2009,** *38*, 34–35.
80. Hamada, T.; Nagase, T.; Watanabe, M.; Watase, S.; Naito, H.; Matsukawa, K. Preparation and Dielectric Property of Photocurable Polysilsesquioxane Hybrids. *J. Photopolym. Sci. Technol.* **2008,** *21*, 319–320.
81. Peng, X. Z.; Horowitz, G.; Fichou, D.; Garnier, F. All-organic Thin-film Transistors Made of Alpha-sexithienyl Semiconducting and Various Polymeric Insulating Layers. *Appl. Phys. Lett.* **1990,** *57*, 2013-1–2013-3.
82. Garnier, F.; Horowitz, G.; Peng, X. Z.; Fichou, D. An All-organic Soft Thin Film Transistor with Very High Carrier Mobility. *Adv. Mater.* **1990,** *2*, 592–594.
83. Bao, Z.; Feng, Y.; Dodabalapur, A.; Raju, V. R.; Lovinger, J. High-performance Plastic Transistors Fabricated by Printing Techniques. *Chem. Mater.* **1997,** *9*, 1299–1301.
84. Lee, H. N.; Lee, Y. G.; Ko, I. H.; Hwang, E. C.; Kang, S. K. Organic Passivation Layers for Pentacene Organic Thin-film Transistors. *Curr. Appl. Phys.* **2008,** *8*, 626–630.
85. Han, S. H.; Kim, J. H.; Son, Y. R.; Jang, J.; Cho, S. M.; Oh, M. H.; Lee, S. H.; Choo, D. J. Passivation of High-performance Pentacene TFT on Plastic. *J. Korean Phys. Soc.* **2006,** *48*, S107–S110.
86. Ahn, T.; Suk, H. J.; Yi, M. H. Polyvinyl Alcohol (PVA)/Layered Silicate Nanocomposite Approaches for Organic Passivation Layers in Organic Thin Film Transistors. *NSTI-Nanotech* **2007,** *1*, 675–676.
87. Kim, W. J.; Koo, W. H.; Jo, S. J.; Kim, C. S.; Baik, H. K. Passivation Effects on the Stability of Pentacene Thin-film Transistors with SnO_2 Prepared by Ion-beam-assisted Deposition. *J. Vacuum Sci. Technol. B* **2005,** *23*, 2357.
88. Han, S. H.; Kim, J. H.; Son, Y. R.; Lee, K. J.; Kim, W. S.; Cho, G. S.; Jang, J.; Lee, S. H.; Choo, D. J. Solvent Effect of the Passivation Layer on Performance of an Organic Thin-film Transistor. *Electrochem. Solid-State Lett.* **2007,** *10*, J68–J70.
89. Stadlober, B.; Zirkl, M.; Beutl, M.; Leising, G. High-mobility Pentacene Organic Field-effect Transistors with a High-dielectric-constant Fluorinated Polymer Film Gate Dielectric. *Appl. Phys. Lett.* **2005,** *86*, 242902.
90. Lee, C. H.; Hsu, C. H.; Chen, I. R.; Wu, W. J.; Lin, C. T. Percolation of Carbon Nanoparticles in Poly(3-Hexylthiophene) Enhancing Carrier Mobility in Organic Thin Film Transistors. *Adv. Mater. Sci. Eng.* **2014.** DOI: 10.1155/2014/878064.
91. Petrovsky, V.; Petrovsky, T.; Kamlapurkar, S.; Dogan, F. Dielectric Constant of Barium Titanate Powders Near Curie Temperature. *J. Am. Chem. Soc.* **2008,** *91*, 3590–3592.
92. Barber, P.; Balasubramanian, S.; Anguchamy, Y.; Gong, S.; Wibowo, A.; Gao, H.; Ploehn, H. J.; zur Loye, H.C. Polymer Composite and Nanocomposite Dielectric Materials for Pulse Power Energy Storage. *Materials* **2009,** *2*, 1697–1733.
93. Ioan, S.; Hulubei, C.; Popovici, D.; Musteata, V. E. Origin of Dielectric Response and Conductivity of Some Alicyclic Polyimides. *Polym Eng. Sci.* **2013,** *53*, 1430–1447.
94. http://web.hep.uiuc.edu/home/serrede/P435/Lecture_Notes/Dielectric_Constants.pdf (accessed Feb 22, 2017)
95. http://www.polymerprocessing.com/polymers/ (accessed Feb 22, 2017)

96. Singh, V. K.; Baquer, M. Impact of Scaling of Dielectric Thickness on Mobility in Top Contact Pentacene Organic Thin Film Transistors. *J. Appl. Phys.* **2012,** *111*, 034905-1–034905-6.
97. Gupta, D.; Hong, Y. Understanding the Effect of Semiconductors Thickness on Device Characteristics in Organic Thin Film Transistors by Way of Two-dimensional Simulations. *Org. Electron.* **2010,** *11*, 127–136.
98. Li, F.; Nathan, A.; Wu, Y.; Ong, B. S. *Organic Thin Film Transistor Integration: A Hybrid Approach;* Wiley: UK, 2011.
99. Majewski, L. A.; Kingsley, J. W.; Balocco, C.; Song, A. M. Influence of Processing Conditions on the Stability of Poly(3-hexylthiophene)-based Field-effect Transistors. *Appl. Phys. Lett.* **2006,** *88*, 222108.
100. Owen, J. W.; Azarova, N. A.; Loth, M. A.; Paradinas, M.; Coll, M.; Ocal, C.; Anthony, J. E.; Jurchescu, O. D. Effect of Processing Parameters on Performance of Spray-deposited Organic Thin-Film Transistors. *J. Nanotechnol.* **2011.** DOI:10.1155/2011/914510.
101. Someya, T.; Kawaguchi, H.; Sakurai, T. In *Cut-and-paste Organic FET Customized ICs for Application to Artificial Skin*, Solid-State Circuits Conference, 2004. Digest of Technical Papers. ISSCC. 2004 IEEE International, 2004; Vol. 1, pp 288–529.
102. Scharnberg, M.; Zaporojtchenko, V.; Adelung, R.; Faupel, F.; Pannemann, C.; Diekmann, T.; Hilleringmann, U. Tuning the Threshold Voltage of Organic Field-effect Transistors by an Electret Encapsulating Layer. *Appl. Phys. Lett.* **2007,** *90*, 013501-3.
103. Feng, L.; Tang, W.; Zhao, J.; Yang, R.; Qiaofeng Li, W. H.; Wang, R.; Guo, X. Unencapsulated Air-stable Organic Field Effect Transistor by All Solution Processes for Low Power Vapor Sensing. *Sci. Rep.* **2016,** *6*, 20671-1–20671-9.
104. Singh, V. K.; Mazhari, B. Impact of Scaling of Dielectric Thickness on Mobility in Top-contact Pentacene Organic Thin Film Transistors. *J. Appl. Phys.* **2012,** *111*, 034905.
105. Zojer, K.; Zojer, E.; Fernandez, A. F.; Gruber, M. Impact of the Capacitance of the Dielectric on the Contact Resistance of Organic Thin-film Transistors. *Phys. Rev. Appl.* **2015,** *4*, 044002.
106. Mittal, P.; Negi, Y. S.; Singh, R. K. Impact of Source and Drain Contact Thickness on the Performance of Organic Thin Film Transistors. *J. Semiconductors* **2014,** *35* (12), 124002-1–124002-7.
107. Kahn, B. E. *Developments in Printable Organic Transistors*; Pira International: UK, 2005.
108. Reese, C.; Roberts, M.; Ling, M. M.; Bao, Z. Organic Thin Film Transistors. *Mater. Today* **2004,** *7*, 20–27.
109. Torsi, L.; Cioffi, N.; Di Franco, C.; Sabbatini, L.; Zambonin, P. G.; Bleve-Zacheo, T. Organic Thin Film Transistors: From Active Materials to Novel Applications. *Solid-State Electron.* **2001,** *45*, 1479–1485.
110. Löffler, S.; Libberton, B.; Richter-Dahlfors, A. Organic Bioelectronic Tools for Biomedical Applications. *Electronics* **2015,** *4*, 879–908.
111. Bhaskar, S. A. Applications of the Organic Thin Film Transistor: Biological Sensing and Fragile X Syndrome. https://www.seas.upenn.edu/sunfest/docs/papers/Bhaskar07.pdf
112. Kuribara, K.; Wang, H.; Uchiyama, N., et al. Organic Transistors with High Thermal Stability for Medical Applications. *Nature Commun.* **2012,** *3*, 1–7.

CHAPTER 7

ORGANIC ELECTRONICS: FOR A BETTER TOMORROW

AJITH JAMES JOSE*, GRETA MARY THOMAS, S. PARVATHY, and ANN TREESSA WILSON

Research and Postgraduate Department of Chemistry, St. Berchmans College, Changanacherry, Kerala 686101, India

**Corresponding author. E-mail: ajithjamesjose@gmail.com*

ABSTRACT

Use of organic materials to build electronic devices holds the promise that future electronic manufacturing will rely on fewer, safer, and more abundant raw materials. Chemical scientists use different types of organic materials such as small molecules and polymers, fullerenes, nanotubes, graphene and carbon-based molecular structures, ensemble of molecule and molecular structures, and hybrid materials for the researches in electronics. These materials are used to build electronic structures and integrate those structures into those electronic devices. For example, small molecules and polymers are used in the manufacturing of organic light-emitting diodes displays, solar cells, transistors, etc. Carbon-based materials are being researched and developed mostly to create bendable or rollable electronic displays and other flexible devices. Organic electronics has the potential to make electronic production, use, and dispose more environmentally sustainable.

7.1 INTRODUCTION

Chemists are synthesizing a wealth of new organic materials for use in electronic devices that create novel properties impossible to replicate with silicon. These materials hold tremendous promise to expand our electronic landscape in ways that will radically change the way society interacts with technology. They use these materials to build electronic structures and then integrate those structures into electronic devices. Many of these devices are early-stage prototypes, with major scientific and engineering challenges still to be surmounted before the prototypes can become real-world products. But others are already commercial realities, some being used on a widespread basis. Hence advanced research and studies conducted in the field of organic electronics are of great importance.[1]

Organic electronics is a field of materials science concerning the design, synthesis, characterization, and application of organic small molecules or polymers that show desirable electronic properties such as conductivity. Scientific work in organic electronics is highly interdisciplinary and involves the design, synthesis, and processing of functional organic and inorganic materials, the development of advanced micro- and nanofabrication techniques, circuit design, and device characterization.

Different types of organic materials used in the research on electronic include small molecules and polymers; fullerenes, nanotubes, graphene, and other carbon-based molecular structures; ensembles of molecules and molecular structures; and hybrid materials. Polymer electronic materials in particular are one of the most active areas of organic electronic research. Carbon-based materials hold tremendous promise for the field of organic electronics because carbon comes in so many different forms, with a wealth of chemistries associated with those different forms.[2]

7.2 MATERIALS USED IN ORGANIC ELECTRONICS

Organic electronics relies on electrically active materials that are based on conjugated organic compounds whose molecules contain carbon and hydrogen elements. The materials are classified in this section as following: (1) semiconductors, (2) conductors, (3) dielectrics, (4) passivation, and (5) substrates. Each material in every class has its advantages and limitations, where often the process conditions as well as the interplay of the material with other layers have a large influence on the device performance. Therefore, the selection of the materials has to be carefully done to meet application and technology parameters such as electrical performance, environmental stability, mechanical and optical properties, and reliability of the devices.

7.2.1 SEMICONDUCTORS

Organic semiconductors are traditionally classified as small molecules or polymers. Polymers often have excellent solubility, which makes them amenable to mass printing processes such as flexographic and gravure printing. On the other hand, small molecules are usually deposited by vacuum sublimation; nevertheless, recent advancements enabled some of the semiconducting small molecules to be processed in solution or dispersion. The carrier mobility is commonly used as a figure of merit to characterize the performance of materials. It is found that the carrier mobility in organic semiconductors varies greatly depending on the choice of material, its chemical purity, and its microstructure. Amorphous films of solution-processed semiconducting polymers usually have mobilities in the range of 10^{-6} to 10^{-3} cm^2/V s.[3] Small-molecule organic semiconductors, on the other hand, are often forming polycrystalline films when deposited by vacuum sublimation, which results in carrier mobilities as large as about 6 cm^2/V s. Unlike inorganic semiconductors, organic semiconductors are not atomic solids but they are π-conjugated materials for which the charge transport mechanism is based on hopping between the individual conjugated molecules. In this case, the mobility is mainly limited by trapping of charges in localized states.[5] As for inorganic media, in a different manner, defaults such as traps, along with molecular and macromolecular structural irregularities, have a crucial impact on the charge transport. In general, the on-going development of organic semiconductors is not limited only to the performance measures but is also extended to other essential issues such as lifetime in real-world environmental conditions, matching over large areas, reproducibility, production yield, and operation voltage.

7.2.2 CONDUCTORS

The need for conductive traces in all electronic products is indispensable. As each conducting material has its own properties, the choice of the material strongly depends on the application. Conductive inks, which are typically consisting of micron-seized conducting flake particles, organic resins, solvents, and rheology modifiers offer promising properties. For applications that demand highly conductive features, silver ink that can have electrical conductivity as large as 10^4 S/cm is a preferable choice.

Another favorable choice for less demanding applications is conductive carbon inks. For the OTFTs, material properties of the conducting layers, especially for the source and drain contacts are very critical as they affect significantly the devices performance. The choice of the material in this case depends on the architecture employed by the OTFT, that is, the order of which the device layers are deposited. The typically used materials are aluminum (Al) or chromium (Cr) for the gate electrode, and gold (Au) for the source and drain contacts. For the design of an all-polymer OTFT, conductive polymers such as polyaniline (PANI) or PEDOT:PSS are also suitable for the gate, source, and drain electrodes.

7.2.3 DIELECTRICS

Dielectrics are used in both active and passive devices such as OTFTs and capacitors, respectively. The majority of OTFTs to date have used inorganic dielectrics, mostly silicon oxide. In fact, the performance of the device depends strongly on the quality, physical properties, and chemical nature of the insulator–semiconductor interface. For instance, trapping states at the mentioned interface immobilize the carrier charges in the channel and correspondingly limit the performance.[6] Significant improvements can be achieved as demonstrated in literature just by inserting few nanometers of organic single layer between the insulator and the semiconductor.[7] In order to take advantage of the complementary design features while not increasing the production cost, the challenge is to ensure that the dielectric material functions well with both *p*- and *n*-channel OTFTs.

7.2.4 PASSIVATION

Passivation materials (encapsulation) are used to protect the devices against environmental influences such as scratches and degradation due to the presence of the water, oxygen, or light. In some applications, the use of encapsulation is highly necessary to ensure an adequate lifetime for the devices. For example, OTFTs that are developed for medical applications can be encapsulated with poly(chloro-p-xylylene) (parylene) and gold layers to protect them against water.[8]

7.2.5 SUBSTRATES

Finally, the substrate is the base material onto which the devices are manufactured. Key material parameters for choosing the substrate material are: optical transmittance, dimensional stability, surface smoothness, durability, barrier capability, temperature tolerance, and mechanical properties (bending radius, deformation, and hysteresis behavior). Nowadays, the majority of applications are using glass (also thin and flexible glass) or stainless steel substrates, as well as polymer substrates such as poly(ethylene terephthalate) (PET) or poly(ethylene-2,6-naphthalate) (PEN). In addition, paper (cellulose) or textile substrates are sometimes used.

7.3 ORGANIC ELECTRONIC DEVICES

The organic materials can be combined to a number of active electronic components such as transistors, light-emitting diodes (LEDs), photovoltaic cells, and various types of sensors, memories, or batteries, as well as passive devices such as conductive traces, antennas, resistors, capacitors, or inductors.

7.3.1 ORGANIC LEDS

The most established and largest sector within the organic electronics industry is organic light-emitting diode (OLED). OLEDs are built from one or more layers of organic and hybrid material (either small molecules or polymers) sandwiched between two electrodes (e.g., indium tin oxide, ITO), all on a plastic or other substrate. OLEDs are built from one or more layers of organic and hybrid material (either small molecules or polymers) sandwiched between two electrodes (e.g., ITO), all on a plastic or other substrate.[9]

The basic device structure of an OLED is shown in Figure 7.1. The structure comprises two organic semiconducting layers, which are sandwiched by anode and cathode electrodes laying on a transparent substrate (e.g., glass). Any type of OLED consists of the following components:

- *Substrate*—the substrate is used to support the OLED. The most commonly used substrate may be a plastic, foil, or even glass.

OLED devices are classified as bottom emission devices if emitted light passes through the transparent substrate on which the panel was manufactured.

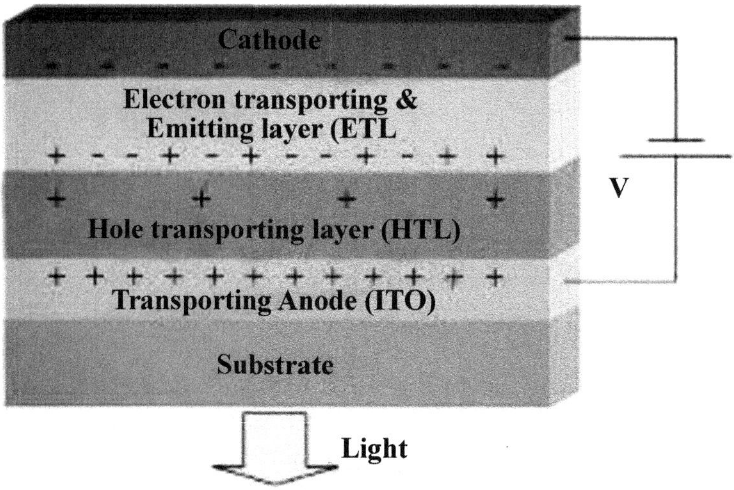

FIGURE 7.1 Basic device structure of an OLED.

- *Anode*—the anode component usually used is ITO. This material is transparent to visible light and is sufficiently conductor and has a high work function which promotes injection of holes into the highest occupied molecular orbital (HOMO) level of the organic layer. A typical conductive layer behaving as a transparent electrode that replaces the traditionally used ITO consists of PEDOT:PSS polymer or poly(3,4-ethylenedioxythiophene):poly(styrenesulfonate) as the HOMO level of this material generally lies between the work function of ITO and the HOMO of other commonly used polymers, reducing the energy barriers for hole injection. Another anode based on graphene yields to performance comparable to ITO transparent anodes.[10]
- *Cathode*—the cathode component depends on the type of OLED required. Noteworthy, even a transparent cathode can be used. Usually metals such as barium, calcium, and aluminum are used as a cathode because they have lesser work functions than anodes which help in injecting electrons into the lowest unoccupied molecular orbital (LUMO) level of the different layers.[11]

- *Electron transport layer*—commonly used components in electrons transport layer are: 2-(4-biphenyl)-5-(4-*t*-butylphenyl)-1,3,4-oxadiazole (PBD), tris(8-hydroxyquinoline) aluminium (Alq3), 1,3,5-tris(*N*-phenylbenzimidizol-2-yl)benzene (TPBI), and bathocuprene.
- *Holes transport layer*—commonly used components in holes transport layer are: *N,N′*-biphenyl-*N,N′*-bis(3-methylphenyl)-1,1-biphenyl-4,4-diamine (TPD) and 1,4-bis(1-naphthylphenylamino) biphenyl (NPB).[12]

Depending on the transparency of the anode and cathode electrodes, the light is transmitted either from top, bottom, or both directions of the device. This basic device structure is called heterostructure OLED, or sometimes referred to as bilayer structure. In this structure, the two organic semiconductors function as hole-transporting and light-emitting layers. The LEDs, regardless whether organic or inorganic, are principally operated by applying an external voltage across the *p–n* junction to accelerate charge carriers of opposite polarities, namely, electrons and holes, from the cathode and anode contacts, respectively. The carriers are driven toward the so called recombination region, which is located at the space charge region of the *p–n* junction and there the carriers form a neutral bound state, or exciton. It is called a recombination region because this is where the electrons recombine with the holes by falling into a lower energy level and realizing energy in the form of a photon. The wavelength (color) of the emitted light depends on the bandgap energy of the materials forming the *p–n* junction. The recombination region, where the luminescent molecular excited states are generated, is typically very small in the single heterostructure LED and is located at the boundary between the two semiconductors. Therefore, to increase the probability of electron–hole recombination and improve the internal quantum efficiency of the device, an additional third semiconducting layer is exploited in a double heterostructure OLED. In this case, the three (organic) semiconductors function as electron-transporting, light-emitting, and hole-transporting layers.[13]

The organic semiconductors can be made of small molecules or polymers. Depending on the materials used, the devices differ mainly in three criteria, namely, fabrication technique and process controllability, operating voltage, and efficiency.[13] Small-molecule thin organic layers are mostly deposited by vacuum evaporation or sublimation, while polymer layers are usually processed in the liquid state by spinning and solidification by heating. Control of the thickness of the organic thin films in a spin-on technique is relatively harder than in vapor deposition. However, polymer-based OLEDs can usually operate at lower power than that of small molecule-based OLEDs. This is owed to the high conductivity of organic polymers. The operation supply voltage of polymer-based OLEDs is in the range of 2–5 V, which is about 1–2 V less than that of small molecule-based OLEDs. Furthermore, the efficiency of polymer-based OLEDs is typically higher than that of the small molecule-based OLEDs. Display as well as lighting industries is currently focusing on different technical approaches for both solution- and vacuum-processable organic materials to develop cost-effective OLEDs.[14–16]

- OLEDs are already widely commercialized in many Samsung and other smartphone models. Additionally, Samsung and LG Electronics have both announced forthcoming launches of large-screen OLED TVs.
- Solid-state lightings (SSLs), including light emitting diode (LED), and OLED lighting, are soon replacing the conventional lighting techniques such as incandescent combustion (candles and incandescent lamps) and gas discharge (fluorescent and induction lamps). SSLs have been promising on account of their superior energy efficiency, absence of hazardous metals, flexible form factors, durability, and their surface emission for design features.
- OLEDs are used to create digital displays in devices such as television screens and computer monitors and portable systems such as mobile phones, digital media players, car radios, digital cameras, car lighting, handheld games consoles, and (personal digital assistants) PDAs. Such portable applications favor the high light output of OLEDs for readability in sunlight and their low power drain.
- Top-emitting OLEDs are better suited for active-matrix applications as they can be more easily integrated with a nontransparent transistor backplane. Manufacturers may use top-emitting OLED displays in smart cards.
- White OLEDs have the true-color qualities of incandescent lighting and emit white light that is brighter, more uniform, and more energy efficient than that emitted by fluorescent lights and incandescent bulbs. As white OLEDs can be manufactured in large sheets, are cost-effective, and also consume less power, they can replace fluorescent lamps and could potentially reduce energy costs for lighting.

- Due to their potentially high level of energy efficiency, even when compared to other OLEDs, phosphorescent organic light emitting diode (PHOLED)s are being used in large-screen displays such as computer monitors or television screens, as well as general lighting needs. PHOLED panels are much more efficient and the colors are stunning. One potential use of PHOLEDs as lighting devices is to cover walls with large area PHOLED light panels.

7.3.2 ORGANIC RADICAL BATTERIES

Organic electronics has become one of the most developing fields. The necessity for the efficient electronic devices has also increased as a result of the increased usage of electronic devices. Therefore, batteries for the next generation should possess the features such as short charging time, long life cycle, high power and energy densities, and environmental friendliness.[14] Organic radical batteries are a new type of radical batteries that use the electrochemical reaction of organic radical polymers. Organic radical batteries use organic radical polymers which are flexible plastics. Due to high reactivity and reversibility of radical reactions, organic radical batteries possess high charge capacity, good cycle ability, and shorter charging time. Organic radical batteries are nontoxic and nonflammable, that is, they are environment-friendly batteries. They are high power alternative to lithium ion batteries.[15] Organic radical batteries consist of cathode and anode electrodes separated by a porous film submerged in an electrolytic solution. The most common subunit in organic radical batteries is the nitroxide radical in (2,2,6,6-tetramethylpiperidin-1-yl)oxyl (TEMPO) which is a stable oxygen-centered molecular radical stabilized by steric or resonance effects. (2,2,6,6-tetramethyl piperidinyloxy-4-yl) methacrylate (PTMA) is obtained by attaching TEMPO radical to the polymer chains. PTMA-based organic radical batteries have high charge density.[16] Nitroxide radicals have small molecular weight per active site among the redox-active organic molecules. The nitroxide radicals have good electron mobility due to their amorphous, solvated, and slightly swollen structure. Nitroxide polymers on burning give carbon dioxide, water, and nitrogen oxide without ash or odor.[17]

The nitroxide radicals possess two redox couples. The positive terminal consists of an organic radical polymer carrying a TEMPO unit and the negative terminal is made up of carbon (graphite). At the anode, nitroxide radical is oxidized to form oxoammonium cation which leads to *p*-type doping of the material. At the cathode, the nitroxide radical is reduced to the hydroxylamine anion which leads to *n*-type doping of the material. When the battery discharges, the nitroxide radical is reduced to hydroxylamine anion and when the battery charges, the hydroxylamine anion is oxidized to nitroxide radical. An oxidation potential of −0.11 V is created. Likewise at the anode, when the battery discharges, the nitroxide radical is oxidized to oxoammonium cation and when the battery charges, the oxoammonium cation is reduced back to nitroxide radical. A reduction potential of +0.87 V is generated.[18]

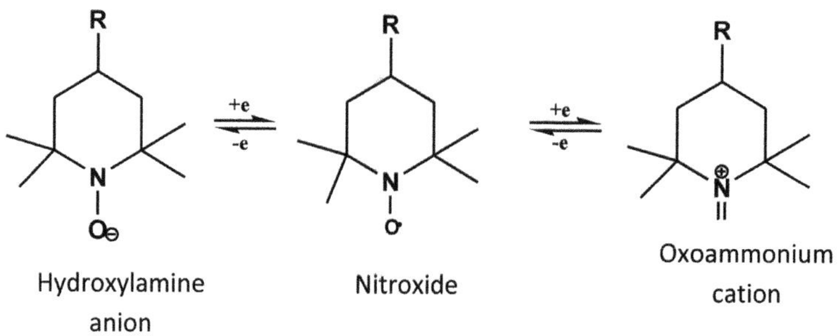

Organic radical batteries are used as an emerging power source and the built-in type organic radical batteries are used to protect information technology (IT) equipment such as desktop personal computer (PC) from losing data during power supply interruption. Organic radical batteries also provide power source for smart card sensors, intelligent papers, radiofrequency identification tags, and micro-sized devices.[19]

7.3.3 ORGANIC SOLAR CELLS

Solar energy is one of the most important renewable energy sources. Earth receives about 1.75 × 1017 W of energy from the sun. The most promising tool for making use of solar energy is its conversion into electrical energy. A photovoltaic cell is a device for conversion of light energy into electrical energy. The most widely used material for solar cell is silicon.[20] The use of organic materials such as conjugated polymers and small molecules makes them particularly attractive compared to the silicon-based solar cells. They are beneficial for commercialization because of ease of processing, mechanical flexibility, transparent, potential for low-cost fabrication of large areas, ecological and economic advantages, etc.[21]

The small molecules and polymers have conjugated π-electron systems, that is, carbon atoms having alternate single and double bonds. The p_Z orbital of these carbon atoms delocalizes to form bonding π orbital and an antibonding π^* orbital. The delocalized π orbital is the HOMO or valence band and the delocalized π^* orbital is the LUMO or conduction band. The energy gap between these energy levels is about 1–4 eV.[22] Light of energy greater than the band gap is absorbed and an exciton (an electron–hole pair bound by electrostatic attraction) is formed. The exciton diffuses to the active interface. By application of effective electric field, the excitons are broken and the charges move to respective electrodes. As a result of charge transfer, an electric current is generated.[23]

In organic photovoltaic, the dissociation of exciton takes place at heterojunctions; an electron donor material and an electron acceptor material are stacked in layers, as shown in Figure 7.2.

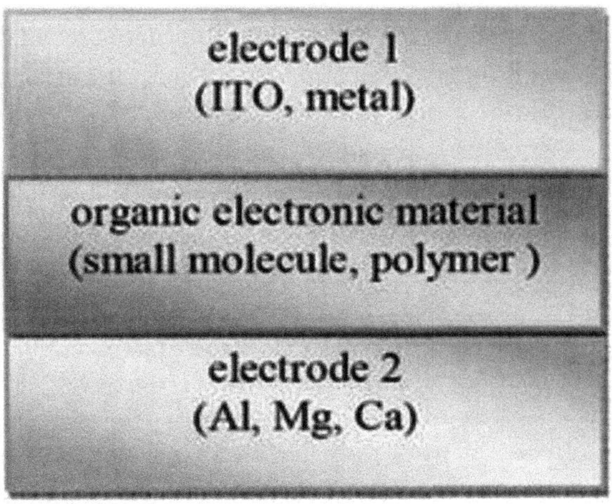

FIGURE 7.2 Schematic representation of a single layer organic photovoltaic.

Single layer junction: It is a metal–insulator–metal (MIM) diode with metal electrodes of different work functions. The organic material is placed between the high work function electrode (ITO) and low work function electrode (metals such as aluminum, magnesium, and calcium). The organic material absorbs photon, and creates an exciton. The potential created due to difference in work function of the electrode separates the exciton and the charges are transported to the respective electrodes.[24]

Bilayer heterojunctions: In bilayer junction, there is a mixture of two undoped semiconductors in between the cathode and anode. The charge separation takes place between the donor–acceptor interfaces. The electron acceptor layers have higher electron affinity and ionization potential compared to the electron donor layer. This junction is also called planar donor–acceptor bilayer.[25]

Mechanical flexibility enables to use organic photovoltaics in clothing, bags, power wearable, and portable electronic devices. High optical absorption coefficient offers possibility for production of very thin solar cells.[26] Different kinds of organic materials including small molecules and polymers such as fullerenes, nanotubes, graphene, other carbon-based molecular structures, and hybrid materials are used to build electronic structures and they are integrated into electronic devices.[27] Organic molecules possess the potential to develop large-scale power generation. Organic solar cells can be manufactured easily and cover large thin-film surfaces. The production method like roll-to-roll technology requires low energy and temperature that makes it cost effective and solution process is widely used in printing techniques.

Organic solar cells have applications in recharging surfaces for laptops, mobile phones, clothes, and packages, in supplying power for small portable devices such as cell phones and MP3 players. Recent developments show that organic solar cells can be used in soldier tents to generate electricity and supply power to military equipments such as night vision scopes, global positioning receivers (GPS), etc.[28]

7.3.4 ORGANIC FIELD-EFFECT TRANSISTORS

Organic field-effect transistors (OFETs) are one of the most promising areas in the field of organic electronics. An OFET is a field effect transistor using an organic semiconductor in its channel. In recent years, the research into the applications of organic semiconductors has increased rapidly; one of their application is in OFETs.[29] In circuitry, transistors are used as either signal amplifies or on/off switches.

In OFET, there are three terminals—the source, drain, and gate. Schematic representation of OFET is shown in Figure 7.3. Also there is a semiconductor layer and an insulating layer between the semiconductor and gate.[30,31] OFETs have several device geometries. Among them, bottom gate with top drain and source electrodes is the most commonly used device geometry. This is because of the similarity of the geometry with thin-field transistors (TFT) models, which allows the devices to use less conductive materials in their design. Field effect transistors behave as a capacitor with a conducting channel between sources and drain electrode. Among the two capacitor plates, one of them is a metal electrode, called gate electrode and the other is a semiconductor. Applied voltage on the gate electrode controls the amount of charge carriers flowing through the system. The direction of charge movement is from source to the drain.

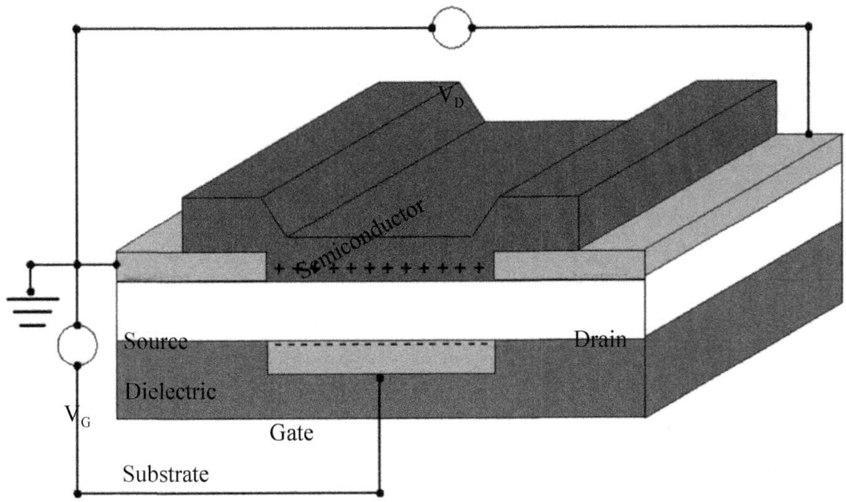

FIGURE 7.3 Schematic representation of OFET.

OFET usually works in accumulation mode. When we apply a negative (positive) voltage between gate and the source electrodes, an electric field induced in semiconductor. This electric field attracts positive (negative) charge carries at the semiconductors/insulator interface between source and drain electrode and overlapping with the gate. By applying a negative (positive) voltage between source and drain electrodes, it is possible to drive the positive (negative) charge carriers across the channel area. The number of charge carriers accumulated in the channel will reduce until the channel is fully depleted of free carriers when we keep increasing gate voltage to positive (negative). The device will be in its off state below the threshold voltage. At this time, there will be no free charge carriers in the channel and there will be no current flow across the channel. If we apply a gate voltage larger than the threshold voltage for a small applied source drain voltage, the gate-induced filed is almost uniformly distributed along the conducting channel. Thus a uniform charge distribution is induced in the channel, so the device is working in the linear region. For larger drain voltages, the gate field at the drain contact is zero. As a result, a depleted area with no induced free charge carriers is present. This phenomenon is called pinch-off.[32] The current flowing across the channel saturated beyond this point further increase in the applied source/drain voltage will produce no significant effect on the measured current.

Researchers are working on OFETs in order to optimize carrier mobility and the on/off current ratio. Performances of the OFET device depend on carrier mobility due to its direct proportionality relationship with semiconductor conductivity. The fairly good range of carrier mobility is 0.1–1.0 cm^2/V s.[33] The on/off current ratio is the ratio of the saturation current when V_{GS} is zero. For the switch-like behavior of OFETs, increasing this ratio is important.[34]

Organic semiconductors having aromatic rings or conjugated π systems are used in OFETs. The alternate single and double bond facilitates their electron delocalization and allows them to conduct electricity. Because of the planar structure of aromatic systems, generally the molecules used in active semiconductor layers are linear or two-dimensional structures. A type of conjugated polymer, thiophene polymer molecules, was used in first reported OFETs. They were able to conduct charge and were of a low cost. Generally used organic semiconductors are linearly fused benzene rings, linearly fused thiophenes, etc. Among this, linearly fused benzene rings such as tetracene and pentacene are commonly used in OFETs. Their structures are simple and their performance has been highly researched.[35]

Hole mobility value of tetracene is up to 0.15 cm^2/V s while hole mobility in OFETs using pentacene molecule is 2.2 cm^2/V s.The higher carrier mobility in pentacene is due to its greater molecular length, which provides more space for carrier movement. Pentacene also has high on/off current ratio. Linear thiophene oligomers and thiophene derivatives constitute another class of materials used in OFETs. Among them, sexithiophene has carrier mobility around 0.1 cm^2/V s. Introduction of alkyl chains to the end of sexithiophene increases both its the mobility and the on/off ratio. Carbazoles are another class of materials currently being investigated for use in OFETs.

Although researches taking place in field of *p*-type OFETs have resulted in promising carrier mobilities and on/off current ratios, there are some problems requiring further research. One of them is the less effectiveness of *n*-type OFETs due the poor electrons mobility in organic semiconductors. So developing an effective *n*-type OFET is an important area for future research. Another problem faced by OFETs today is the anisotropy inherent in most of the organic semiconductors due to the linear nature. This has to be rectified by excelled control of molecule orientation.

OFETs have applications in optical communication system, advance display technologies, SSL, and electronic materials that mimic human skin.[36]

7.4 FUTURE PROSPECTIVES

OLEDs in lamps, smartphone displays and the latest televisions, and radio frequency identification (RFID) tags or sensors used to test blood sugar levels are good examples here. Roll-on displays, solar cells on wallpaper, or medical diagnostic devices that are worn close to the skin have been talked about in the media for years too. Connected systems such as "smart" textiles, fitness bracelets, and even household appliances and cars with the organic electronic devices are of future importance. The most important thing here is to develop extremely low-cost multifunctional chips which include the sensors, electronic systems, wireless modules, and energy sources. In future we can also think about transistors in which all the parts are composed of organic materials, that is, a fully OFET. Such devices will have high potential application.

KEYWORDS

- **organic electronics**
- **solar cells**
- **light-emitting diode**
- **applications**
- **graphene**

REFERENCES

1. Clemens, W.; Lupo, D.; Hecker, K.; Breitung, S.; Eds. OE-A Roadmap for Organic and Printed Electronics, White Paper, 4th ed.; Organic Electronics Association: Frankfurt 2009 (White Paper).

2. Klauk, H., Ed. *Organic Electronics: Materials, Manufacturing and Applications*, 2nd ed.; Wiley-VCH: Weinheim, 2008.
3. Klauk, H. Organic Thin-film Transistors. *Chem. Soc. Rev.* **2010,** *39* (7), 2643–2666.
4. Choi, C. G.; Bae, B.-S. Effects of Hydroxyl Groups in Gate Dielectrics on the Hysteresis of Organic Thin Film Transistors. *Electrochem. Solid State Lett.* **2007,** *10* (11), H347–H350.
5. Moliton, A.; Hiorns, R. C. The Origin and Development of (Plastic) Organic Electronics. *Polym. Int.* **2012,** *61* (3), 337–341.
6. Heremans, P.; Dehaene, W.; Steyaert, M.; Myny, K.; Marien, H.; Genoe, J.; Gelinck, G.; van Veenendaal, E. In *Circuit Design in Organic Semiconductor Technologies*, Proceedings of the European Solid-state Device Research Conference, Helsinki, Finland, Sept 12–16, 2011; pp 5–12.
7. Zschieschang, U.; Ante, F.; Kälblein, D.; Yamamoto, T.; Takimiya, K.; Kuwabara, H.; Ikeda, M.; Sekitani, T.; Someya, T.; Blochwitz-Nimoth, J.; Klauk, H. Dinaphtho[2,3-b:2',3'-f]Thieno[3,2-b]Thiophene (DNTT) Thin-film Transistors with Improved Performance and Stability. *Org. Electron.* **2011,** *12* (8), 1370–1375.
8. Kuribara, K.; Wang, H.; Uchiyama, N.; Fukuda, K.; Tokota, T.; Zschieschang, U.; Jaye, C.; Fischer, D.; Klauk, H.; Yamamoto, T.; Takimiya, K.; Ikeda, M.; Kuwabara, H.; Sekitani, T.; Loo, Y.-L.; Someya, T. Organic Transistors with High Thermal Stability for Medical Applications. *Nat. Commun.* **2012,** *3*, 7231–7237.
9. Tang, C. W.; VanSlyke, S. A. Organic Electroluminescent Diodes. *Appl. Phys. Lett.* **1987,** *51* (12), 913–915.
10. Wu J.; Agrawal, M.; Becerril, H. A.; Bao, Z.; Liu, Z.; Chen, Y.; Peumans, P. Organic Light-emitting Diodes on Solution-processed Graphene Transparent Electrodes. *ACS Nano* **2010,** *4*, 43–48.
11. Friend, R. H.; Gymer, R. W.; Holmes, A. B.; Burroughes, J. H.; Marks, R. N.; Taliani, C.; Bradley, D. D. C.; Dos Santos, D. A.; Brédas, J. L.; Lögdlund, M.; Salaneck, W. R. Electroluminescence in Conjugated Polymers. *Nature* **1999,** *397*, 121–128.
12. Bellmann, E.; Shaheen, S. E.; Thayumanavan, S.; Barlow, S.; Grubbs, R. H.; Marder, S. R.; Kippelen, B.; Peyghambarian, N. New Triarylamine-containing Polymers as Hole Transport Materials in Organic Light-emitting Diodes: Effect of Polymer Structure and Cross-linking on Device Characteristics. *Chem. Mater.* **1998,** *10*, 1668–1676.
13. Forrest, S.; Burrows, P.; Thompson, M. The Dawn of Organic Electronics. *IEEE Spectrum* **2000,** *37* (8), 29–34.
14. Halls, J. J. M.; Walsh, C. A.; Greenham, N. C.; Marseglia, E. A.; Friend, R. H.; Moratti, S. C.; Holmes, A. B. Efficient Photodiodes from Interpenetrating Polymer Networks. *Nature* **1990,** *376*, 498–500.
15. Foley, D. NEC Develops New Ultra-thin Flexible, Rechargeable Battery Boasting Super-fast Charging Capacity. *NEC Corporation*, October 30, 2012.
16. Stoddart, A. *Flexible Battery Power*; October 30, 2012.
17. Nakahara, K.; Iriyama, J.; Iwasa, S.; Suguro, M.; Satoh, M.; Cairns, E. J. *J. Power Sources* **2017,** *167*, 870.
18. Nishide, H.; Oyaizu, K. Materials Science. Toward Flexible Batteries. *Science* **2008,** *319*, 737–738.
19. Oyama, N.; Tatsuma, T.; Sato, T.; Sotomura, T. Dimercaptan–Polyaniline Composite Electrodes for Lithium Batteries with High Energy Density. *Nature* **1995,** *373*, 598–600.

20. Gustafsson, G.; Cao, Y.; Treacy, G. M.; Klavetten, K.; Colaneri, N.; Heeger, A. J. Flexible Light Emitting Diodes Made from Soluble Conducting Polymers. *Nature* **1992,** *357*, 477.
21. Green, M. A.; Emerg, K.; King, D. L.; Hishikaw, Y.; Warta, W. Solar Cell Efficiency Tables. *Prog. Photovoltaics* **2006,** *14*, 45–51.
22. Moule, A. J.; Meerholz, K. Minimizing Optical Losses in Bulk Heterojunction Polymer Solar Cells. *Appl. Phys.: Laser Optics* **2007,** *86*, 721–727.
23. Macinnes, D. Jr.; Druy, M. A.; Nigrey, P. J.; Nairns, D. P.; MacDiarmid, A. G.; Heeger, A. J. Organic Batteries: Reversible n- and p-type Electrochemical Doping of Polyacetylene, (CH)x. *Chem. Commun.* **1981,** 317–319.
24. Brabee, C.; Scherf, U.; Dyakonov, V.; Ed. *Organic Photovoltaics: Materials, Device Physics and Manufacturing Technologies*; Wiley, 2007; pp 978–988.
25. Geim, A. K.; Novoselov, K. S. The Rise of Graphene. *Nat. Mater.* **2007,** *6*, 183–191.
26. Dyakanov, V. Electrical Aspects of Operation of Polymers: Fullerene Solar Cells. *Thin Solid Films* **2004,** *451–452*, 493.
27. Burroughes, J. H.; Bradley, D. D. C.; Brown, A. R.; Marelks, R. N.; MacKay, K.; Friend, R. H.; Burn, P. L.; Holmes, A. B. Light-emitting Diodes Based on Conjugated Polymers. *Nature* **1990,** *347*, 539–541.
28. Tvingstedt, K.; Inganas, O. Electrode Grids for ITO Free Organic Photovoltaic Devices. *Adv. Mater.* **2007,** *19*, 2893–2897.
29. Salleo, A; Chabinyc, M. L.; Yang, M. S.; Street, R. A. Polymer Thin-film Transistors with Chemically Modified Dielectric Interfaces. *Appl. Phys. Lett. IEEE* **2002,** *81* (23), 4383–4385.
30. Horowitz, G. Organic Field-effect Transistors. *Adv. Mater.* **1998,** *10*, 365–377.
31. Pierret, R.F. *Semiconductor Device Fundamentals*; Addison Wesley Longman: Reading, MA, 1996; pp 525–732.
32. Brown, A. R.; Jarrett, C. P.; De Leeuw, D. M.; Matters, M. Field-effect Transistors Made from Solution-processed Organic Semiconductors. *Synthetic Metals* **1997,** *88*, 37–55.
33. Roncali, J.; Leriche, P.; Cravino, A. From One- to Three-dimensional Organic Semiconductors: In Search of the Organic Silicon? *Adv. Matter.* **2007,** *19*, 2045–2060.
34. Sirringhanus, H.; Friend, R. H.; Li, X. C.; Moratti, S. C.; Holmes, A. B.; Feeder, N. Bis(dithienothiophene) Organic Field Effect Transistors with a High ON/OFF Ratio. *Appl. Phys. Lett.* **1997,** *71*, 26.
35. Roichman, Y.; Tessler, N. Charge Transport in Conjugated Polymers: The Influence of Charge Concentration. *Synthetic Metals* **2003,** *135–136*, 443–444.
36. Facchetti, A.; Yoon, M.-H.; Marks, T. J. Gate Dielectrics for Organic Field-effect Transistors: New Opportunities for Organic Electronics. *Adv. Matter.* **2005,** *17*, 1705–1725.

CHAPTER 8

BASIC PRINCIPLE OF FLUORESCENCE AND DESIGN OF SENSORS FOR IONS

ASHISH KUMAR* and SWAPAN DEY

Department of Applied Chemistry, Indian Institute of Technology (ISM), Dhanbad 826004, Jharkhand, India

Corresponding author. E-mail: ashishkumarrana1988@gmail.com

ABSTRACT

The basic principle of fluorescence emission process and designing of chemosensors is described in this chapter. The concept of fluorescence emission and quenching is explained on the basis of Jablonski diagram. The design of fluorescent chemosensors is a combination of fluorophore and ionophore moieties. A brief discussion on Stokes shift, fluorescence lifetime, fluorescence quenching, and quantum yield calculation has also been provided. All the mentioned sensing is based on intramolecular charge transfer and photo-induced electron transfer mechanisms in the recognition of selective ions.

8.1 INTRODUCTION

The term *sensing* is applicable for selective identification of analytes by a specific device, sensor. Sensors are applied in daily life such as medicinal (monitoring blood pressure, heart rates, glucose levels, etc.), environmental (e.g., CO_2, UV rays, toxic ions, sound, temperature, pressure, etc.), and industrial (e.g., electronics, automobiles, and waste management). Therefore, the design and development of new sensors has been of great importance of research in 21st century.

In supramolecular chemistry, the monitoring of ions (cations and anions) and neutral (amino acids) species using synthetic molecular devices (i.e., sensors) with responsible transformation of signals have drawn reputation. This chapter has paid attention on the development of such molecular system that can monitor a definite analyte in solution by transformation of fluorescence properties. Such type of molecular system is famous as *fluorescent sensors* or *fluorescent probes* or *chemosensors*.

The objective of discussion of this chapter is based on particular points of fluorescent chemosensors, with basic principle of fluorescence sensing and photochemistry, Jablonski diagram, Stokes shift, fluorescence lifetime, fluorescence quenching and quantum yield of fluorescence, and examples of some fluorescence chemosensors for ions detection.

8.2 BASIC CONCEPT OF FLUORESCENCE SENSING

The design and development of ligands (sensors) for the selective identification and complexation of appropriate guests such as cations, anions, and neutral species are important objectives of supramolecular chemistry.[1–3] A number of examples of supramolecular chemistry are presented in chemical and biological systems, the monitoring of concentration of ions, such as cations (Na^+, K^+, Ca^{2+}, Fe^{3+}, Cu^{2+}, and Zn^{2+}), anions (carboxylates, halides, and phosphates), and molecules (e.g., oxygen, carbon monoxide, or glucose, amino acid, uric acid, etc.).[4,5] In addition to that, in vitro and in vivo analyses are very important, because these ions are available in many biological processes such as muscle contraction, nerve impulses, cell activity regulation including apoptosis, and so on. The uncontrolled concentration of biological species generated some diseases. The monitoring of small ions, medium size molecules, and large size molecules has vital importance in medical diagnoses as well as in monitoring drug levels.[4] Besides, the exact monitoring of environmental samples has also strained attention in current research, where lifestyle changes have been increasing the environmental pollutions in addition to releasing of unconsumed ions and other materials from industries.[6–8] So, the development of fluorescent sensors is highly useful in wide range of industries for continuous, multipurpose and single monitoring.

Historically, the monitoring of ions and molecules from different media such as biological tissue, blood, cells, environmental samples, etc. has the main drawback of inaccurate analysis and time taking process. The use of sensors for selectively targeting various analytes and molecules overcomes these drawbacks. In supramolecular chemistry, many fluorescence sensors have been synthesized for selective sensing of environmentally and biologically important ions and molecules. On the basis of these reasons, fluorescent chemosensors overcome all these problems, and the field of supramolecular chemistry has gained more reputation in past two decades.[4]

8.2.1 BASIC IDEOLOGIES OF FLUORESCENT SENSOR

A sensor has ability to provide quantifiable signal when it comes in contact with a suitable analyte. The emission of signal in the form of appropriate energy by sensor in presence of analytes is known as signal transduction. Fluorescence emission is based on the ideal signal transduction mechanism, which is applied as potential in sensing applications. It can be used as an extremely sensitive technique, because fluorescent sensors emit observable signal at very low concentration of analytes ($>10^{-6}$ M).

8.2.2 BUILDING BLOCKS OF FLUORESCENT SENSOR

The application of luminescence for sensing is very imperative because luminescence have high multipurpose utilization in many spectroscopy methods. It can be applied in real time, application for monitoring of low concentration of analyte such as ions and molecules using μM concentration of sensors. In presence of analyte, sensors emit signal, which can be easily detected through low-cost instrument.[9–11]

Hence, the design and development of fluorescent sensors depend on two key moieties such as fluorophore (signaling moiety) and ionophore (binding moiety). Fluorophore is a signaling moiety which acts as a signal transducer and provides signals in absence or presence of analyte. Ionophore is responsible for binding of appropriate analytes in selective manner, such as size of ions (cations or anions), availability of coordination atoms, hardness or softness, etc. (Fig. 8.1). Beyond this, the identification of analytes also depends on various conditions such as pH, polarity of solvents, ionic strength, etc.

In the case of fluorescent sensing, an active pathway has to be opened in between fluorophore and ionophore to pass the message. When an analyte is recognized by sensor through ionophore, it modulates the photophysical properties of fluorophore. The sensors may be synthesized by space through connectivity of two moieties, fluorophore and ionophore. The directly connected sensor means ionophore and fluorophore are both present in same molecule. The ionophore and fluorophore are attached through covalent bond, either in conjugation or non-conjugation. Figure 8.2a,b represents schematic diagram of sensor design.

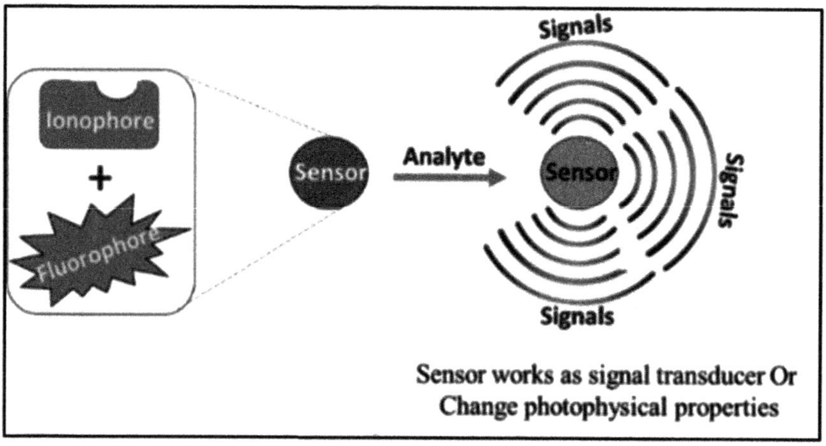

FIGURE 8.1 Representation of a fluorescent sensor designed.

FIGURE 8.2 (a) An integrated model sensor, (b) space model sensor.

The principle of structural design of sensor is an understanding of the basic photophysical properties for selected sensor. The changes of such photophysical behavior are monitored for the determination of potential analytical recognition. These aspects are applicable in the field of photochemistry[12,13] and will be explained in detail in the following section.

8.3 PRINCIPLES OF PHOTOCHEMISTRY

8.3.1 INTRODUCTION

As discussed in the above section, the luminescence emission is the key of fluorescence sensing. In general, luminescence is light emission phenomena from electronically excited state of any substrate such as ions and molecules and they are divided into two main classes, fluorescence and phosphorescence. Fluorescence emission arises from singlet excited state (*S*) and this type of feature in a molecule is referred as fluorophore. Phosphorescence (phosphors) is also an emission process which occurs from triplet excited state (*T*). In both the processes either singlet (*S*) or triplet (*T*) state is promoting electron from ground state or ground state orbital, which is called the highest occupied molecular orbital (HOMO) to excited state or higher energy orbital, which is called the lowest unoccupied molecular orbital (LUMO) by applying light irradiation (Fig. 8.3). The HOMO and LUMO define the transfer of electron from different orbitals such as n, π, and π^*. The singlet and triplet excited states are defined with respect to ground state orbital electron. We know that when two electrons are present in one orbital with reverse spin, then it is called ground state (S_0). If, the light excitation promotes electron from ground state to excited state (higher energy) without change in spin, then it is called singlet excited state, while, a triplet excited state is developed by spin conversion or spin flipped or same spin with respect of ground state and is energetically unfavorable process. This spin alteration process occurred through intersystem crossing (ISC) can be accelerated by heavy atom such as bromide or iodide group present in fluorophore moiety. In addition to that, the enhancement of ISC has also been found in presence of paramagnetic species in solution.[14] The formation of singlet excited state is a spin-allowed transition, while the formation of triplet state is spin-forbidden transition. As it is an energetically unfavorable process, it cannot be normally occurred by excitation of an electron. The excited state electron (either singlet or triplet state) return to the ground state (an excited state is deactivated) by releasing energy in the form of light is called fluorescence or phosphorescence. However, the excited state electron also comes to the ground state by other processes rather than luminescence, is called fluorescence quenching. This process occurs by thermal deactivation, for example, electron transfer or energy transfer with other molecules or collision with solvent molecules. The electron transfer process is attracting more attention in fluorescence sensing and will be discussed later. The requirement of energy for electron transfer or excitation depends on gap between ground state and excited state energy level, and energy gap changed with changing molecules. The size of energy gap is the main reason of fluorescence happening which is occurred at high energy, such as for shorter wavelength (found to below 430 nm), within visible region (430–800 nm, long emission), and within the near-infrared (NIR) region (800–1600 nm, NIR emission).

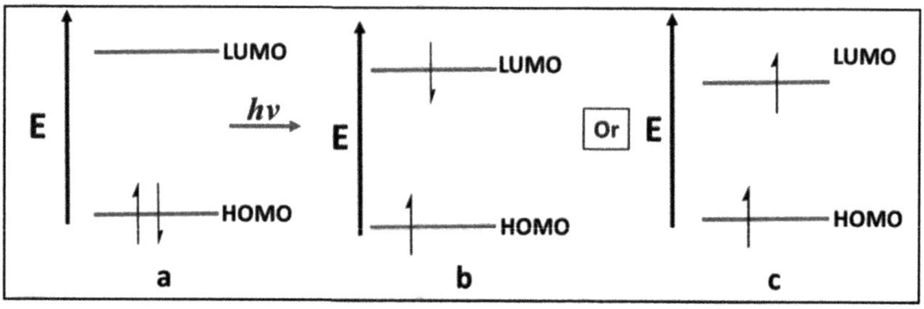

FIGURE 8.3 Molecular orbital diagram is showing the promotion of an electron from HOMO to LUMO energy level upon excitation of light: (a) ground state, (b) singlet excited state, and (c) triplet excited state.

8.3.2 JABLONSKI DIAGRAM

The process of absorption and emission of light has been frequently illustrated by Jablonski diagram.[15] It is used as a conversation of starting point of absorption and emission of light. It exists in various forms of energy level to illustrate many molecular processes.

In 1935, Jablonski understood many facts about the electronic energy levels and proposed a diagram that can explain all possible disperse excited states of energy, which is well known as Jablonski diagram, Figure 8.4a. This diagram shows that a molecule absorbs light from ground state (S_0) of vibrational level to a range of singlet first excited state (S_1) of vibrational level. This is very fast process and occurs within 10^{-15} s, which means the time of displacement of nuclei is too short (known as the Franck–Condon principle). The excitation process can be observed in absorption or emission spectrum of the sensors, which is recorded by UV–visible spectrophotometer. Byron and Werner[16] studied and reported typical absorption spectra of anthracene that depicted several peaks, which stipulated different transitions from ground state (S_0) vibrational level to different excited state (S_n) vibrational level. For example, the emission spectrum of anthracene shows more than one peak and its appearance is like mirror image or a symmetrical nature of absorption spectra. It is found to be longer wavelength and indicates that the absorption and emission occur at same transition state (Fig. 8.4b). This type of observation was also found in various anthracene-based sensors.[17] Most of the fluorophores or chromophores have shown broad absorption spectra as in rhodamine and naphthalimide chromophore.[17–19] According to Jablonski diagram (Fig. 8.4a), the absorption wavelength appeared at shorter wavelength and fluorescence emission appeared at higher wavelength. This phenomenon clearly indicates that the loss of vibrational energy via internal conversion occur in excited state. The difference between maximum absorption and emission wavelength is known as Stoke shift (Fig. 8.4c). Several researchers devoted to synthesize chromophores or fluorescence sensors that exhibited large Stoke shift.[20–22]

8.3.3 STOKES SHIFT

Investigation of the Jablonski diagram (Fig. 8.4a) reveals that the energy of emission is less than that of absorption. Hence, fluorescence is characteristically found in longer wavelength or lower energy. This phenomenon was first observed in 1852 by Sir George Gabriel Stokes.[23] The difference between absorption and emission wavelength of fluorophore molecules is called Stokes shift.[24–26] Many of fluorescence molecules have been synthesized and reported on the basis of Stokes shift.[26–28] The decrease or increase of Stokes shift depends on solvent and its environmental effects. The Stokes shift mostly depends on polarity of solvent, besides this, other factors also affect the fluorescence emission spectra such as viscosity of solvent, rate of solvent relaxation, conformational changes of chemosensors, inflexibility of local environment, proton transfer and excited state reactions, sensor–sensor interaction, and change in radiation and nonradiative decay rate. All these parameters affect the emission maxima. The excited state of fluorophore is stabilized by polar solvent molecules. Usually, fluorophore has large dipole moment in the excited state (μ_e) as compared to ground state (μ_g). The polarity of solvent generally modulates emission spectra of fluorescence polar compounds. Nonpolar compounds (unsubstituted

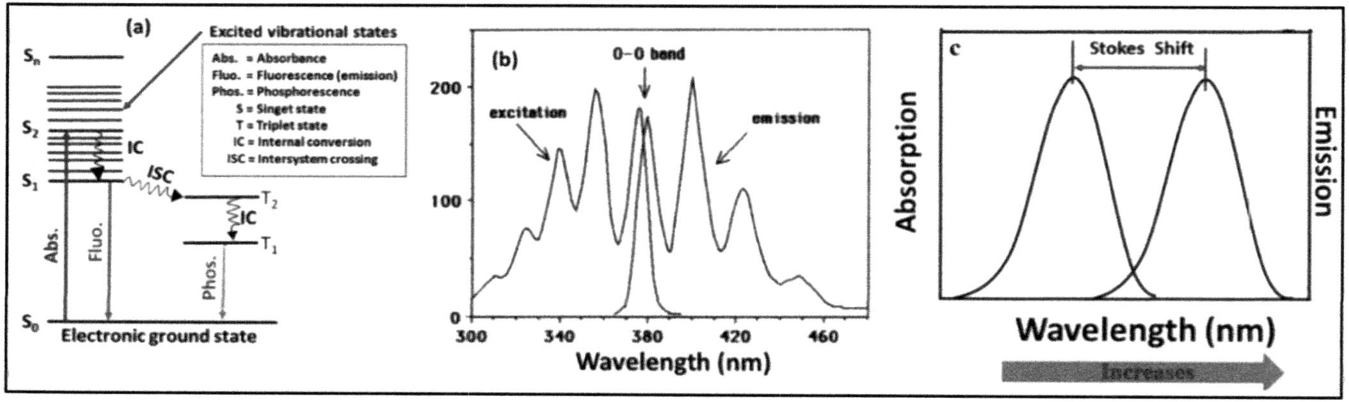

FIGURE 8.4 (See color insert.) (a) Jablonski diagram, represented various singlet states excited vibration level and photophysical process occur in different way such as singlet and triplet state, (b) excitation and emission spectra of anthracene, represent that small Stokes shift in highly symmetrical spectra, and (c) absorption and emission spectra of molecules or atoms.

Source: Part b—Adapted with permission from Ref. [16].

aromatic) are less sensitive toward polar solvents. The shifting of emission spectra in presence of different polar and nonpolar solvents is demonstrated by Jablonski diagram (Fig. 8.5).

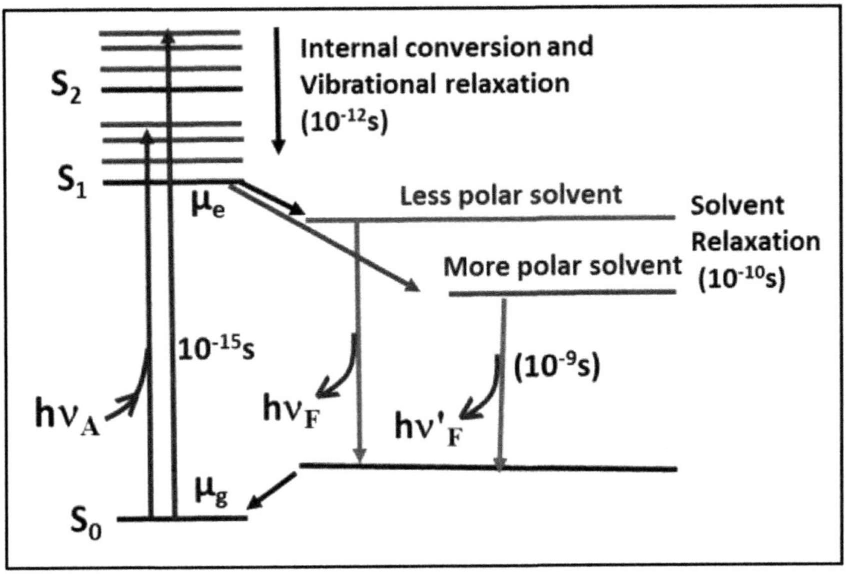

FIGURE 8.5 Jablonski diagram for solvent relaxation of fluorescence compound.

A new compound, trans-4-dimethylamino-4′-(1-oxobutyl)-stilbene (DOS) was synthesized by Safarzadeh-Amiri et al.[29] which contains two functional groups; one is dimethyl amino that acts as electron donor and other is C=O (carbonyl) group that acts as electron acceptor. The DOS showed different emission spectra ranging from blue to red region, that is, shift toward longer wavelength due to solvent polarity and viscosity effects. This showed the stabilization of excited intramolecular charge transfer (ICT) state. The presence of electron donating and withdrawing groups in same molecules such as DOS show more solvent effects, similar to fluorophore such as 4-dimethylamino-4′-nitrostilbene, PNBD [30] and BPNBD,[30] 6-anilino-2-naphthalenesulfonic acid,[31] etc. The dotted line shows that dipalmitoyl-L-α-phosphatidylcholine (DPPC) is present in DOS, is nearly polar to ethylacetate which indicated that the polarity of DPPC is greater than cyclohexane and less than butanol.

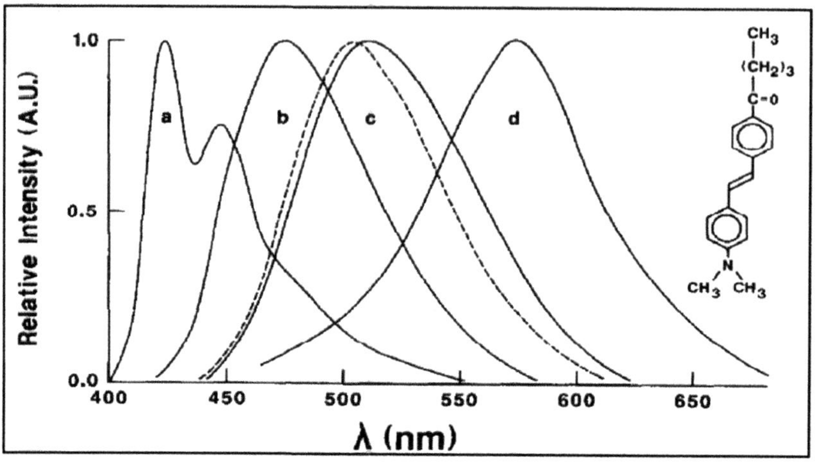

FIGURE 8.6 Fluorescence emission spectra of trans-4-dimethylamino-4′-(1-oxobutyl)-stilbene (DOS) in different solvents (a) cyclohexane, (b) toluene, (c) ethyl acetate, and (d) butanol, dotted line represents that emission spectra of DOS from DPPC vesicles.

Source: Adapted with permission from Ref. [29].

On the basis of solvents polarity effect, recently reported fluorescence chemosensors PNBD and BPNBD by Liu et al. represent that the presence of electron withdrawing group and donating group in the same molecule shows solvent effect. These types of molecules are frequently used in sensing.[25,27]

8.3.4 FLUORESCENCE LIFETIME

The fluorescence lifetime (τ) is the most important phenomenon of fluorophore. It is the time spend by molecules in excited states, to attain the ground state vibrational level (S_1) from the highest excited state vibrational energy level by dissipating vibration energy. According to Jablonski diagram, a molecule absorbs light and emits photons. The emission of light gives the information of lifetime available in a molecule. The lifetime is generally explained on the basis of average time spend by molecules in excited state to return to the ground state. This process is very fast and happens in nanoseconds, but the increasing of lifetime to microsecond and milliseconds has been observed in triplet excited state (Fig. 8.4a). The lifetime in nanoseconds has been measured by spectrometers possessing single photon counting for single excited state. For longer lifetime, nearly microsecond time scale spectrometers (bench-top) are employed, which can operate in two modes: steady state and time resolved.

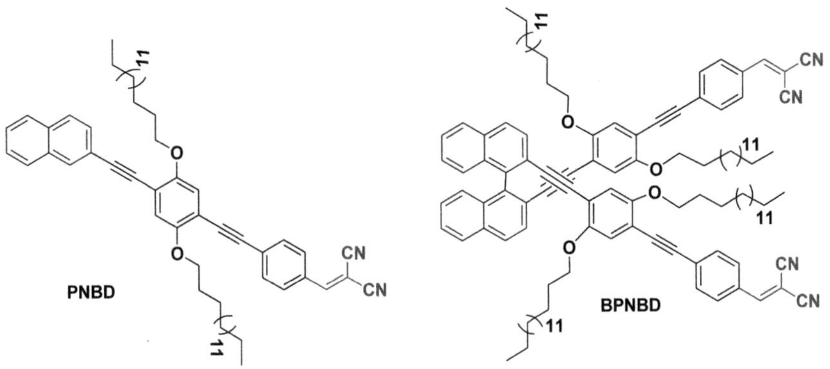

FIGURE 8.7 Structure of PNBD and BPNBD have donating and withdrawing group shows solvent polarity effect.

The lifetime measurement is an important phenomenon used in sensing technique. It gives information of lifetime changed by its local environment that is effected in a sensor. For example, sensors for measuring gases, generally O_2 is used for lifetime measurements.[32] Such sensors do not have specific binding position for these gases, but the gas itself can interact or interfere the photophysical properties of the sensors, often raised as collision quenching.[33] For example, O_2 sensing can modulate the lifetime of the exited state in the presence of the gas; this effectively works in sensors based on transition metal complexes. The emission arising from the metal complex, due to metal-to-ligand charge transfer (MLCT) affected by O_2, is commonly quenched emission of MLCT transition.[34,35]

8.3.5 FLUORESCENCE QUENCHING

The decrease in fluorescence intensity is called quenching. Fluorescence quenching processes may occur in extensive variety of mechanisms, such as contact quenching, collision quenching, and fluorescence resonance energy transfer (FRET), which are diagrammatically represented below (Fig. 8.8) and also explained through Jablonski diagram (Fig. 8.8d).[36] The emission of fluorophore can be influenced by interaction with other fluorescent or nonfluorescent molecules, which are present in same solution system that can "quench" excited state fluorophore.

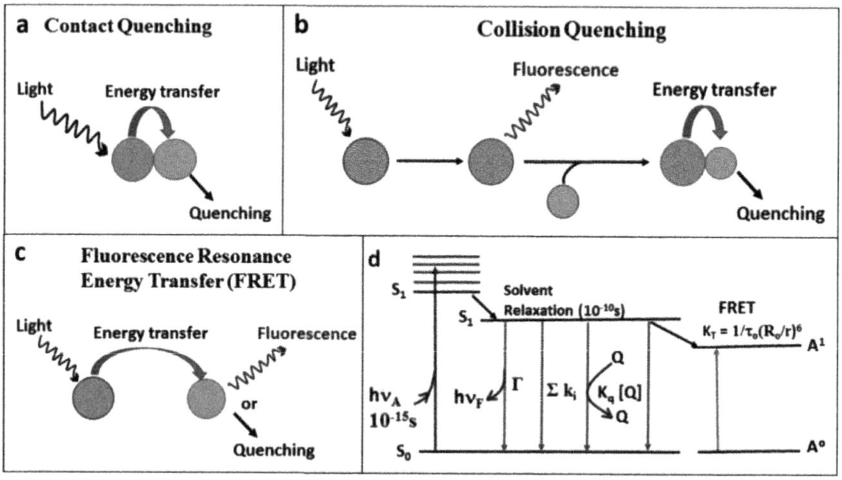

FIGURE 8.8 Fluorescence quenching process is based on (a) contact quenching, (b) collision quenching, (c) fluorescence resonance energy transfer (FRET), and (d) simplified Jablonski diagram for collision quenching and FRET, where ΣK_i is nonradiative paths, R_0 is Forester distance in resonance energy transfer, K_T is transfer rate in resonance energy transfer, Γ is radiative decay rate, r is anisotropy, and τ_0 is unquenched lifetime.

Contact quenching process occurs when fluorophore molecules form complex before excitation with quenching molecule. Immediately, transfer of excited state energy occurs from one fluorophore molecule to other molecule, which deactivates the excited state energy as heat. *Collision quenching* occurs in excited state interaction of fluorophore molecules with other molecule in presence of solution. They deactivate excited state fluorophore molecules, called fluorophore quenchers. The excitation energy of fluorophore immediately transfers to the contact or quencher molecules and the relaxation of excited fluorophore takes place. Collisional quenching is illustrated by simplified Jablonski diagram in Figure 8.8d. The collisional quenching is extensively applied in sensing, the decrease in intensity is determined by using Stern–Volmer equation (eq 8.1).

$$\frac{I_0}{I} = (1 + K[Q]) = (1 + k_q \tau_0 [Q]) \tag{8.1}$$

In this equation, K is Stern–Volmer quenching constant, ($K_{sv} = k_q\tau_0$), k_q is the bimolecular quenching constant, τ_0 is the lifetime in absence of quencher, [Q] is concentration of quencher, where I_0 and I are the fluorescence emission in the absence and presence of the quencher [Q]. An enormous number of molecules act as collisional quenchers, such molecules are oxygen, halogen, amines, and electron-deficient molecules such as acrylamide.

In case of *FRET*, the emission arises by excitation of donor fluorophore to emit photon. An acceptor molecule, which may be fluorescent or non-fluorescent, is designed approximately with the donor fluorophore to absorb the emitted photon, can effectively quench the emitted light in the case of nonfluorescent molecules. The discharged photon is also decreased when fluorophore is used as quencher. The different quenching processes are mentioned below.

8.3.6 QUANTUM YIELD OF FLUORESCENCE

The capability of fluorescence emission of molecules is pronounced as a fluorescence quantum yield (Φ).[12,13] It is dependent on ratio of number of photons emitted related to the number of photons absorbed. The largest quantum yield of fluorescence molecules is 1.0 (100%), which is possible only when each photon is absorbed and emitted. The quantum yield of any fluorescent compound which has greater than 0.10 or 10% is considered as fluorescent. An alternative way to describe the fluorescence quantum yield is on the basis of rate of excited state decay or nonradiative decay to S_0 (K_{nr}). The fluorescence quantum yield ratio is dependent on rate constant of emission rate of the fluorophore (Γ) and on the nonradiative decay to S_0 (K_{nr}), both depopulate the excited state. So, the fraction of fluorophore which decay through emission and the quantum yield is given by the following equation (eq 8.2).

$$\Phi = \frac{\Gamma}{\Gamma + K_{nr}} \tag{8.2}$$

Thus, we know that quantum yield can be equal to unity (1.0), that is, the radiationless rate decay is much less than the radiative rate decay, ($K_{nr} << \Gamma$). The quantum yield can be calculated by applying several methods, which is compared with well-known standard quinine sulfate ($\Phi_F = 0.542$ in 0.1N H_2SO_4), frequently used as primary standard references.[37,38] But other than this, many compounds can also be used as a standard such as 9,10-dimethyl- or pheynylanthracene, in addition to various transition metal complexes. Here, we mentioned one method to determine quantum yield in the field of sensing, where the collection of data with different concentration is often used. In this method, a number of solutions ranging from different concentrations, 0.10–0.02, can be used by maintaining their linearity. The identical absorbance of test sample and reference solution can be used and excited at the same wavelength that absorbs same number of photons. The calculation of quantum yield by comparing integrated data of emission spectra is done using eq 8.3.

$$\Phi = \Phi_r \times \frac{m_x}{m_r} \times \frac{(\eta_x)^2}{(\eta_r)^2} \tag{8.3}$$

where x, r, and η denote the test sample, reference standard, and refractive index of solvent, respectively. Firstly, plotting the integrated area of emission spectra of each standard and tested samples, a straight line with the slop m is obtained, from which the quantum yield is calculated using the above mentioned eq 8.3.

The quantum yield of fluorescence compounds also depends on different conditions such as organic solvent,[39,40] temperature,[41] and UV-radiation.[37,42] In this regard, rhodamine dyes have high fluorescence quantum yield, which can be calculated by using, as reference, quinine sulfate dehydrate in presence of sulfuric acid.[38,41]

8.4 EXAMPLES OF FLUORESCENCE CHEMOSENSORS FOR IONS DETECTION

In field of sensing, fluorescence compound is a highly delicate technique for identification of selective molecules and can be detected at the molecular level. Commercially, existing spectrometer for sensing purpose is not allowed to detect single molecules. Therefore, in the field of molecular recognition, the developed sensors can identify ions with alteration of wavelength or fluorescence emission. The sensing strength of sensors is monitored on the basis of binding constant calculation of host–guest (sensing) interaction, using absorption or emission spectra. The subsequent section discusses the synthesis of sensors and their designing principles for ions sensing. Here, we mention some selective examples which are reported by several research groups for ions recognition and their photophysical properties (Fig. 8.1). The sensing methods were frequently explained with various mechanisms, such as ICT,[43,44] photo-induced electron transfer (PET),[45,46] FRET,[47] through-bond energy transfer (TBET),[48] excimer,[45,49] Excited state intramolecular proton transfer (ESIPT),[50] etc. However, we only focus and discuss on the mechanism and examples of ICT and PET processes.

8.4.1 SENSORS POSSESSING ICT

In ICT process, communication pathway in-between fluorophore and ionophore is directly connected through π–π or n–π conjugation, which provides signals from ionophore to fluorophore (Fig. 8.2a).

An example of such type of sensor **1** (*N-butyl*-4,5-di[(pyridin-2-ylmethyl)amino]-1,8-naphthalimide)[51] is designed and synthesized using fluorescence moiety of 1,8-naphthalic anhydride. This sensor is based on ICT mechanism, where, two groups of 2-picolylamine are attached directly at four and fifth position of fluorophore of naphthalimide. Sensor **1** was directly attached to two amines which acted as electron donor, that is, the lone pair of electron was delocalized all over the aromatic ring and it exhibited ICT mechanism. During photophysical study, sensor **1** showed green fluorescence at 525 nm, when it coordinated to Cu^{2+} ion, the solution turned into blue fluorescence and peak was shifted to 475 nm. The available free electrons were coordinated and resulted blocked ICT. This fluorescence change was observed through long-range UV light.

Ketoaminocoumarin was used for signal transducer moiety because it has high emission and photostability properties. The quantum yield (Φ) of sensors **3** (7-diethylamino-3-(4-dimethylaminobenzoyl)coumarin, Φ = 0.001) and **4** (3-benzoyl-7-diethylaminocoumarin, Φ = 0.35) has been calculated by Spech et al. in 1982.[52] Sensors **2** and **3** did not show any significant changes in absorption and emission wavelength with monovalent metal ions, while red shift in absorptions (15 nm and 63 nm, respectively) was obtained in presence of Pb^{2+} ion. The large bathochromic shifting of 3•Pb^{2+} was observed for strong stabilization of ICT, due to complexation of metal ions via chelation of coumarin carbonyl group. Sensor **2** exhibited fluorescence enhancements with Pb^{2+}, Ba^{2+}, and Cu^{2+} corresponds to 40, 12, and 18-fold, respectively. Nevertheless, sensor **3** only responds to Cu^{2+} ion with 26-fold fluorescence enhancement. The binding mode of complex 2•Pb^{2+} was explained on the basis of Job's plot, infrared and proton spectra.

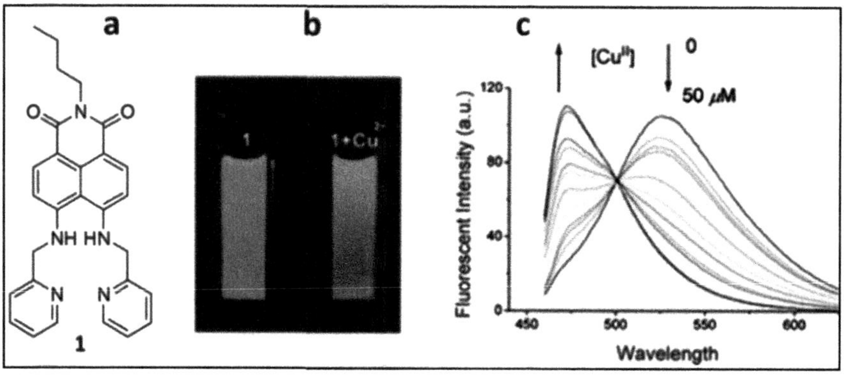

FIGURE 8.9 **(See color insert.)** (a) Structure of sensor **1**, (b) fluorescence emission of sensor **1** and 1 + Cu^{2+}, and (c) fluorescence titration spectra of sensor **1** with different concentration of Cu^{2+} ion.

Ferrocene-aminonaphthalate derivatives (sensor **5**) were synthesized by Bhatta et al.[54] for selective recognition of Hg^{2+} with turn-on properties. Sensor **5** itself showed an weak fluorescence in CH_3CN/H_2O, upon addition of Hg^{2+}, dramatic enhancement in fluorescence quantum yield was observed up to 58-fold with 68 nm red shift in emission spectrum. The binding position of imine C=N and triazole group was directly attached and it displayed weak fluorescence because of C=N bond isomerization-induced quenching effect. Upon addition of Hg^{2+} ion, imine and triazole nitrogen get coordinated and form five-membered ring. The conjugation of electron increases, and extended ICT mechanism was occurred. In acidic pH, sensor **5** was protonated on imine nitrogen and it took part in hydrogen bonding with more feasible triazole nitrogen atom with enhancement of fluorescence for annihilation of the C=N isomerization route. In presence of strong base, fluorescence property was decreased

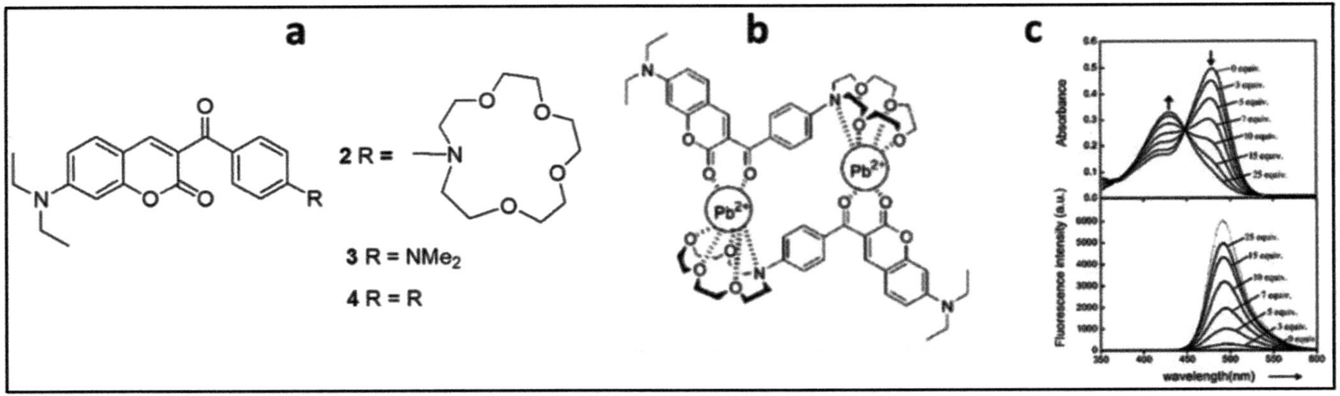

FIGURE 8.10 (a) Structure of sensors **2**, **3**, and **4**, (b) possible mode of binding of sensor **2** with Pb^{2+}, and (c) absorption and emission fluorescence titration spectra of sensor **2** with Pb^{2+}.

Source: Adapted with permission from Ref. [52].

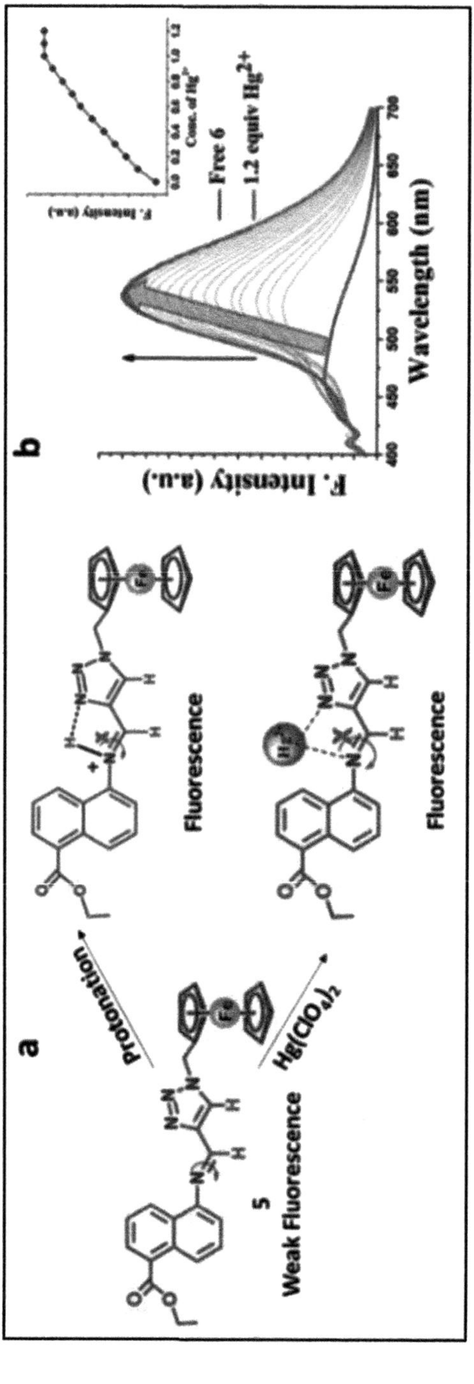

FIGURE 8.11 **(See color insert.)** (a) Possible mode of binding of sensor **5** and enhancement of ICT and (b) fluorescence titration spectra of sensor **5** with Hg^{2+}, inset change intensity plot.

Source: Reprinted with permission from Ref. [54]. © 2017 by American Chemical Society.

progressively, resulting *off–on–off* fluorescence switching probes. The hydrogen bonding process was studied in density functional theory and was found the enhancement of ICT process. Sensor **5** was the first example of triazole and imine-conjugated ferrocene-naphtholate system which modulated fluorescence and ICT in presence of Hg^{2+} and H^{+}. In the field of sensing, the limit of detection is very important and it was found very high up to nanomolar concentration (2.7×10^{-9} M).

The ICT mechanism is not only applicable for cations or protonation but it also works as anion sensor. Here, we mention some examples of anion sensors (**6**, **7**, and **8**) that were based on naphthalimide attached to amidourea (sensor **6**)[55] or thiourea (sensor **7**)[56] and urea moiety (sensor **8**).[57] The significant color changes can be observed through naked eye (Fig. 8.12). The photophysical titration of sensor **6** was carried out in dimethyl sulfoxide (DMSO) solution; a broad absorption peak was appeared at 414 nm and smaller shoulder at 560 nm. Upon addition of AcO^{-} ion, peak at 560 nm was increased and peak at 414 nm was decreased, progressively. Besides this, one new peak at 350 nm and two isosbestic points were appeared at 465 and 380 nm, respectively. The change of color from yellow to deep purple was clearly observed through naked eye

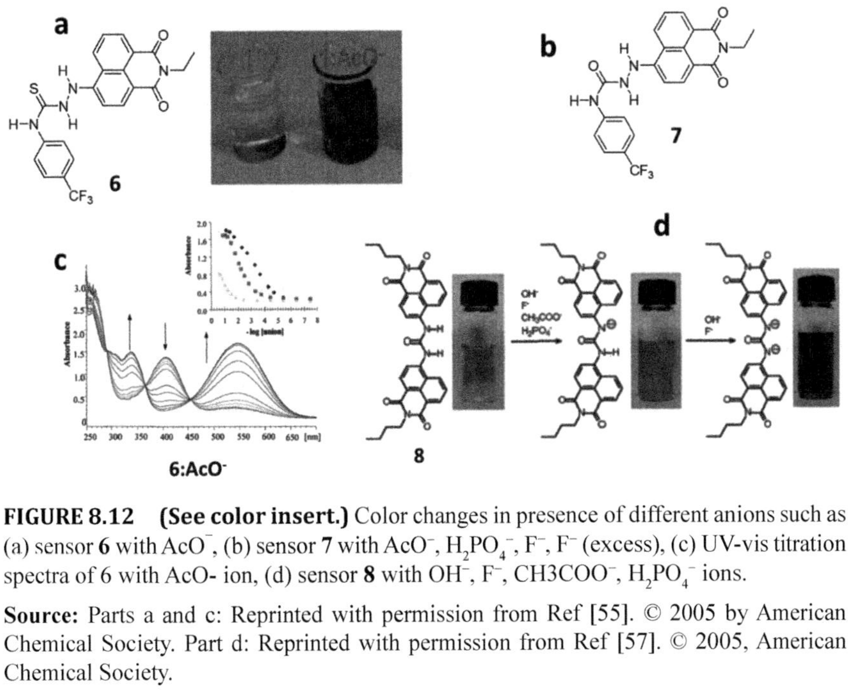

FIGURE 8.12 **(See color insert.)** Color changes in presence of different anions such as (a) sensor **6** with AcO^{-}, (b) sensor **7** with AcO^{-}, $H_2PO_4^{-}$, F^{-}, F^{-} (excess), (c) UV-vis titration spectra of 6 with AcO- ion, (d) sensor **8** with OH^{-}, F^{-}, $CH3COO^{-}$, $H_2PO_4^{-}$ ions.

Source: Parts a and c: Reprinted with permission from Ref [55]. © 2005 by American Chemical Society. Part d: Reprinted with permission from Ref [57]. © 2005, American Chemical Society.

in absence or presence of AcO^- ions in DMSO, respectively. The color change depends on anion binding with thiourea moiety through hydrogen bonding. It took into account to improve ICT character with related color changes. The numbers of chemosensors based on ICT mechanism for ions recognition are listed below in Table 8.1.

8.4.2 SENSORS DESIGNED ON PET

The developing fluorescence sensors are commonly used strategies on PET mechanism. Few PET sensors were designed and developed by de Silva et al. for various ions and neutral spaces.[64,65] This method is based on space model, in which a space will be in-between fluorophore and ionophore or receptor molecules. In PET sensors, spacer was used as a covalent bond which minimized $n–\pi^*$ or $\pi–\pi^*$ interaction. In the time of recognition, sensor showed some changes in fluorescence emission spectra with binding of selective analytes. These changes were detected through naked eye under UV light and fluorescence emission instrument. In PET sensors, receptor generally contains nonbonding electron pair with high energy. After excitation, the electron in the HOMO is transferred to the LUMO, which occurs from HOMO of the receptor to the LUMO excited fluorophore. So, this is deactivation process of the excited state through nonradiative path and results in fluorescence quenching effect. When ionophore is coordinated to suitable analytes, the energy of HOMO is reduced as compared to the fluorophore and results in reduced electron transfer progression for redox potential of ionophore and PET process is reduced and generates fluorescence emission.[66] PET mechanism is represented as schematic in Figure 8.13.

8.4.3 EXAMPLES OF PET SENSORS

PET-based sensors **16** and **17** were reported by de Silva et al. in 1985,[67,68] which were based on anthracene moiety for sensing of pH and K^+ ion, respectively in Figure 8.14. In sensor **16**, anthracene fluorophore is attached with -$N(Me)_2$ group in presence of methylene as a spacer, in similar way sensor **17** has been synthesized, but a crown ether moiety is attached in the place of -$N(Me)_2$ group. The emission spectrum of sensor 16 was highly pH dependent as the amine group is capable to transfer electron in

TABLE 8.1 List of Chemosensors for Identification of Ion Supports with ICT Mechanism.

Chemosensors	ICT	Ions	F. intensity	LOD (M)	Reference
9. 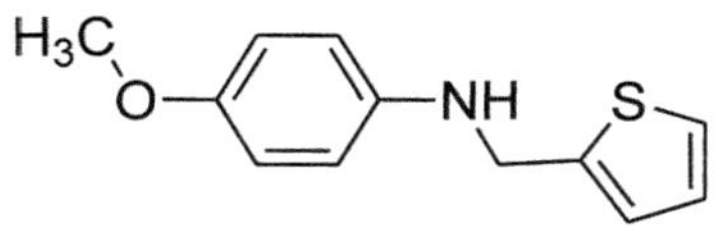	Enhanced	Ag^+	Increase	5.0×10^{-6}	[43]
10. 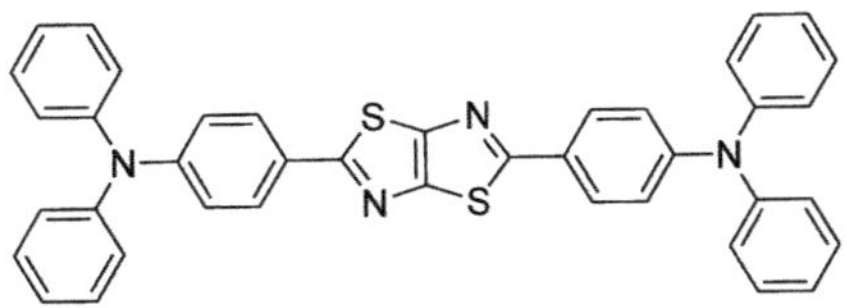	Enhanced	Acids	Changes	NA	[58]
11. 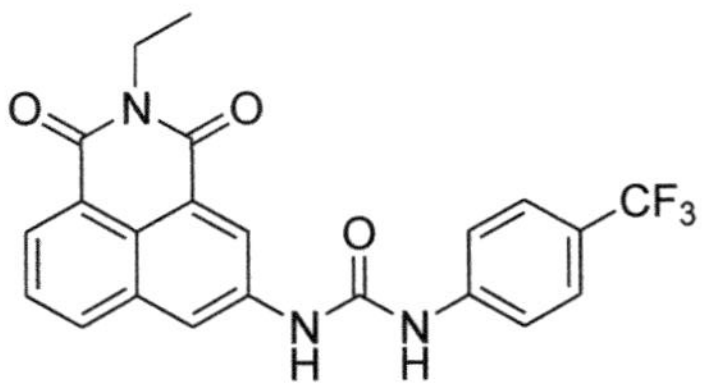	Enhanced	F^-	Increases	NA	[59]
12. 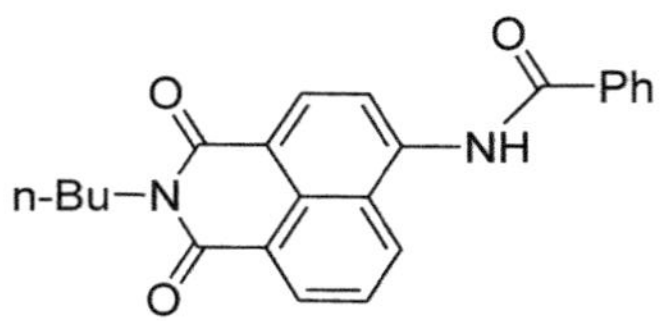	Enhanced	F^-	Decreases and new peak	NA	[60]

TABLE 8.1 *(Continued)*

Chemosensors	ICT	Ions	F. intensity	LOD (M)	Reference
13. 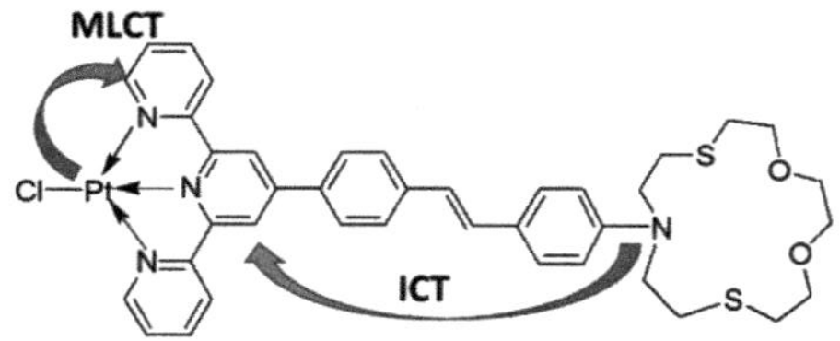	Block	Hg^{2+}	Colorimetric response	μM	[61]
14. 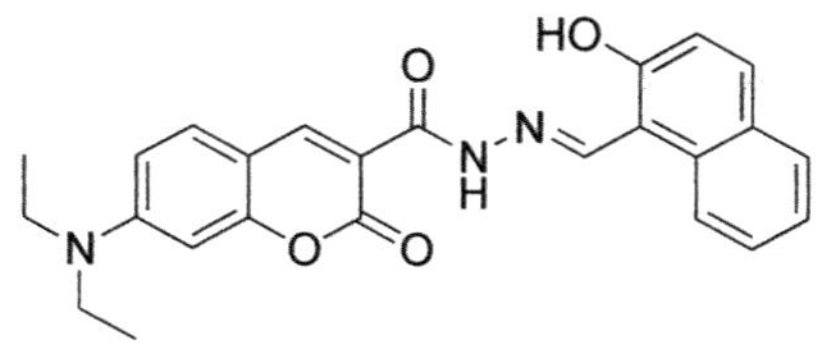	Inhibit	Al^{3+}	Increases	3.3×10^{-6}	[62]
15. 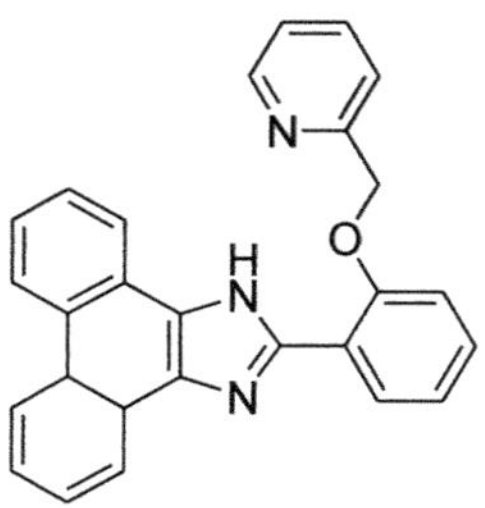	Off	Cr^{3+}	Increases	1.7×10^{-8}	[63]

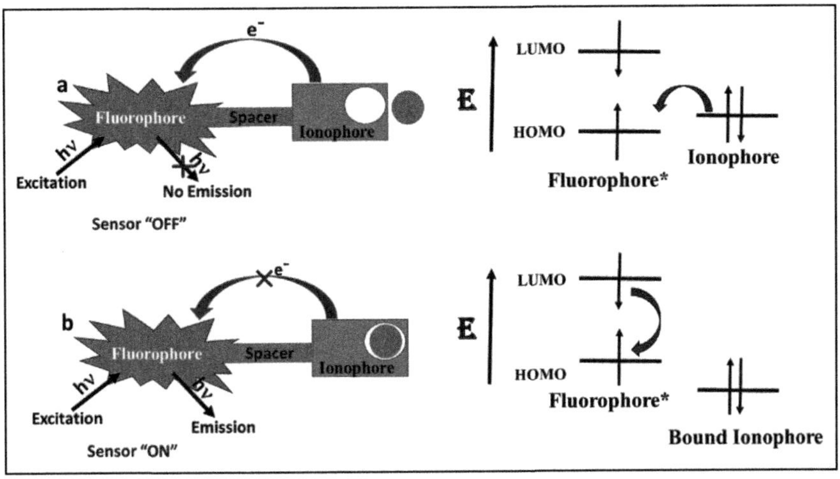

FIGURE 8.13 Schematic representation of PET off (reduced emission) and PET on (enhanced emission).

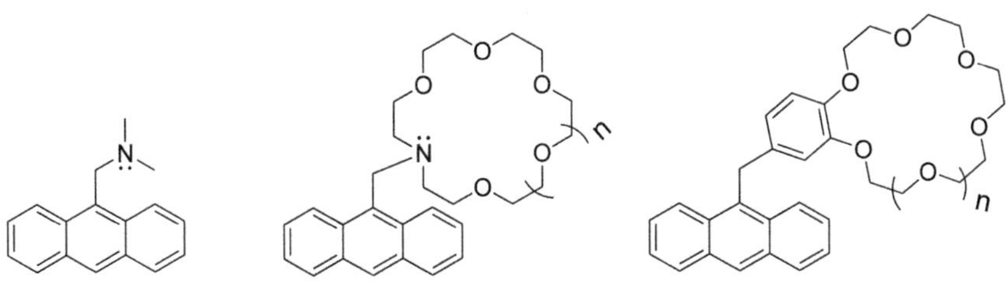

16 17 n = 0, 1 18 n = 0, 1

FIGURE 8.14 PET sensors reported by de Silva et al.: sensor **16** for H^+, sensors **17** and **18** for both Na^+ and K^+ ions.

Source: Adapted with permission from Refs. [67,68,69].

basic conditions, which led to the quenching of fluorescence properties. In acidic conditions, the electron transfer from lone pair of N-atom was blocked and hence PET process was strongly inhibited, resulting *switch on*. On other side, sensor **16** was H^+ ion sensitive with enhancement of fluorescence intensity and quantum yield. The reversibility of the sensor was shown with simple protonating and deprotonating amine and enhancement of PET became active in de-protonation system. Sensors **17** and **18** have same binding site *N*-(9-anthrylmethyl)monoaza-18-crown-6, anthracene moiety as a fluorophore, and methylene as a spacer. Sensor **17** has weak emission in methanol in presence of benzyltrimethylammonium hydroxide, and PET mechanism was enhanced or electron transfer process from amine to excited state fluorophore anthracene occurred. Upon addition of Na^+ and K^+ ion, the electron transfer system of lone pair nitrogen atom is blocked due to coordination with ions and PET process is diminished. As a result, fluorescence intensity or quantum yield was increased. The reversibility changes also happened in absence and presence of proton just like sensor **16**.

TABLE 8.2 The effect of cavity size in sensing and quantum yield.[69]

	Metal-free (Φ_F)	**Na^+ (Φ_F)**	**K^+ (Φ_F)**	**H^+ (Φ_F)**
$n = 1$	0.003	0.053	0.14	0.38
$n = 0$	0.009	0.057	0.019	0.41

Sensor **18** was similar to sensor **17** and showed almost same mechanism and properties toward H^+, Na^+, and K^+ binding, where Φ_F was fluorescence quantum yield.

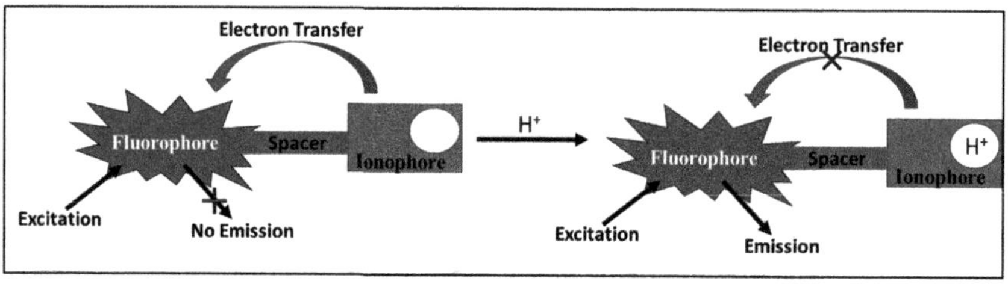

FIGURE 8.15 Schematic representation of pH sensor through PET.

The diagram shows that the emission was not generated or switched off, when ionophore transferred electron to the excited state of fluorophore. Upon addition of proton, the free ionophore amine was protonated and oxidation potential was increased with inhibition of electron transfer process and *switched on* mechanism was resulted. The reversibility of PET *off* and PET *on* showed in presence and absence of proton, respectively.

TABLE 8.3 List of Chemosensors for Identification of Ion Supports with PET Mechanism.

Chemosensors	PET	Ions	F. intensity	LOD (M)	Reference
19. 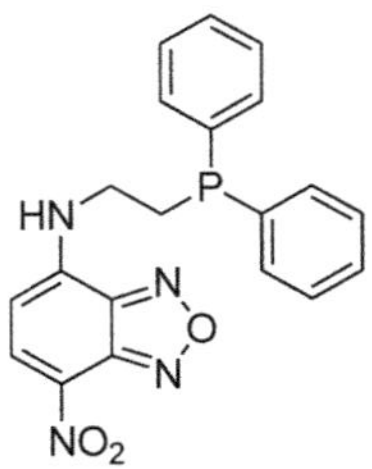	Block	H_2O_2	Increase	NA	[70]
20. 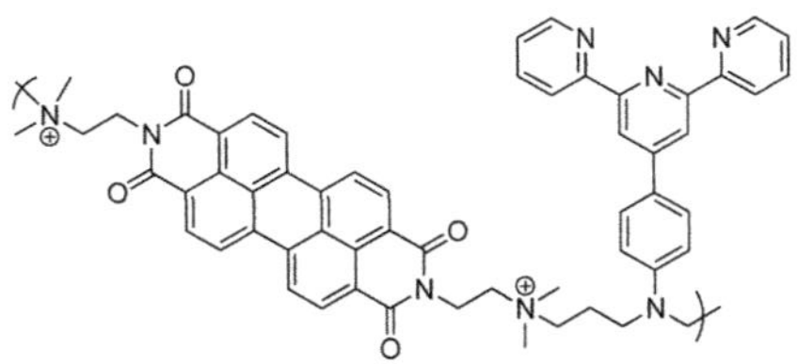	Block	Fe^{3+}	Increase	9.2×10^{-7}	[71]
21. 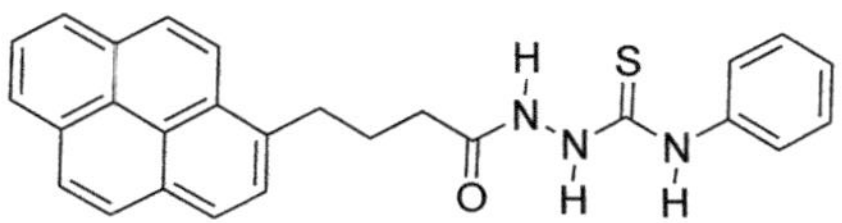	Enhanced	Anions	Quenched	NA	[72]

TABLE 8.3 *(Continued)*

Chemosensors	PET	Ions	F. intensity	LOD (M)	Reference
22. 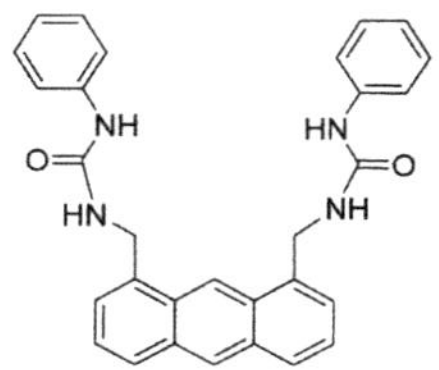	Enhanced	F^-	Quenched	NA	[46]
23. 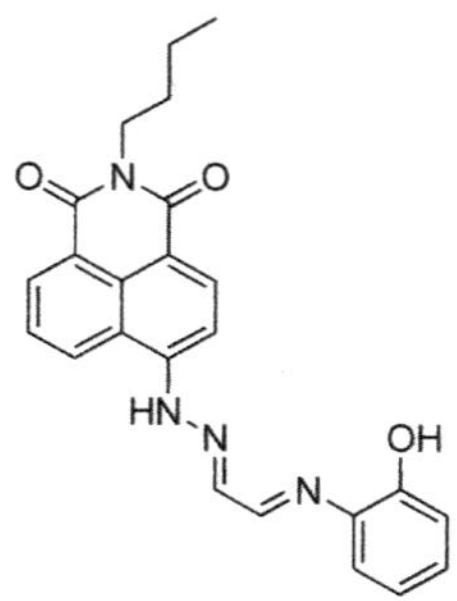	Block	Cu^{2+}	Increase	0.15 μM	[73]
24. 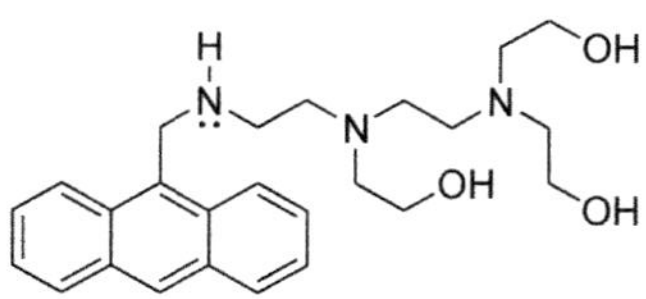	Inhibit	Cd^{2+}	Increases	3.3×10^{-6}	[74]

Following are a few chemosensors based on PET mechanism reported by different research groups (Table 8.3).

8.5 CONCLUSION

This chapter explains the principle of fluorescence sensing and design of sensors by using various routes. The mentioned chemosensors are based on various fluorescence moieties which have binding site for selective identification of targeting ions. The reported chemosensors have both fluorophore and ionophore groups, which are connected through a spacer. Here, photophysical properties of ICT and PET mechanism-based chemosensor are also discussed and some chemosensors are listed for ions recognition.

KEYWORDS

- **emission**
- **Jablonski diagram**
- **ICT**
- **PET**
- **sensing**

REFERENCES

1. Anslyn, E. V. *J. Biol. Chem.* **2007,** *72*, 687–699.
2. Jung, J.; Jo, J.; Dinescu, A. *Organ. Process Res. Develop.* **2017**, *21*, 1689–1693.
3. Zhou, X.; Lee, S.; Xu, Z.; Yoon, J. *Chem. Rev.* **2015**, *115*, 7944–8000.
4. de Silva, A. P.; Tecilla, P. *See Special Issue on Chemical Sensing: J. Mater. Chem.* **2005,** *15*, 2617–2976.
5. Spichiger-Keller; Ursula, E. *Chemical Sensors and Biosensors for Medical and Biological Applications;* U.S. Spichiger-Keller, Wiley-VCH: Weinheim, New York 1998.
6. dos Santos, C. M. G.; Harte, A. J.; Quinn, S. J.; Gunnlaugsson, T. *Coord. Chem. Rev.* **2008,** *252*, 2512–2527.
7. Gale, P. A.; Garcia-Garrido, S. E.; Garric, J. *Chem. Soc. Rev.* **2008,** *37*, 151–190.
8. Caltagirone, C.; Gale, P. A. *Chem. Soc. Rev.*, **2009,** *38*, 520–563.

9. Sessler, J. L.; Gale, P.; Cho, W.-S. *Anion Receptor Chemistry;* Cambridge, UK: Royal Society of Chemistry, 2006.
10. Callan, J. F.; de Silva, A. P.; Magri, D. C. *Tetrahedron* **2005,** *61*, 8551–8588.
11. Dong, X.; Wang, S.; Gui, C.; Shi, H.; Cheng, F.; Tang, B. Z. *Tetrahedron* **2018,** *74*, 497–505.
12. Turro, N. J. *Modern Molecular Photochemistry*; University Science Books: California, USA, 1991.
13. Gilbert, A.; Baggott, J. *Essentials of Molecular Photochemistry;* Blackwell Science Ltd: England, 1991.
14. Skoog, D. A.; Holler, F. J.; Nieman, T. A. *Principles of Instrumental Analysis*, 5th ed.; Brooks/Cole, 1998.
15. Jabłoński, A. *Zeitschrift für Physik;* Belmont (Calif.) : Brooks/Cole, **1935,** *94*, 38–46.
16. Byron, C. M.; Werner, T. C. Experiments in Synchronous Fluorescence Spectroscopy for the Undergraduate Instrumental Chemistry Course. *J. Chem. Edu.* **1991,** *68*, 433.
17. Dong, X.; Zhou, Y.; Song, Y.; Qu, J. *J. Fluorine Chem.* **2015,** *178*, 61–67.
18. Kumar, A.; Kumari, C.; Sain, D.; Hira, S. K.; Manna, P. P.; Dey, S. *Chem. Select* **2017,** *2*, 2969–2974.
19. Prasanna de Silva, A.; Rice, T. E. *Chem. Commun.*, **1999,** 163–164. DOI: 10.1039/A809119F.
20. Mukherjee, K.; Chio, T. I.; Gu, H.; Banerjee, A.; Sorrentino, A. M.; Sackett, D. L.; Bane, S. L. *ACS Sensors* **2017,** *2*, 128–134.
21. Bukowska, P.; Piechowska, J.; Loska, R. *Dyes Pigments* **2017,** *137*, 312–321.
22. Loudet, A.; Burgess, K. *Chem. Rev.* **2007,** *107*, 4891–4932.
23. Stokes, G. G. *Phil. Trans. R. Soc. London* **1852,** *142*, 463–562.
24. Piatkevich, K. D.; Malashkevich, V. N.; Morozova, K. S.; Nemkovich, N. A.; Almo, S. C.; Verkhusha, V. V. *Sci. Rep.* **2013,** *3*, 1847.
25. Abeywickrama, C. S.; Wijesinghe, K. J.; Stahelin, R. V.; Pang, Y. *Chem. Commun.* **2017,** *53*, 5886–5889.
26. Raj, P. J.; Bahulayan, D. *Tetrahedron Lett.* **2017,** *58*, 2122–2126.
27. Santiago-González, B.; Vázquez-Vázquez, C.; Blanco-Varela, M. C.; Gaspar Martinho, J. M.; Ramallo-López, J. M.; Requejo, F. G.; López-Quintela, M. A. *J. Colloid Interface Sci.* **2015,** *455*, 154–162.
28. Bukowska, P.; Piechowska, J.; Loska, R. *Dyes Pigments* **2017,** *137*, 312–321.
29. Safarzadeh-Amiri, A.; Thompson, M.; Krull, U. J. *J. Photochem. Photobiol. A: Chem.* **1989,** *47*, 299–308.
30. Liu, H.; Xu, X.; Peng, H.; Chang, X.; Fu, X.; Li, Q.; Yin, S.; Blanchard, G. J.; Fang, Y. *Phys. Chem. Chem. Phys.* **2016,** *18*, 25210–25220.
31. Sueishi, Y.; Fujita, T.; Nakatani, S.; Inazumi, N.; Osawa, Y. *Spectrochim. Acta Part A: Mol. Biomol. Spectrosc.* **2013,** *114*, 344–349.
32. Vanderkooi, J. M.; Maniara, G.; Green, T. J.; Wilson, D. F. *J. Biol. Chem.* **1987,** *262*, 5476–5482.
33. Dmitriev, R. I.; Papkovsky, D. B. *Cell. Mol. Life Sci.* **2012,** *69*, 2025–2039.
34. McGee, K. A.; Mann, K. R. *J. Am. Chem. Soc.* **2009,** *131*, 1896–1902.
35. Ramos, L. D.; da Cruz, H. M.; Morelli Frin, K. P. *Photochem. Photobiol. Sci.* **2017,** *16*, 459–466.

36. Lakowicz, J. R. *Principles of Fluorescence Spectroscopy*; Kluwer Academic/Plenum, USA 1999.
37. Sarkar, S.; Gandla, D.; Venkatesh, Y.; Bangal, P. R.; Ghosh, S.; Yang, Y.; Misra, S. *Phys. Chem. Chem. Phys.* **2016,** *18*, 21278–21287.
38. Chen, X.; Wang, F.; Hyun, J. Y.; Wei, T.; Qiang, J.; Ren, X.; Shin, I.; Yoon, J. *Chem. Soc. Rev.* **2016,** *45*, 2976–3016.
39. Losev, A. P.; Byteva, I. M.; Gurinovich, G. P. *Chem. Phys. Lett.* **1988,** *143*, 127–129.
40. Moreira, L. M.; de Melo, M. M.; Martins, P. A.; Lyon, J. P.; Romani, A. P.; Codognoto, L.; Santos, S. C. d.; Oliveira, H. P. M. d. *J. Braz. Chem. Soc.* **2014,** *25*, 873–881.
41. Kubin, R. F.; Fletcher, A. N. *J. Luminescence* **1982,** *27*, 455–462.
42. Tiwari, D.; Tanaka, S.-I.; Inouye, Y.; Yoshizawa, K.; Watanabe, T.; Jin, T. *Sensors* **2009,** *9*, 9332.
43. Tharmaraj, V.; Devi, S.; Pitchumani, K. *Analyst* **2012,** *137*, 5320–5324.
44. Li, J.; Yim, D.; Jang, W.-D.; Yoon, J. *Chem. Soc. Rev.* **2017,** *46*, 2437–2458.
45. Guan, J.; Zhang, P.; Wei, T.-B.; Lin, Q.; Yao, H.; Zhang, Y.-M. *RSC Advan.* **2014,** *4*, 35797–35802.
46. Kim, S. K.; Yoon, J. *Chem. Commun.* **2002,** 770–771. DOI: 10.1039/B110139K.
47. Wang, C.; Liu, Y.; Cheng, J.; Song, J.; Zhao, Y.; Ye, Y. *J. Luminescence* **2015,** *157*, 143–148.
48. Kumari, C.; Sain, D.; Kumar, A.; Debnath, S.; Saha, P.; Dey, S. *RSC Advan.* **2016,** *6*, 62990–62998.
49. Kraskouskaya, D.; Cabral, A. D.; Fong, R.; Bancerz, M.; Toutah, K.; Rosa, D.; Gardiner, J. E.; de Araujo, E. D.; Duodu, E.; Armstrong, D.; Fekl, U.; Gunning, P. T. *Analyst* **2017,** *142*, 2451–2459.
50. He, L.; Dong, B.; Liu, Y.; Lin, W. *Chem. Soc. Rev.* **2016,** *45*, 6449–6461.
51. Xu, Z.; Xiao, Y.; Qian, X.; Cui, J.; Cui, D. *Org. Lett.* **2005,** *7*, 889–892.
52. Specht, D. P.; Martic, P. A.; Farid, S. *Tetrahedron* **1982,** *38*, 1203–1211.
53. Chen, C.-T.; Huang, W.-P. A Highly Selective Fluorescent Chemosensor for Lead Ions, *J. Am. Chem. Soc.* **2002,** *124*, 6246–6247.
54. Bhatta, S. R.; Mondal, B.; Vijaykumar, G.; Thakur, A. ICT–Isomerization-Induced Turn-On Fluorescence Probe with a Large Emission Shift for Mercury Ion: Application in Combinational Molecular Logic. *Inorg. Chem.* **2017,** *56*, 11577–11590.
55. Gunnlaugsson, T.; Kruger, P. E.; Jensen, P.; Tierney, J.; Ali, H. D. P.; Hussey, G. M. Colorimetric "Naked Eye" Sensing of Anions in Aqueous Solution. *J. Org. Chem.* **2005,** *70*, 10875–10878.
56. Ali, H. D. P.; Kruger, P. E.; Gunnlaugsson, T. *N. J. Chem.* **2008,** *32*, 1153–1161.
57. Esteban–Gómez, D.; Fabbrizzi, L.; Licchelli, M. Why, on Interaction of Urea-Based Receptors with Fluoride, Beautiful Colors Develop. *J. Org. Chem.* **2005,** 70, 5717–5720.
58. Wang, K.; Huang, S.; Zhang, Y.; Zhao, S.; Zhang, H.; Wang, Y. *Chem. Sci.* **2013,** *4*, 3288–3293.
59. Duke, R. M.; Gunnlaugsson, T. *Tetrahedron Lett.* **2011,** *52*, 1503–1505.
60. Liu, B.; Tian, H. *J. Mater. Chem.* **2005,** *15*, 2681–2686.
61. Chung, S.-K.; Tseng, Y.-R.; Chen, C.-Y.; Sun, S.-S. *Inorg. Chem.* **2011,** *50*, 2711–2713.
62. Qin, J.-C.; Yang, Z.-Y. *Anal. Meth.*, **2015,** *7*, 2036–2040.
63. Fan, C.; Huang, X.; Black, C. A.; Shen, X.; Qi, J.; Yi, Y.; Lu, Z.; Nie, Y.; Sun, G. *RSC Adv.* **2015,** *5*, 70302–70308.

64. de Silva, A. P.; Gunaratne, H. Q. N.; Gunnlaugsson, T.; Huxley, A. J. M.; McCoy, C. P.; Rademacher, J. T.; Rice, T. E. *Chem. Rev.* **1997,** *97*, 1515–1566.
65. Bissell, R. A.; de Silva, A. P.; Gunaratne, H. Q. N.; Lynch, P. L. M.; Maguire, G. E. M.; McCoy, C. P.; Sandanayake, K. R. A. S. *Topics Curr. Chem.* **1993,** *168*, 223.
66. Culzoni, M. J.; Munoz de la Pena, A.; Machuca, A.; Goicoechea, H. C.; Babiano, R. *Anal. Meth.* **2013,** *5*, 30–49.
67. de Silva, A. P.; Rupasinghe, R. A. D. D. *J. Chem. Soc. Chem. Commun.* **1985,** 1669–1670. DOI: 10.1039/C39850001669.
68. de Silva, A. P.; Sandanayake, K. R. A. S. *Tetrahedron Lett.* **1991,** *32*, 421–424.
69. de Silva, A. P.; de Silva, S. A. *J. Chem. Soc. Chem. Commun.* **1986**, 1709–1710. DOI: 10.1039/C39860001709.
70. Onoda, M.; Uchiyama, S.; Endo, A.; Tokuyama, H.; Santa, T.; Imai, K. *Org. Lett.* **2003,** *5*, 1459–1461.
71. Jin, L.; Liu, C.; An, N.; Zhang, Q.; Wang, J.; Zhao, L.; Lu, Y. *RSC Adv.* **2016,** *6*, 58394–58400.
72. Liu, W.-X.; Jiang, Y.-B. *Org. Biomol. Chem.* **2007,** *5*, 1771–1775.
73. Chen, Z.; Wang, L.; Zou, G.; Tang, J.; Cai, X.; Teng, M.; Chen, L. *Spectrochim. Acta Part A: Mol. Biomol. Spectrosc.* **2013,** *105*, 57–61.
74. Tsukamoto, K.; Iwasaki, S.; Isaji, M.; Maeda, H. *Tetrahedron Lett.* **2013,** *54*, 5971–5973.

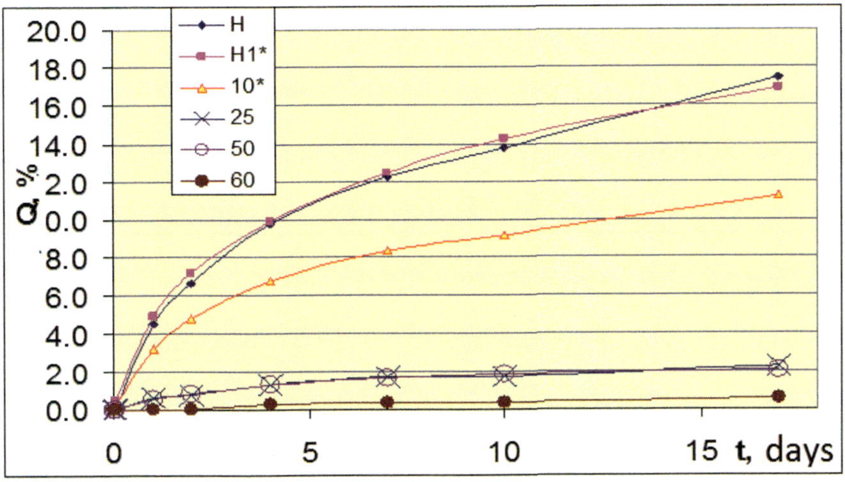

FIGURE 2.12 Swelling in ink-solvent (TM Inkwin) of composites without filler (H and H1) and with 10–60 wt% of gypsum.

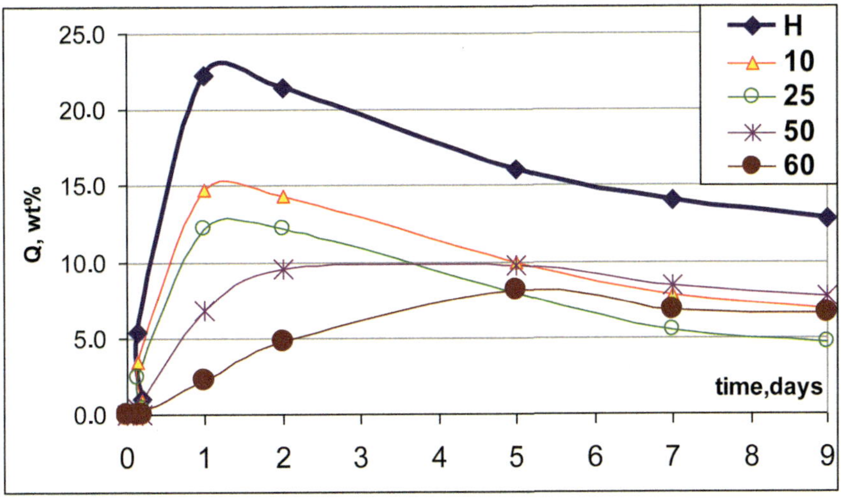

FIGURE 2.13 Curves of swelling in acetone of composites without filler (H) and with 10–60 wt% of gypsum.

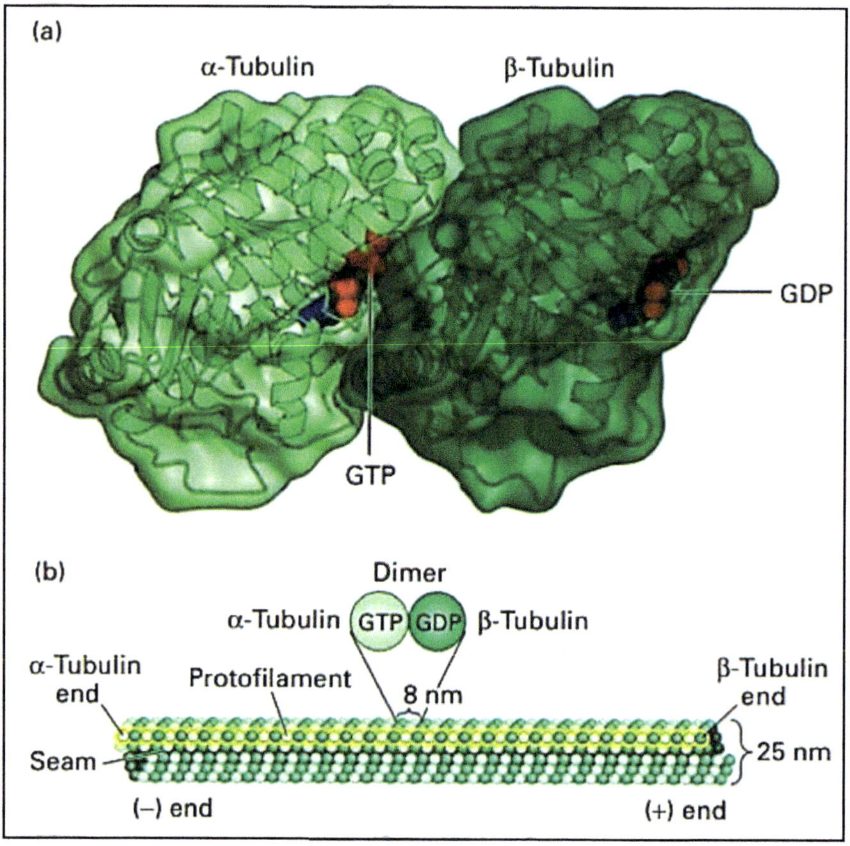

FIGURE 4.3 Details of MT's substructure, α- and β-tubulin.

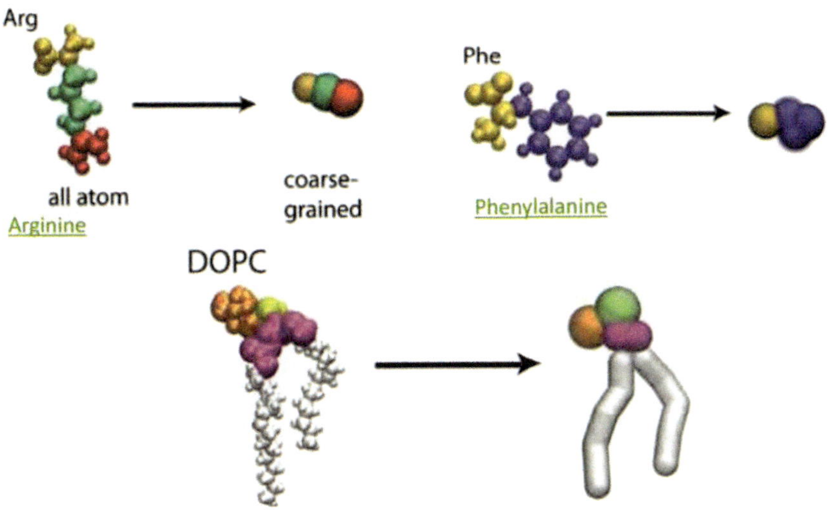

FIGURE 4.5 Atomic categorization based on the RBCG method.

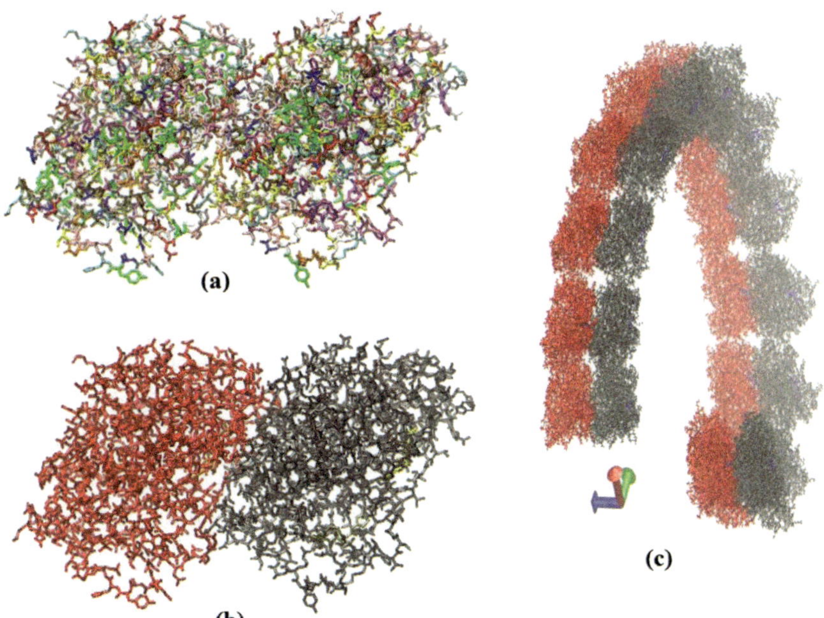

FIGURE 4.6 (a) Aminoacids in α- and β-tubulin, (b) α- and β-tubulin as a substructure of one loop, and (c) all-atom model of one loop.

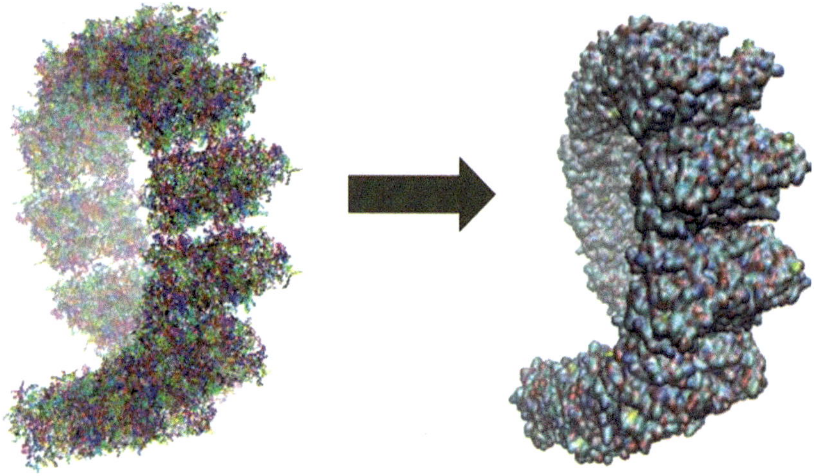

FIGURE 4.7 All-atom loop (left) and coarse-grained loop (right).

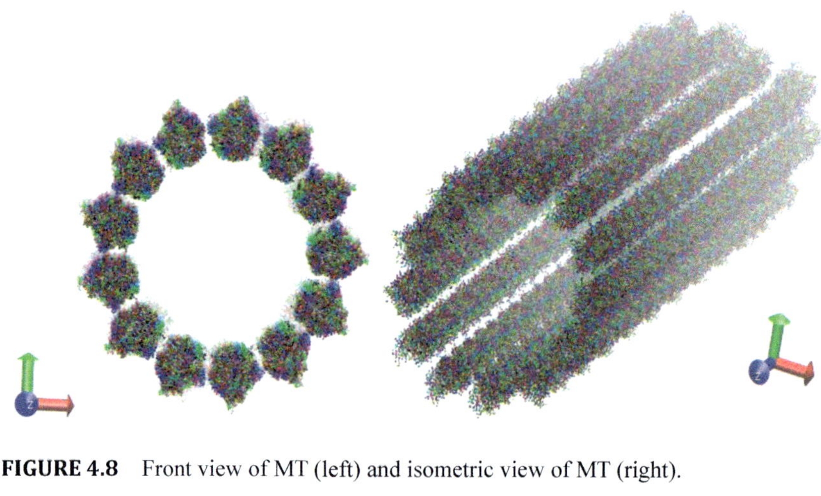

FIGURE 4.8 Front view of MT (left) and isometric view of MT (right).

FIGURE 4.9 Solvated MT with water.

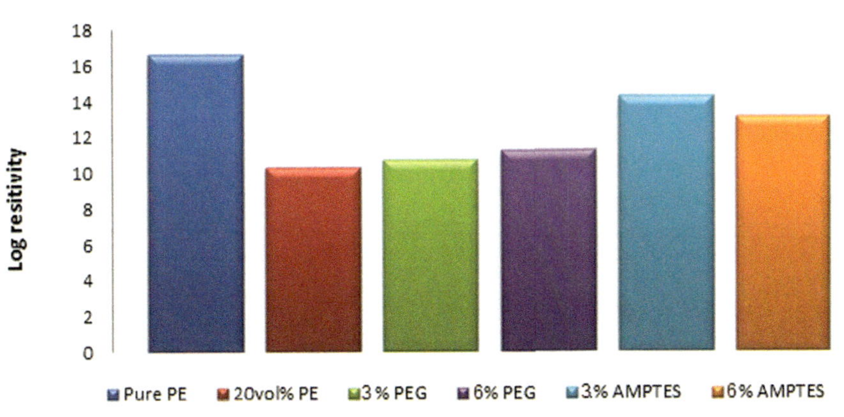

FIGURE 5.13 Log resistivity comparison of LLDPE, 20 vol % ZnO/LLDPE composites.

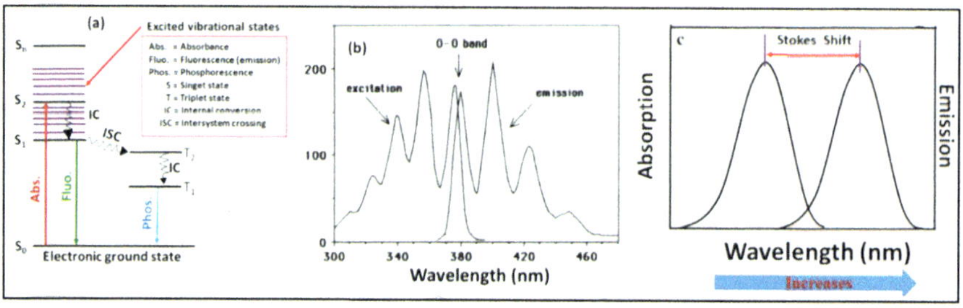

FIGURE 8.4 (a) Jablonski diagram, represented various singlet states excited vibration level and photophysical process occur in different way such as singlet and triplet state, (b) excitation and emission spectra of anthracene, represent that small Stokes shift in highly symmetrical spectra, and (c) absorption and emission spectra of molecules or atoms.

Source: Part b—Adapted with permission from Ref. [16].

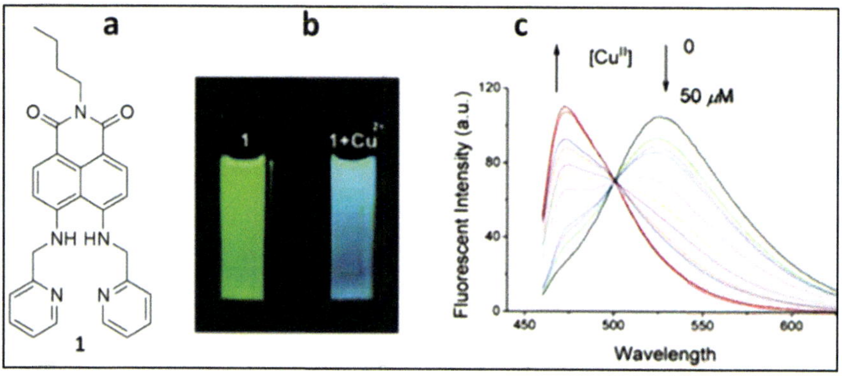

FIGURE 8.9 (a) Structure of sensor **1**, (b) fluorescence emission of sensor **1** and 1 + Cu^{2+}, and (c) fluorescence titration spectra of sensor **1** with different concentration of Cu^{2+} ion.

Source: Reprinted with permission from Ref. 51.

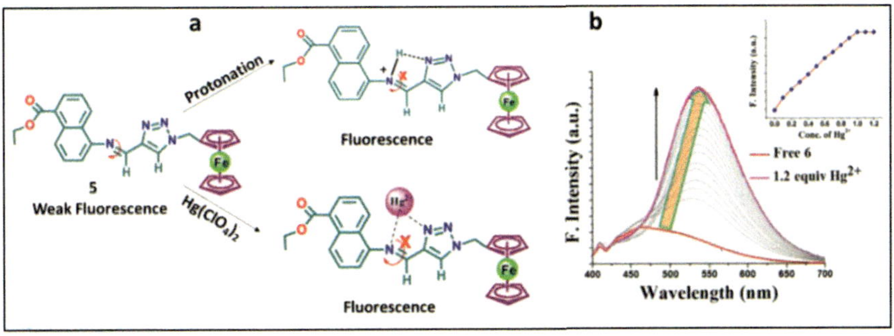

FIGURE 8.11 (a) Possible mode of binding of sensor **5** and enhancement of ICT and (b) fluorescence titration spectra of sensor **5** with Hg^{2+}, inset change intensity plot.

Source: Reprinted with permission from Ref. [54]. © 2017 by American Chemical Society.

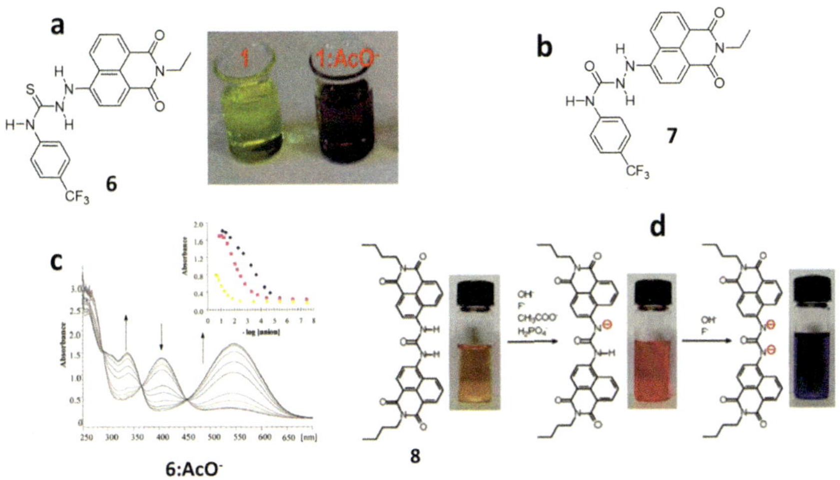

FIGURE 8.12 Color changes in presence of different anions such as (a) sensor **6** with AcO^-, (b) sensor **7** with AcO^-, $H_2PO_4^-$, F^-, F^- (excess), (c) sensor **8** with OH^-, F^-, CH3COO$^-$, $H_2PO_4^-$ ions.

Source: Parts a and c: Reprinted with permission from Ref. [55]. © 2005 by American Chemical Society. Part d: Reprinted with permission from Ref. [57]. © 2005, American Chemical Society.

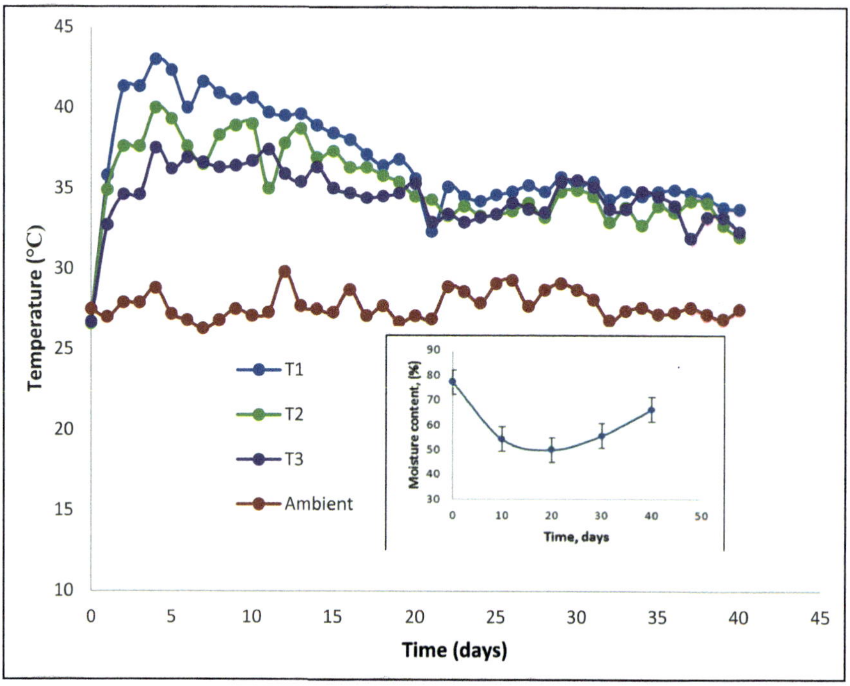

FIGURE 14.3 Temperature trend during the first 40 days of the composting process. Small figure shows the moisture content profile in the compost mixture (standard errors (SE) in vertical bars).

CHAPTER 9

SUPERCONDUCTIVITY AND QUANTUM COMPUTING VIA MAGNETIC MOLECULES

FRANCISCO TORRENS[1,*] and GLORIA CASTELLANO[2]

[1]*Institut Universitari de Ciència Molecular, Universitat de València, Edifici d'Instituts de Paterna, P.O. Box 22085, E-46071 València, Spain*

[2]*Departamento de Ciencias Experimentales y Matemáticas, Facultad de Veterinaria y Ciencias Experimentales, Universidad Católica de Valencia San Vicente Mártir, Guillem de Castro-94, E-46001 València, Spain*

[*]*Corresponding author. E-mail: torrens@uv.es*

ABSTRACT

People live in a world conditioned by an enormous ability to acquire and process information, and the advances were possible, greatly, thanks to the continuous miniaturization of electronic devices. This chapter reviews superconductivity (SC) and magnetism coexistence by chemical design, critical approach to pseudosciences as a didactic exercise, SC, quantum computing via magnetic molecules, quantum simulations and tensor nets in condensed matter and high energy, quantum fluctuations of the space–time geometry in spectral scheme, and how quantized space and time explain properties of relativity and dark matter. This study wants to establish a dialogue between science and pseudosciences from an educational viewpoint. The goal of this revision is to show the current state of the field and how to make use of the available technology to obtain meaningful results in a prequantum-computing era.

9.1 INTRODUCTION

The goal of this chapter is to show the current state of the field and how to make use of the available technology to obtain meaningful results in a prequantum-computing era.[1]

In earlier publications, fractal hybrid-orbital analysis[2,3]; resonance[4]; molecular diversity[5]; periodic table of the elements[6,7]; law, property, information entropy, molecular classification, simulators[8–15] labor risk prevention, preventive healthcare at work with nanomaterials[16–18]; and developing sustainability via nanosystems and devices[19] were reviewed.

The present chapter deals with superconductivity (SC) and quantum computing via magnetic molecules. It reviews SC and magnetism coexistence by chemical design. It wants to establish a dialogue between science and pseudosciences from an educational viewpoint. It revises SC. The aim of this chapter is to initiate a debate by suggesting a number of questions (Q), which can arise when addressing subjects of quantum computing via magnetic molecules, and quantum simulations and tensor nets in condensed matter and high energy, and provide, when possible, answers (A) and hypotheses (H). It discusses quantum fluctuations of the space–time geometry in spectral scheme. Finally, it analyzes how quantized space and time explain properties of relativity and dark matter.

9.2 SC AND MAGNETISM COEXISTENCE BY CHEMICAL DESIGN

Although SC and ferromagnetism coexistence in one compound is rare, some examples do exist. Methods to prepare physically hybrid structures with both competing phases are known, which rely on the nanofabrication of alternating conducting layers. Chemical methods of building up hybrid materials with organic molecules (SC layers) and metal complexes (magnetic layers) provided examples of SC with magnetic properties but not fully ordered. Coronado group reported a chemical design strategy via the self-assembly in solution of macromolecular nanosheet building blocks to engineer SC and magnetism coexistence in $[Ni_{0.66}Al_{0.33}(OH)_2][TaS_2]$ at 4 K.[20] The method was shown in isostructural $[Ni_{0.66}Fe_{0.33}(OH)_2][TaS_2]$, in which magnetic ordering was shifted from 4 to 16 K.

9.3 CRITICAL APPROACH TO PSEUDOSCIENCES AS A DIDACTIC EXERCISE

Gaita-Ariño group proposed three teaching activities.[21] They ranged in complexity and were adequate for implementation in classrooms and laboratories from secondary education to university level. The *leitmotiv* for the activities was series dilutions, and the concepts of substance quantity, molecules, and concentration. Transversally, this was used to point out the pseudoscientific fraud of homeopathy, which in recent years was alarmingly popular. An important benefit of a basic understanding of chemical concepts was the ability to detect the kind of hocus-pocus. One of the problems dealt with how many dilutions one should do to reach an only Au nanoparticle (NP). Taking into account the rise experienced by nanoscience and nanotechnology field, the fact of working with pupils the synthesis and characterization of NPs of interest in a number of fields became a tool useful from the motivational aspect. For that, a simple synthesis procedure of Au NPs was incorporated. Au NPs can be related to experiences in informal contexts.

9.4 SUPERCONDUCTIVITY

The intense magnetic fields that must be generated in some plants are induced via enormous electric currents.[22] If it were not for SC, Joule effect would make that; due to losses and heating, the technology was not feasible. This is a phenomenon that appears in certain materials and produces that material resistivity to the passage of electric current be null, eliminating Joule effect. Despite SC being discovered at the beginning of 20th century, nowadays no complete satisfactory theory exists on it that explain all observed phenomena. However, some approximations explain many of them, for example, Bardeen–Cooper–Schrieffer (BCS) and Ginzburg–Landau theories. *BCS theory*: This theory of 1957 points out that electrons in a superconductor material forms the so-called Cooper pairs, in which movement via a suitable crystalline solid presents, because of quantum effects, no resistance. *Ginzburg–Landau theory*: The theory focuses on the problem from a macroscopic viewpoint. It is based on thermodynamic variable *Helmholtz function* and its minimization, due to electrons mobility in the superconductor. The theory of 1950 is able to predict better SC in heterogeneous materials.

9.5 QUANTUM COMPUTING VIA MAGNETIC MOLECULES

Gaita-Ariño group proposed hypotheses/questions/answer (H/Q/A) on quantum computing.[23]

H1 (Feynman, 1982). To apply quantum laws to the information subject.[24,25]
Q1. What will people do via a quantum computer?
Q2. How to construct it?
Q3. Why to construct it?
A3. To construct a quantum computer, it requires overcoming challenges in physics and technology.
H2 (Hund, 1927). His rules determine the *term symbol* corresponding to ground state of a *multielectron atom.*

9.6 QUANTUM SIMULATIONS/TENSOR NETS IN CONDENSED MATTER/HIGH ENERGY

Quantum simulation uses a different system to emulate the behavior of the problem that many-body quantum systems are hard to describe (cf. Fig. 9.1). Cirac proposed Q/H/A on quantum simulation descriptions of high-energy physics (HEP).

Q1. How does quantum physics tackle the problem of complexity?
H1 (Feynman, 1982). Simulating physics with computers.
Q2 (Feynman, 1982). How good can computers compute people's world?
A2. Number of parameters required to describe quantum systems grows exponentially with number of particles, and so on.
H2 (Feynman, 1982). Take a lattice, discretize time, and put a quantum bit (*qubit*) in every vertex.
H3 (Feynman, 1982). If you want to make a simulation of nature, you should better make it quantum mechanical.
Q3. Challenge: How to tune the Hamiltonian?
Q4. Experiments?
He proposed the following questions and answers on tensor network descriptions of HEP.
Q5. How much *entangled* are the states?
A5. For local Hamiltonians $H = \Sigma_i h_i$, little.

Q6. What do we learn?
A6. The experimental Hamiltonian scales polynomially.

FIGURE 9.1 A quantum simulator: optical network.

9.7 QUANTUM FLUCTUATIONS OF THE SPACE–TIME GEOMETRY IN SPECTRAL SCHEME

The spectral scheme for space–time geometry is a totally new framework for quantitatively describing the space–time geometry in terms of the spectra of a certain elliptic operator (typically Laplacian one) on the space in question. The central idea of the framework can be symbolically stated as *Let us* hear *the shape of the Universe!* A number of advantages of the framework exist compared to the traditional geometrical description in terms of Riemannian metric. After sketching the basics of the spectral scheme, Seriu gave a formula for the Einstein–Hilbert action, which is a central quantity for the general theory of relativity (GTR), in terms of the spectral scheme.[26] He paid attention to its application to the quantum universe and saw how the quantum fluctuations of space–time could be effectively described in terms of the spectral scheme.

9.8 QUANTIZED SPACE/TIME EXPLAIN PROPERTIES OF RELATIVITY/DARK MATTER

Some thought experiments are used to quantize space and time.[27] The model adopts the idea that space and time are separate but related entities. Space is quantized into a unit called the *speck*. When specks are near fermions, the interaction increases the rate of gauge boson production near the fermions, which is related to the apparent rate of the passage of time. Specks exist in a standard state; they contribute greatly to gauge boson production; however, once the speck produces a gauge boson, it shrinks in size and becomes a low-energy (LE) one. The LE specks present a reduced ability to form gauge bosons, but when they do, they shrink in size with every produced gauge boson. The properties and others are used to generate a model for quantized space and time, which is then used to explain some of the properties associated with special theory of relativity and GTR (e.g., time dilation, relativistic changes in mass and length). The model provides some insight into how some dark matter observations arose. Consider two issues: it is entirely possible to have an incorrect grammatical description of a phenomenon but a correct mathematical description; it is easier to prove that a mathematical description is correct than the grammatical description.

9.9 DISCUSSION

A critical approach to pseudosciences is necessary, especially due to their didactic and educational perspectives. How to treat pseudosciences in the lecture room? How to teach a student of physician to deal with patient's pseudosciences? This theme is related to that of social controversy in science, and history, philosophy and sociology of science, for example, Kuhn–Popper controversy.[28-32]

Initiating a debate by suggesting a number of questions, which can arise when addressing subjects of quantum computing via magnetic molecules, and quantum simulations and tensor nets in condensed matter and high energy, would provide, when possible, answers and hypotheses.

9.10 FINAL REMARKS

From the present results and discussion, the following final remarks can be drawn:

1. In a world conditioned by an enormous ability to acquire and process information, the advances were possible, greatly, thanks to the continuous miniaturization of electronic devices.
2. A critical approach to pseudosciences is necessary, especially due to their didactic and educational perspectives. How to treat pseudosciences in the lecture room? How to teach a student of physician to deal with patient's pseudosciences? This theme is related to that of social controversy in science and history of science.
3. It is interesting to be up to date in SC, quantum computing via magnetic molecules, quantum simulations and tensor nets in condensed matter and high energy, quantum fluctuations of the space–time geometry in spectral scheme, and how quantized space and time explain properties of relativity and dark matter.
4. Showing the current state of the field, it is needed to know how to make use of the available technology to obtain meaningful results in a prequantum-computing era.
5. Initiating a debate by suggesting a number of questions, which can arise when addressing subjects of quantum computing via magnetic molecules, and quantum simulations and tensor nets in condensed matter and high energy, would provide, when possible, answers and hypotheses.

ACKNOWLEDGMENTS

The authors thank support from Generalitat Valenciana (Project No. PROMETEO/2016/094) and Universidad Católica de Valencia *San Vicente Mártir* (Project Nos. PRUCV/2015/617 and UCV.PRO.17-18.AIV.03).

KEYWORDS

- **space–time structure**
- **quantized space**
- **quantized time**
- **quantum computation**
- **quantum simulation**
- **superconducting qubit**

REFERENCES

1. Nouailhat, A. *An Introduction to Nanosciences and Nanotechnology*; Wiley: Hoboken, NJ, 2010.
2. Torrens, F. Fractals for Hybrid Orbitals in Protein Models. *Complex. Int.* **2001,** *8*, 1–13.
3. Torrens, F. Fractal Hybrid-orbital Analysis of the Protein Tertiary Structure. *Molecules* **2002,** *7*. DOI: 10.3390/70100026.
4. Torrens, F.; Castellano, G. Resonance in Interacting Induced-Dipole Polarizing Force Fields: Application to Force-field Derivatives. *Algorithms* **2009,** *2*, 437–447.
5. Torrens, F.; Castellano, G. Molecular Diversity Classification Via Information Theory: A Review. *ICST Trans. Complex Syst.* **2012,** *12* (10–12), e4.
6. Torrens, F.; Castellano, G. Reflections on the Nature of the Periodic Table of the Elements: Implications in Chemical Education. In *Synthetic Organic Chemistry*; Seijas, J. A., Vázquez Tato, M. P., Lin, S. K., Eds.; MDPI: Basel, Switzerland, 2015; Vol. 18, pp 1–15.
7. Putz, M. V.; Ed. *The Explicative Dictionary of Nanochemistry*; Apple Academic–CRC: Waretown, NJ (In Press).
8. Torrens, F.; Castellano, G. Reflections on the Cultural History of Nanominiaturization and Quantum Simulators (Computers). In *Sensors and Molecular Recognition*; Laguarda Miró, N., Masot Peris, R., Brun Sánchez, E., Eds.; Universidad Politécnica de Valencia: València, Spain, 2015; Vol. 9, pp 1–7.
9. Torrens, F.; Castellano, G. Ideas in the History of Nano/Miniaturization and (Quantum) Simulators: Feynman, Education and Research Reorientation in Translational Science. In *Synthetic Organic Chemistry*; Seijas, J. A., Vázquez Tato, M. P., Lin, S. K., Eds.; MDPI: Basel, Switzerland, 2016; Vol. 19, pp 1–16.
10. Torrens, F.; Castellano, G. Nanominiaturization and Quantum Computing. In *Sensors and Molecular Recognition*; Costero Nieto, A. M., Parra Álvarez, M., Gaviña Costero, P., Gil Grau, S., Eds.; Universitat de València: València, Spain, 2016; Vol. 10, pp 31-1–31-5.
11. Torrens, F.; Castellano, G. Nanominiaturization, Classical/Quantum Computers/ Simulators, Superconductivity and Universe. In *Methodologies and Applications for Analytical and Physical Chemistry*; Haghi, A. K., Thomas, S., Palit, S., Main, P., Eds.; Apple Academic–CRC: Waretown, NJ (In Press), 2018; pp 27–44.
12. Torrens, F.; Castellano, G. Superconductors, Superconductivity, BCS Theory and Entangled Photons for Quantum Computing. In *Physical Chemistry for Engineering and Applied Sciences: Theoretical and Methodological Implication*; Haghi, A. K., Aguilar, C. N., Thomas, S., Praveen, K. M., Eds.; Apple Academic–CRC: Waretown, NJ (In Press) 2018; pp 379-387.
13. Torrens, F.; Castellano, G. EPR Paradox, Quantum Decoherence, Qubits, Goals and Opportunities in Quantum Simulation. In *Theoretical Models and Experimental Approaches in Physical Chemistry: Research Methodology and Practical Methods*; Haghi, A. K., Ed. (In Press) 2019; Vol. 5, pp 319-336.
14. Torrens, F.; Castellano, G. Nanomaterials, Molecular Ion Magnets, Ultrastrong, Spin–Orbit Couplings in Quantum Materials. In *Physical Chemistry for Chemists and Chemical Engineers: Multidisciplinary Research Perspectives*; Vakhrushev,

A. V., Haghi, R., de Julián-Ortiz, J. V., Allahyari, E., Eds.; Apple Academic–CRC: Waretown, NJ (In Press) 2019; pp 181-190.

15. Torrens, F.; Castellano, G. Nanodevices and Organization of Single Ion Magnets and Spin Qubits. In *Chemical Science and Engineering Technology: Perspectives on Interdisciplinary Research*; Haghi, A. K., Ed.; Apple Academic–CRC: Waretown, NJ, 2018 (In Press) 2018; pp 3-26.
16. Torrens, F.; Castellano, G. *Book of Abstracts, Certamen Integral de la Prevención y el Bienestar Laboral, València, Spain, September 28–29, 2016*; Generalitat Valenciana–INVASSAT: València, Spain, 2016; p 3.
17. Torrens, F.; Castellano, G. Nanoscience: From a Two-Dimensional to a Three-dimensional Periodic Table of the Elements. In *Methodologies and Applications for Analytical and Physical Chemistry*; Haghi, A. K., Thomas, S., Palit, S., Main, P., Eds.; Apple Academic–CRC: Waretown, NJ (In Press).
18. Torrens, F.; Castellano, G. El Trabajo con Nanomateriales: Historia Cultural, Filosofía Reduccionista/Positivista y Ética. In *Tecnología, Ciencia y Sociedad*; Gherab-Martín, K. J., Ed., Global Knowledge Academics: València, Spain (In Press) 2018; pp 11-35.
19. Torrens, F.; Castellano, G. Developing Sustainability via Nanosystems and Devices: Science–Ethics. In *Chemical Science and Engineering Technology: Perspectives on Interdisciplinary Research*; Balköse, D., Ribeiro, A. C. F., Haghi, A. K., Ameta, S. C., Chakraborty, T., Eds.; Apple Academic–CRC: Waretown, NJ (In Press).
20. Coronado, E.; Martí-Gastaldo, C.; Navarro-Moratalla, E.; Ribera, A.; Blundell, S. J.; Baker, P. J. Coexistence of Superconductivity and Magnetism by Chemical Design. *Nat. Chem.* **2010,** *2*, 1031–1036.
21. Abellán, G.; Rosaleny, L. E.; Carnicer, J.; Baldoví, J. J.; Gaita-Ariño, A. La Aproximación Crítica a las Pseudociencias Como Ejercicio Didáctico: Homeopatía y Diluciones Sucesivas. *An. Quím.* **2014,** *110*, 211–217.
22. Fernández-Cosials, K.; Barbas Espa, A., Eds. *Curso Básico de Fusión Nuclear*; Jóvenes Nucleares–Sociedad Nuclear Española: Madrid, Spain, 2017.
23. Aromí, G.; Gaita-Ariño, A.; Luis, F. Computación Cuántica con Moléculas Magnéticas. *Rev. Esp. Fís.* **2016,** *30* (3), 21–24.
24. Feynman, R. P. There is Plenty of Room at the Bottom. *Caltech Eng. Sci.* **1960,** *23*, 22–36.
25. Feynman, R. P. Simulating Physics with Computers. *Int. J. Theor. Phys.* **1982,** *21*, 467–488.
26. Seriu, M. Quantum Fluctuations of the Space–Time Geometry in the Spectral Scheme. *Sci. Res. Essays* **2016,** *11* (23), 247–254.
27. Remillard, R. B. J. Quantized Space and Time Explain Properties of Relativity and Dark Matter. *J. Appl. Theor. Phys. Res.* **2017,** *1* (3), 6–10.
28. Kuhn, T. S. *The Structure of Scientific Revolutions*; University of Chicago: Chicago, IL, 1962.
29. Popper, K. *All Life Is Problem Solving*; Routledge: London, 1999.
30. Bunge, M. *La Investigación Científica: Su Estrategia y su Filosofía*; Ariel: Barcelona, Spain, 1985.
31. Bunge, M. *El Problema Mente–Cerebro*; Tecnos: Madrid, Spain, 1985.
32. Bunge, M. *Ciencia, Técnica y Desarrolo*; Laetoli: Pamplona, Spain, 2014.

SECTION III

Prospects and Challenges of Composites

CHAPTER 10

NOVEL SEPARATION PROCESSES, WATER SCIENCE, AND COMPOSITE MEMBRANES: A FAR-REACHING REVIEW AND A VISION FOR THE FUTURE

SUKANCHAN PALIT[1,2,*]

[1]*Department of Chemical Engineering, University of Petroleum and Energy Studies, Energy Acres, P.O. Bidholi via Premnagar, Dehradun 248007, Uttarakhand, India*

[2]*43, Judges Bagan, P.O. Haridevpur, Kolkata 700082, India*

[*]*E-mail: sukanchan68@gmail.com; sukanchan92@gmail.com*

ABSTRACT

The world of environmental engineering and water science are veritably moving from one visionary paradigm toward another. Environmental engineering science and chemical process engineering stand in the midst of scientific vision and deep comprehension. Human civilization's immense scientific prowess, the scientific grit and determination, and the futuristic vision of chemical engineering will all lead a long and visionary way in the true emancipation of science and engineering globally. Novel separation processes in chemical process engineering include membrane separation processes. Nontraditional environmental engineering tools include advanced oxidation processes. In this chapter, the author deeply elucidates the scientific potential, the scientific success, and the immense scientific ingenuity in the field of application of novel separation processes in environmental engineering. Water purification, industrial wastewater treatment, and drinking water treatment are the utmost needs of human society today. Water science and water technology today are integrated with membrane science and desalination science. This chapter unfolds the scientific vision and the scientific stewardship of the world of membrane science in water purification domain. The challenges and the vision of membrane science will all lead a long and visionary way in the true realization of wastewater treatment and water science in decades to come.

10.1 INTRODUCTION

The world of environmental engineering and chemical process engineering are today in the midst of deep scientific vision and vast scientific farsightedness. Global climate change, the depletion of fossil fuel resources, and the frequent environmental catastrophes are urging the scientific domain to gear forward toward newer innovation and scientific paradigm. Novel separation processes such as membrane science are changing the vast scientific firmament of environmental engineering science. In today's modern civilization, technology and engineering science are moving ahead at a rapid pace from visionary paradigm toward another. Scientific vision and scientific profundity in the field of membrane separation processes are veritably far-reaching and surpassing vast and versatile scientific frontiers. In this chapter, the author deeply elucidates the need of innovations, the vision, and the scientific determination in research pursuit in membrane separation techniques and water science and technology. The chapter unfolds the scientific intricacies and the vast technical knowhow in the field of both membrane science as well as membrane technology. Drinking water treatment in the similar manner needs to be readdressed as global water shortage challenges the vast scientific firmament. Provision of clean drinking water and the provision of basic human needs are the needs of human civilization today. Technology and engineering science has few answers to the ever-growing concerns for global climate change, global warming, and depletion of fossil fuel resources. In such a crucial juncture of human scientific history and time, water purification techniques, desalination, and the greater link with chemical process engineering are the utmost need of human scientific progress today. The author deeply ponders on the scientific success, the vast scientific profundity, and the scientific ingenuity in the path toward environmental sustainability and environmental protection science.

10.2 THE VISION OF THIS STUDY

Global climate change and frequent environmental catastrophes are the challenges and the vision of science and engineering in modern human civilization today. Industrial wastewater treatment, drinking water treatment, and environmental protection today stand in the midst of scientific profundity and deep scientific forbearance. The vision of this study is to target the needs of water purification technologies and wastewater treatment tools to human society and human scientific progress. The author pointedly focuses on the different novel separation processes, water purification techniques, and the vast world of chemical process engineering. Chemical engineering and novel separation processes are two opposite sides of the visionary coin today. Human scientific regeneration in the field of water science and technology is in a state of immense scientific cauldron and in a state of disaster. Environmental engineering catastrophes are challenging the scientific firmament today. Research and development initiatives in water technologies are the veritable need of the hour. This chapter throws a bright light on the scientific ingenuity, the vast scientific farsightedness, and the technological validation behind novel separation processes. Success of conventional environmental engineering techniques is scarce. Thus the immediate need of nonconventional environmental engineering techniques such as membrane science and advanced oxidation techniques. This chapter highlights the vast scientific needs of environmental engineering and chemical process engineering toward a greater emancipation of science and technology.

10.3 WHAT DO YOU MEAN BY NOVEL SEPARATION PROCESSES?

Novel separation techniques, environmental engineering science and chemical process engineering, are veritably linked with each other in modern science and present-day human civilization. Novel separation processes include membrane science. It has today applications in all areas of water science and technology. Technological and scientific validations are the necessities of science and engineering today. Global water crisis has today urged the scientists and engineers to delve deep into the world of membrane science and water science. Human scientific research pursuit thus today needs to be envisioned and reframed as science and engineering move forward.

10.3.1 WHAT DO YOU MEAN BY COMPOSITE MEMBRANES?

Membrane science is today in the path of newer scientific rejuvenation. Composite membranes are the next generation innovative materials. Novel separation processes such as membrane science needs to be reenvisioned and reframed as regards the effectivity and efficiency of the separation phenomenon. Thin-film composite membranes are semipermeable membranes manufactured primarily for use in water purification and water desalination membranes. They also have immense use in chemical applications such as batteries and fuel cells. A thin-film composite membrane can be considered as a molecular sieve constructed in the form of a film from two or more layered materials. Thin-film composite materials are the smart materials of tomorrow. Scientific profundity, scientific ingenuity, and deep scientific vision stand as major pillars in the scientific pursuit in membrane science today. Today environmental engineering science and chemical process engineering are interconnected by a visionary umbilical cord. Global water scenario is extremely dismal and steps should be taken to mitigate drinking water crisis. In such a crucial juncture of human history and time, thin-film composite membranes assume immense importance. Thin-film composite membranes are commonly classified as nanofiltration and reverse osmosis membranes. Here comes the vast importance of composite membranes. Thin-film composites membranes typically suffer from compaction effects under high pressure. Thin-film composite membranes are used (1) in water purification, (2) as a chemical reaction buffer, and (3) in industrial gas separations. Thus, composite membranes have vast and versatile use in drinking water treatment, industrial wastewater treatment, and water purification.

10.4 WHAT DO YOU MEAN BY AOPs?

Advanced oxidation processes (AOPs) and integrated AOPs are witnessing dramatic challenges in modern science and present-day human civilization. Technology and science of environmental engineering need to be reenvisioned and reorganized with the passage of scientific history and time. Advanced oxidation techniques today stands in the midst of deep scientific vision and vast scientific ingenuity. Nonconventional environmental engineering tools are changing the face of human scientific research pursuit in environmental engineering. Reverse osmosis, nanofiltration, and ultrafiltration are challenging the vast scientific firmament of vision and determination.[1] AOPs, in a broad sense, are a set of chemical treatment tools designed to remove organics (and sometimes inorganic) materials in wastewater by oxidation through reactions with hydroxyl radicals. Mainly AOPs refer to chemical processes that employ ozone, hydrogen peroxide, and ultraviolet/light.[1]

10.5 SCIENTIFIC VISION, THE DEEP SCIENTIFIC CONSCIENCE, AND THE INTROSPECTION BEHIND MEMBRANE SCIENCE

The deep scientific vision and conscience of membrane science are changing the vast face of research pursuit in environmental engineering science. The needs of industrial wastewater treatment, the vast vision of drinking water treatment, and the world of water purification are the veritable forerunners toward a newer eon in the field of chemical process engineering. Human technological vision, scientific grit, and determination are the necessities of scientific research pursuit in water treatment today. Reverse osmosis, nanofiltration, ultrafiltration, and microfiltration are the needs of the world of water purification today. The deep scientific conscience needs to be reenvisioned as membrane science and novel separation processes enter into a new era of scientific grit and determination. Membrane science is the need of water science and purification in modern science and present-day human civilization. The challenge and the vision lie in the hands of effectiveness and the efficiency of the environmental engineering treatment procedure.[1,2] Technology and engineering science are in the midst of deep vision and thoughtful scientific regeneration. In this entire treatise, the author deeply elucidates the success and vision of wastewater treatment and industrial pollution control in the furtherance of science and technology. As science and engineering move forward toward a newer age of vision and might, novel separation techniques and environmental engineering tools assume immense importance. The author pointedly focuses on vision, targets, and the scientific fortitude behind novel separation techniques. Membrane science and AOPs are the areas of newer scientific regeneration. A deep scientific introspection and vast scientific envisioning are needed as regards application of novel separation processes to water and environment. Human scientific research pursuit, the needs of water science and technology and the world of scientific and engineering challenges will lead a long and visionary way in the true realization of environmental sustainability today.

10.6 THE SCIENCE OF WATER PURIFICATION

The science of water purification is today entering into a new era of vast scientific farsightedness and scientific foresight. The global need is for eradication of greenhouse warming and global climate change. In the similar vein, depletion of fossil fuel resources is changing the face of scientific research pursuit in petroleum engineering science and chemical process engineering. Technology and engineering science thus stand in the midst of deep scientific forbearance and adjudication. In this chapter, the author deeply proclaims the need for water purification in the proliferation of global science and technology initiatives. The challenge for global research and development initiatives in water science and technology needs to be widened and reenvisioned with the passage of scientific history and the visionary timeframe.

10.7 THE VAST WORLD OF NOVEL SEPARATION PROCESSES

The vast world of novel separation processes is changing the face of human civilization and human scientific endeavor. The provision of clean drinking water is a veritable scientific imperative. Novel separation processes and its applications today stand in the midst of deep scientific forbearance and an unending vision. The vast world of novel separation processes and nonconventional environmental engineering tools are witnessing immense scientific envisioning and vast scientific forbearance. Reverse osmosis, nanofiltration, ultrafiltration, and electrodialysis are the needs of human scientific endeavor in water science and technology today. Chemical process engineering is linked with novel separation processes by an unsevered umbilical cord. Every branch of engineering science and technology are highly challenged today as human civilization treads forward toward a newer visionary eon. Novel separation processes such as membrane science also is moving toward a newer scientific direction and knowledge dimension. The need of membrane science and desalination is immense as environmental engineering science globally faces immense turmoil with the ever-growing concerns for global climate change. In such a crucial juxtaposition of science and engineering, novel separation processes, its vision, and scientific fortitude need to be readdressed.

10.8 SIGNIFICANT SCIENTIFIC ENDEAVOR IN MEMBRANE SCIENCE

Membrane science is the need of the human society today. Separation techniques and chemical process engineering need to be reenvisioned with the passage of scientific history and time. Nanofiltration, ultrafiltration, microfiltration, and reverse osmosis are changing the vast scientific firmament of environmental engineering science. Human factor engineering and environmental engineering science are the necessities of scientific endeavor today. In the similar manner, industrial wastewater treatment and drinking water treatment are the marvels of science and engineering today. In this chapter, the author deeply elucidates the immense necessity of membrane separation processes in the furtherance of science and engineering of water and environment. Scientific research pursuit in water and environment today needs to be more streamlined and envisioned as human civilization moves forward.

Shannon et al.[1] discussed in deep details science and technology for water purification in the coming decades. This chapter is a comprehensive review in the field of water purification. One of the decisive challenges afflicting people throughout the world is absolutely inadequate supply of clean water and sanitation. Human scientific understanding and prudence are in the state of deep distress as water and environment moves toward newer scientific regeneration.[1] Groundwater heavy metal contamination is a disaster to human civilization today. South Asian countries like Bangladesh and India are in a state of immense catastrophe due to groundwater poisoning. Problems with water are expected to grow worse in the coming years, with water scarcity occurring globally, even in regions currently considered water rich. Today, the state of environment is immensely dismal. In this chapter, the author discusses some of the science and technology being developed to improve the disinfection and decontamination of water as well as the areas of reuse of water and efficient desalination of brackish water. The pivotal challenges of this paper are (1) disinfection, (2) decontamination, and (3) reuse and reclamation, and the vast world of desalination. Today, human scientific challenges are immense and groundbreaking. Provision of clean drinking water is under serious threat in modern human civilization. This paper takes up that immense challenge to address and envision the immediate need of innovative water purification technologies.[1]

Giese et al.[2] discussed with wide vision and deep conscience water purification by membranes and the role of polymer science. Human scientific research endeavor in water purification is vastly advancing and crossing visionary scientific endeavor. Now two of the greatest challenges facing the 21st century are the provision of electricity and water at affordable costs.[2] Membrane technology is today at a promising state and is the necessity of water science today. Innovative polymer chemistry and applied physics along with deep discussion of membrane science are the hallmarks of this well-researched treatise. The authors gave a brief background of world water resources, the vast scientific ingenuity of membrane operations. This chapter also delved deeply in the field of membrane manufacture and membrane fabrication.[2] The other areas of this chapter are pore-flow membranes, polyamide thin-film composites, and solution–diffusion transport mechanisms of membranes. Membrane technology is a revolutionary avenue of science and engineering today. This chapter throws immense light on the necessity of membrane science on water purification technologies.[2]

Arnot et al.[3] discussed and deeply comprehended cross-flow and dead-end microfiltration of oily water emulsions and also mechanisms and modeling of the flux decline. Membrane fouling today is in a difficult stage.[3] The focus of this chapter is on the mechanisms and modeling of the flux decline. The chapter opens up a new phase in the field of microfiltration and fouling phenomenon. The science of fouling is still in a latent stage and needs to be investigated in details. In this chapter, the authors target microfiltration membrane functioning and the fouling associated with it.[3]

Sarkar et al.[4] with deep scientific forbearance and insight discussed the Donnan membrane principle and the opportunities for sustainable-engineered processes and materials. The Donnan membrane technique is essentially a specific domain of the second law of thermodynamics dealing solely with completely ionized electrolytes.[4] The targets of this chapter are the thermodynamic approach of membrane-separation principles. Membrane is a revolutionary area of research pursuit today. The goal and mission of this chapter is to present some distinctive environmental applications research forays that utilize the basic tenets of Donnan membrane principle but do not require electricity or transmembrane pressure as the driving force.[4] The authors in this chapter deeply elucidate the scientific discernment, the scientific understanding, and the scientific prudence in membrane separation techniques.[4]

Science and technology of membrane science are today moving from one visionary paradigm toward another. Membrane science and water science are today two opposite sides of the visionary coin. In this entire chapter, the author pointedly focuses on the scientific success of environmental engineering and chemical engineering science in today's scientific landscape.

10.9 RECENT SCIENTIFIC ENDEAVOR IN THE FIELD OF ADVANCED OXIDATION PROCESSES

AOPs and integrated AOPs are witnessing dramatic and drastic challenges in the path toward scientific emancipation today. Water science and technology depend on novel separation processes and nonconventional environmental engineering tools. Human civilization today stands in the midst of deep scientific vision and forbearance. AOPs are the pivots of scientific research pursuit in environmental engineering and environmental management today. Technological validation, scientific serendipity, and the travails of environmental engineering will all lead a long and visionary way in the true emancipation of science and engineering today. AOPs and integrated AOPs are changing the scientific mindset of human scientific endeavor today. Technology and engineering science today have practically no answers to the burning issue of arsenic groundwater contamination and its aftereffects. In this chapter, the author reiterates the scientific success, the scientific ardor, and the vision behind scientific research pursuit in nonconventional environmental engineering processes and its diverse applications.

Munter[5] deeply discussed with immense scientific conscience the current status and prospects AOPs. The paper presents an overview of theoretical basis, efficiency, economics, laboratory and pilot plant testing, design, and modeling of different AOPs (combinations of ozone and hydrogen peroxide with ultraviolet radiation and catalysts). Scientific vision, the vast and wide scientific ingenuity, will all be the forerunners of a greater vision of AOPs and other environmental engineering tools. Hazardous organic wastes from industrial, military, and commercial operations represent one of the greatest human challenges to environmental engineers and scientists. AOPs are alternatives to the incineration of wastes, which are immense advantages as well as disadvantages. Conventional incineration is commonly thought to be an excellent alternative to landfill, but as presently practiced, incineration may bring about serious problems due to release of toxic compounds such as polychlorinated dibenzodioxins and polychlorinated dibenzofurans into the environment via the incinerator off gas emissions and/or fly ash. Human scientific endeavor today is in a state of immense rejuvenation and immense scientific forbearance. The AOPs have proceeded along one of the two routes: (1) oxidation with oxygen in temperature ranges intermediate between ambient conditions and those found in incinerators and wet air oxidation processes in the region of 1–20 MPa and 200–300°C and (2) the use of high-energy oxidants such as ozone and hydrogen peroxide and/or photons that are able to generate highly reactive intermediates that is OH radicals. Human scientific regeneration today is at its helm as science and engineering of water and environment surges ahead toward a newer visionary eon. Technological validation of environmental engineering processes is the need of human civilization today. In the similar manner, AOPs also are in the process of newer scientific overhauling and deep scientific rejuvenation. One brilliant environmental engineer defined AOPs as "near ambient temperature and pressure water treatment processes which involve the generation of hydroxyl radicals in sufficient quantity to effect water purification." The hydroxyl radical (OH) is a powerful, nonselective chemical oxidant which acts very rapidly with most organic compounds. These reaction rate constants vary in quite a wide range from 0.01 to 10^4 M^{-1} s^{-1}. Once generated, the hydroxyl radicals aggressively attack virtually all organic compounds. Depending upon the nature of the organic species, two types of initial attack are possible: the hydroxyl radical can abstract a hydrogen atom from water, as with alkanes or alcohols, or it can add itself to the contaminant, as in the case of olefins or aromatic compounds. The attack by the OH radical, in the presence of oxygen, initiates a complex cascade of oxidative reactions leading to the mineralization of the organic compound. The exact routes of these reactions are still not very clear in today's scientific research pursuit. For example, chlorinated organic compounds are oxidized first to intermediates, such as aldehydes and carboxylic acids and finally to CO_2, H_2O, and the chloride ion. Nitrogen in organic compounds is usually oxidized to nitrate or to free nitrogen, sulfur is oxidized to the sulfate. Cyanide is oxidized to cynate, which is then further oxidized to CO_2 and NO_3^- (or perhaps N_2). Human scientific conscience, the deep scientific doctrine, and the farsightedness of science and technology will all lead a long and visionary way in the true envisioning of novel separation processes today. This paper reviews the entire gamut of scientific doctrine and the scientific profundity in the field of AOPs. In general, the AOPs when applied in the right place, gives a world of vast scientific opportunities to reduce the contaminants' concentration from several hundreds ppm to less than

5 ppb. This is the reason why they are envisioned as the water treatment processes of the 21st century. Technological envisioning, the vast scientific needs of the 21st century, and the world of environmental management are the forerunners toward a newer eon in environmental sustainability today.

Cesaro et al.[6] discussed with vast scientific insight and deep scientific conscience wastewater treatment by combination of AOPs and vast world of conventional biological systems. Scientific conscience, scientific profundity, and the needs for environmental sustainability are all today forerunners toward a newer vision in nonconventional environmental engineering tools such as AOP. One of the most challenging and inspiring issues of the last decades is the presence of recalcitrant compounds in the effluents of wastewater treatment plants due to their vast levels of toxicity on both human health and the environment. The challenges and the vision of AOPs are immense and groundbreaking in today's scientific landscape. In the context of successful environmental engineering paradigm, AOPs, which are oxidation methods relying on the action of highly reactive species such as hydroxyl radicals, are raising immense interest for the removal of these organic compounds not treatable by conventional methods due to their high chemical stability and low biodegradability.[6] Human scientific ardor, the vast scientific serendipity, and the needs of human perseverance are the pallbearers toward a newer knowledge dimension in the field of novel separation tools today. This vast and well-researched treatise aims to discuss the most common AOPs used as a pretreatment of wastewater for its biological and chemical processing, to highlight the enhancement of wastewater biological treatability supplied by different advanced oxidation tools. Biodegradability is the hallmark of vast scientific evolution and scientific profundity today. The targets, the vision, and the immense challenges of biodegradability are deeply discussed in this chapter. The action mechanism of AOPS relies on the formation of highly reactive oxidant species, mainly hydroxyl radicals, which can react with recalcitrant compounds until their mineralization occurs.[6] The vast world of scientific intricacies, the world of scientific validation, and the technological underpinnings are the needs and the torchbearers of a greater emancipation of science of AOPs today. This chapter stands in the forefront of a deeper scientific issue of environmental engineering science.[6]

Tsydenova et al.[7] elucidated with deep scientific vision solar-enhanced AOPs for water treatment and simultaneous removal of pathogens and chemical pollutants. This review investigates the feasibility of simultaneous removal of pathogens and chemical pollutants by solar-enhanced

AOPs.[7] Integrated AOPs, the needs for biological treatment and the vast world of bioremediation will lead a long and visionary way in unraveling the scientific truth of AOPs today. Bioremediation is the scientific marvel of today's scientific endeavor in environmental engineering science. The AOPs are based on in situ generation of reactive oxygen species, most notably hydroxyl radicals that are capable of destroying and decimating both pollutant molecules and pathogen cells. In this treatise, the authors deeply ponder on the solar-enhanced AOPs which are new direction in the field of environmental engineering science.[7] This technique is highly effective in conventional water/wastewater treatment particularly in fields where there is a scarce access to centralized drinking water and sewage treatment facilities.[7] This well-researched treatise targets the needs of simultaneous removal of pathogens and chemical pollutants with the help of solar-enhanced AOPs, titanium dioxide photocatalysis, and photo-Fenton techniques.[7] Human scientific regeneration in the field of environmental engineering separation processes and chemical process engineering is of utmost necessity today. In this chapter, the authors also elucidate on the contribution of catalyst/oxidant concentration, incident radiation flux, and pH in AOP wastewater treatment techniques.[7]

Lamsal et al.[8] with vast scientific foresight deeply discussed and compared AOPs for the removal of natural organic matter (NOM). This study investigated the impact of UV, ozone, AOPs including ozone/ultraviolet, hydrogen peroxide/ultraviolet, hydrogen peroxide/ozone in the change of molecular weight distribution and disinfection by product formation potential. Science of AOPs is immensely advanced today. This technology is today far-reaching and surpassing vast scientific frontiers.[8] Human scientific challenges are immense today. The authors deeply discussed AOPs in a vastly experimental effort.[8]NOM is a complex heterogeneous mixture of diverse compounds with varying molecular weight and science. A common drinking water paradigm is to remove NOM as it is a precursor for unwanted disinfection by-products during chemical disinfection processes.[8] NOM has also found to be shown to contribute to fouling of membranes, the production of biologically unstable water, and other unwanted water-quality issues. The application of AOPs has assumed immense importance in the drinking water industry as an additional technique for removing NOM and minimizing the formation of recalcitrant chemicals in drinking water and industrial wastewater. During advanced oxidation treatment, hydroxyl radicals are formed which act as a strong oxidant and transform the NOM. During advanced oxidation treatment,

NOM is partially oxidized and higher molecular weight compounds are transformed into smaller and more biodegradable compounds such as aldehydes and carboxylic acids. This study deeply evaluated ozone, ultraviolet, and three AOPs including hydrogen peroxide/ozone, hydrogen peroxide/ultraviolet, and ozone/ultraviolet for the removal of NOM and other nondegradable compounds in the wastewater. Scientific vision, the deep scientific prudence, and the vast world of technological validation are the truths of this chapter.[8]

Covinich et al.[9] with lucid and cogent insight in a review paper deeply discussed AOPs for wastewater treatment in the pulp and paper industry. The challenge and the vision of science and technology are deeply illustrated in details in this paper. The effluents of some pulp and paper processes are extremely pollutant, because of their large volume and their refractory nature. Biological tools cannot degrade these compounds. AOPs are vastly characterized by the capability of exploiting the degradation potential and high reactivity of hydroxyl radicals. AOPs can achieve total mineralization, transforming recalcitrant compounds into oxygen compounds (carbon dioxide and hydrogen peroxide), or partial mineralization, transforming them into more biodegradable compounds.[9] Today science and engineering are crossing vast and versatile scientific frontiers. The high reactivity and low selectivity of these radicals are important attributes that make these promising technologies. The authors in this chapter deeply target the application of AOPs in pulp and paper industries. Oxidation is defined as the transfer of one or more electrons from an electron donor (reductant) to an electron acceptor (oxidant) which has a higher affinity of electrons. These electron transfer result in the chemical transformation of both the oxidant and the reductant, producing in some cases chemical species with an odd number of valence electrons.[9] Technological validation and vast scientific motivation in scientific research pursuit are the necessities of human scientific paradigm and vast scientific genre today. This treatise upholds the deep fact of science and technological evolution in application of AOP in pulp and paper industry.[9]

Malhotra et al.[10] discussed with cogent insight degradation of ferrohexacyanide by AOPs. The forbearance and conscience of science are of highest order in today's world of water purification and environmental engineering.[10] Degradation of ferrohexacyanide by AOPs are studied in minute problems. AOPs included ozone and its combination with hydrogen peroxide and ultraviolet radiation. Results demonstrated that ultraviolet radiation alone was not sufficient to degrade cyanide ion but an oxidant was required for complete degradation. The effect of AOPs on the degradation of cyanide ion was studied using potassium ferrocyanide solution at cyanide concentration of 100 mg/L.[10] The vast ingenuity of AOP is discussed in deep details in this paper.

Andreozzi et al.[11] elucidated with immense farsightedness AOPs for water purification and recovery. Advanced oxidation encompasses a common feature: the capability of exploiting the high reactivity of HO radicals in driving oxidation processes which are immensely suitable for achieving the complete abatement and complete mineralization of even less-reactive water pollutants. In the last 25 years, a rather progressive revolution of the research activities devoted to environmental protection has been recorded as the consequence of the special attention paid to the environment by social, political, and legislative domain.[11] The destruction of toxic pollutants is a major scientific imperative and the techniques involve nonbiological technologies as well. These tools involve conventional phase separation techniques (adsorption processes, stripping techniques) and methods which ameliorate contaminants (chemical oxidation, reduction). Chemical oxidation involves mineralization of the contaminants to carbon dioxide, water, and other organics. The authors in this chapter covered Fenton processes, photo-assisted Fenton processes, photocatalysis, and ozone–water system. The other pillars of this article are oxalic acid treatment, ozone treatment, and hydrogen peroxide treatment of industrial wastewater. Technologically, industrial wastewater treatment today stands in the midst of deep scientific vision and vast scientific prudence and ingenuity. This chapter opens up new windows of innovation and scientific instinct in the field of water and environment in decades to come.

10.10 ENVIRONMENTAL SUSTAINABILITY AND THE VISION OF WATER TREATMENT

Sustainable development whether energy or environmental is challenging the vast scientific firmament in modern science and present-day human civilization. Environmental sustainability and global climate research and development initiatives are the two opposite sides of the visionary coin. The challenges and the vision of water treatment and the wide world of water purification are challenging the vast scientific landscape. In this chapter, the author deeply elucidates the intense scientific intricacies, the scientific vision, and the scientific travails in the path toward scientific emancipation in sustainable development today. Holistic sustainable development in the field of environment and energy is the need of human civilization today. Technological reenvisioning and the vast world of scientific validation will surely usher in a new era and a new eon in the field of environmental science. Today, the world of science and technology is witnessing immense overhauling and a deep reformation. Thus, the need of sustainable development in the field of energy and environment. In addition, the vicious and the vibrant world of environmental catastrophes such as large groundwater contamination needs to be eradicated with immediate effect. Here also the vision of environmental sustainability is of utmost importance. Technological overhauling and scientific validation are the cornerstones of scientific endeavor in environmental sustainability today. The need for water treatment and industrial pollution control assumes immense importance with the march of science of sustainability.

10.11 ARSENIC AND HEAVY METAL GROUNDWATER REMEDIATION

Arsenic and heavy metal groundwater remediation are the utmost need of the hour along with energy and environmental sustainability. Provision of clean drinking water today stands in the midst of deep scientific crisis in developing and developed nations around the world. Technology and engineering science has few answers to the monstrous issue of heavy metal groundwater contamination. In this chapter, the author pointedly focuses on the global water shortage issue. Arsenic groundwater contamination is a burning issue in South Asia, developing and developed nations around the world. Human scientific ingenuity is witnessing immense scientific difficulties as well as scientific serendipity with the progress of environmental engineering science. The immense necessity of human scientific progress is the proliferation of research and development initiatives in the field of water science and technology. The scientific underpinnings of environmental engineering, chemical process engineering and water technology needs to be thoroughly reenvisioned and readdressed with the passage of scientific history and the visionary timeframe. Disastrous health effects of the citizens of developing nations due to heavy metal drinking water contamination are challenging the veritable scientific fabric of human civilization today. The imminent need of the hour is the rebuilding and revamping of the huge world of water science and water engineering. Human scientific vision in research and development initiatives in groundwater remediation are today in a state of immense catastrophe. Thus the need of a whole-hearted effort in the true realization of environmental sustainability. Environmental management is similarly a veritable "handrail" of sustainable development. India and Bangladesh today stand in the midst of a deep environmental catastrophe which is the burning issue of groundwater contamination. Science and technology has few answers to the ever-growing concerns of human health due to groundwater heavy metal and arsenic contamination. Thus, the need of a well concerted effort in environmental management. Today, the underpinning of scientific research effort in environmental engineering is environmental management. The underpinnings of technological innovations should also be environmental management. In this chapter, the author deeply proclaims the need of innovations and scientific instinct in the furtherance of environmental engineering emancipation. Arsenic groundwater contamination stands in the midst of deep crisis and unending health issues amongst the citizens of India and Bangladesh. Human scientific regeneration is thus of utmost necessity in the furtherance of science and engineering today.

Hashim et al.[12] in a comprehensive review discussed remediation technologies for heavy metal contaminated groundwater. The contamination of water is due to natural soil sources and anthropogenic sources. Remediation of contaminated groundwater is a research initiative of immense importance. Billions of people around the world use groundwater for drinking purposes. Here comes the need of newer innovations and scientific advancements in water purification technologies. Technology and engineering today has practically no answers to the burning issue of arsenic groundwater contamination in South Asia, particularly India and Bangladesh. In this well-researched paper, 35 approaches for groundwater treatment have been reviewed in details and classified under 3 categories that is (1) chemical, (2) biochemical/biological/biosorption, and (3) physicochemical treatment processes.[12] Keeping environmental sustainability in mind, bioremediation and biosorption are the pillars and cornerstones of research pursuit in environmental protection today. In this chapter, the authors touched upon sources, chemical property, and speciation of heavy metals in groundwater and technologies for treatment of heavy metal contaminated groundwater. Vast scientific evolution, scientific prudence, and deep ingenuity should be the focus of science and engineering today. Biological, biochemical, and biosorptive treatment technologies in industrial and drinking water treatment are the other pivots of this treatise.[12]

Human civilization today stands in the midst of scientific vision and deep scientific doctrines. Science and engineering are two huge colossuses with a definite vision of its own. Global water challenges are changing the vast scientific firmament. The vision, the scientific might, and determination of water science needs to be revamped and reenvisioned as science and technology moves forward. In this chapter, the author repeatedly focuses on the needs of innovations in water science and water technology today.

10.12 FUTURE RECOMMENDATIONS AND THE FUTURE AREAS OF RESEARCH

Future areas of research are groundbreaking in the world of environmental engineering science and chemical process engineering. Human's immense scientific grit and determination is in a dismal state of affairs. Technology needs to be revamped and reformed as regards environmental engineering and chemical engineering. Today global warming and global climate changes are immensely challenged and need to be reenvisioned. Future areas of research work should be in the field of new techniques in water purification and industrial wastewater treatment. Innovations, scientific instincts, and scientific forbearance are the pillars of research pursuit today. Global warming and global climate change are the challenges and the scientific vision of tomorrow. Membrane science and its branches are not new innovations but yet latent areas of science and engineering today. Applications of membrane science in environmental engineering are the necessities and scientific imperatives of modern science.[13–15]

10.13 FUTURE FLOW OF SCIENTIFIC THOUGHTS

Scientific thoughts and scientific vision are today in the midst of scientific grit and determination. Future of science and technology in global scenario is vast and versatile. Industrial and environmental pollution control today stands in the midst of deep crisis as well as vision of science. Future of environmental and chemical engineering science is extremely bright in today's human civilization and scientific progress. Water science and technology should be more envisioned and readdressed with the passage of scientific history and visionary timeframe. Arsenic groundwater contamination is challenging the scientific landscape in developing countries particularly South Asian countries. Human scientific vision is at a dismal state in developing as well as developed countries with the evergrowing concern for heavy metal groundwater contamination and poisoning. Scientific forays in water science and technology and research and development initiatives in groundwater remediation need to be reframed as science and engineering surge forward in the present century. The challenge and the vision of engineering science today lie in the hands of environmental and chemical process engineers. Provision of pure drinking water stands as a major scientific imperative in the path of human civilization. Sustainability, such as environmental, energy, social, and economic, is the challenges of human civilization today. Human stands today in the midst of deep scientific vision and immense scientific degradation with the growing concern of global warming and climate change. This vision and challenge of environmental sustainability needs to be inculcated in the minds of global citizens today. Future flow of scientific thoughts should be directed toward research and development initiatives in the field of water purification and pollution control. Again the challenge and the vision of environmental sustainability lie at the hands of civil society and scientists. Technology and engineering science are surpassing vast and versatile scientific boundaries today. The utmost needs of human scientific progress are sustainable development, energy security, and provision of basic human needs such as water, electricity, food, and shelter. The vision and goal of research initiatives in environmental engineering should be toward true emancipation of water science and technology and novel separation processes today. Science is so much ebullient and replete with splendor and might in today's human civilization. This chapter unravels the immense scientific barriers in water technology and targets the true emancipation of environmental sustainability.[13–15]

10.14 CONCLUSION AND SCIENTIFIC PERSPECTIVES

Science and engineering are in the modern civilization moving from one visionary paradigm toward another. Environmental engineering science and chemical process engineering are today in the path of newer scientific regeneration and ingenuity. Global water shortage and global climate change are veritably changing the face of human scientific endeavor. Drinking water purification, industrial wastewater treatment, and water purification techniques are the utmost needs of the vast area of environmental engineering science. Future scientific perspectives in the field of environmental engineering should be directed toward alleviating global water crisis, heavy metal, and arsenic groundwater contamination. Heavy metal remediation of groundwater or drinking water is a burning global issue in South Asia, as well as many developing and developed nations around the world. The author in this chapter pointedly focuses on the scientific success, the vast scientific acuity, and the scientific vision behind water pollution control and application of novel separation processes. Novel separation techniques involve membrane science while nonconventional environmental engineering tools involve AOPs. The entire chapter unfolds the scientific vision behind membrane science treatment of water pollution control and drinking water treatment. Scientific evolution, deep scientific provenance, and vast technological validation of industrial wastewater treatment are the other salient features of this chapter.

ACKNOWLEDGMENT

The author acknowledges the contributions of his late father Shri Subimal Palit, an eminent textile engineer from India who taught the author rudiments of chemical engineering science.

KEYWORDS

- **water**
- **vision**
- **novel**
- **composite**
- **separation**
- **oxidation**
- **advanced**

REFERENCES

1. Shannon, M. A.; Bohn, P. W.; Elimelech, M.; Georgiadis, J. G.; Marinas, B. J.; Mayes, A. M. *Science and Technology for Water Purification in the Coming Decades*; Nature Publishing Group: USA, 2008; pp 301–310.
2. Giese, G. M.; Lee, H.-M.; Miller, D. J.; Freeman, B. D.; Mcgrath, J. E.; Paul, D. R. Water Purification by Membrane: The Role of Polymer Science. *J. Polym. Sci. B Polym. Phys.* **2010,** *48*, 1685–1710.
3. Arnot, T. C.; Field, R. W.; Koltuniewicz, A. B. Cross Flow and Dead End Microfiltration of Oily-Water Emulsions, Part-II, Mechanisms and Modeling of the Flux Decline. *J. Membr. Sci.* **2000,** *169*, 1–15.
4. Sarkar, S.; Sengupta, A. K.; Prakash, P. The Donnan Membrane Principle: Opportunities for Sustainable Engineered Processes and Materials. *Environ. Sci. Technol.* **2010,** *44*, 1161–1166.
5. Munter, R. Advanced Oxidation Processes: Current Status and Prospects. *Proc. Estonian Acad. Sci.* **2001,** *50* (2), 5–80.
6. Cesaro, A.; Naddeo, V.; Belgiorno, V. Wastewater Treatment by Combination of Advanced Oxidation Processes and Conventional Biological Systems. *Int. J. Bioremed. Biodegrad.* **2013,** *4* (8), 1–8.
7. Tsydenova, O.; Batoev, V.; Batoeva, A. Solar Enhanced Advanced Oxidation Processes for Water Treatment: Simultaneous Removal of Pathogens and Chemical Pollutants. *Int. J. Environ. Res. Public Health* **2015,** *12*, 9542–9561.
8. Lamsal, R.; Walsh, M. E.; Gagnon, G. A. Comparison of Advanced Oxidation Processes for the Removal of Natural Organic Matter. *Water Res.* **2011,** *45*, 3263–3269.
9. Covinich, L. G.; Bengoechea, D. I.; Fenoglio, R. J.; Area, M. C. *Am. J. Environ. Eng.* **2014,** *4* (3), 56–70.
10. Malhotra, S.; Pandit, M.; Tyagi, D. K. Degradation of Ferrohexacyanide by Advanced Oxidation Processes. *Indian J. Chem. Technol.* **2005,** *12*, 19–24.
11. Andreozzi, R.; Caprio, V.; Insola, A.; Marotta, R. Advanced Oxidation Processes (AOP) for Water Purification and Recovery. *Catal. Today* **1999,** *53*, 51–59.
12. Hashim, M. A.; Mukhopadhyay, S.; Sahu, J. N.; Sengupta, B. Remediation Technologies for Heavy Metal Contaminated Groundwater. *J. Environ. Manag.* **2011,** *92*, 2355–2388.
13. Cheryan, M. *Ultrafiltration and Microfiltration Handbook*; Technomic Publishing Company, Inc.: USA, 1998.
14. Palit, S. Filtration: Frontiers of the Engineering and Science of Nanofiltration: A Far-reaching Review. In *CRC Concise Encyclopedia of Nanotechnology*; Ubaldo, O.-M., Kharissova, O. V., Kharisov, B. I., Eds.; CRC Press: Taylor and Francis, Boca Raton, Florida, USA 2016; pp 205–214.
15. Palit, S. Advanced Oxidation Processes, Nanofiltration, and Application of Bubble Column Reactor. In *Nanomaterials for Environmental Protection*; Kharisov, B. I., Kharissova, O. V., Rasika Dias, H. V., Eds.; Wiley: USA, 2015; pp 207–215.

WEB REFERENCES

https://en.wikipedia.org/wiki/Membrane_technology.
https://www.sciencedirect.com/science/article/pii/S2214993715300105.
https://www.sswm.info/content/membrane-filtration.
https://www.suezwaterhandbook.com/water.../fundamental...water-treatment/membran.
www.fountainmagazine.com/.../Drinking-Water-from-the-Sea-Polymeric-Membranes.
onlinelibrary.wiley.com/doi/10.1002/9783527631407.ch1/summary.
https://www.ncbi.nlm.nih.gov/pmc/articles/PMC4021920/.
www.veoliawatertechnologies.co.za/water.../membrane.../reverse-osmosis-RO-water-tr.
https://www.ncbi.nlm.nih.gov/pubmed/7456867.
https://www.wwdmag.com/desalination/membrane-filtration-water-and-wastewater.
https://www.usbr.gov/research/dwpr/reportpdfs/report029.pdf.

CHAPTER 11

GENETIC ALGORITHM, FLUIDIZATION ENGINEERING, AND THE FUTURISTIC VISION OF COMPOSITES AND CHEMICAL ENGINEERING

SUKANCHAN PALIT*

Department of Chemical Engineering, University of Petroleum and Energy Studies, Energy Acres, P.O. Bidholi via Premnagar, Dehradun 248007, Uttarakhand, India

E-mail: sukanchan68@gmail.com, sukanchan92@gmail.com

ABSTRACT

Human scientific endeavor in the field of chemical engineering is witnessing drastic and dramatic challenges. Scientific vision, scientific cognizance, and deep scientific knowledge are all leading a long and effective way in the true emancipation of chemical process engineering, applied mathematics, and the world of engineering science. Energy sustainability and engineering science are today the two opposite sides of the visionary coin. Global water crisis, the depletion of fossil fuel resources, and the needs of energy sustainability are leading human civilization to a newer era of holistic sustainability. Technology and engineering today have few answers to the vicious problem of global water crisis and depletion of fossil fuel resources. Energy engineering, chemical process engineering, and environmental engineering science lie in the midst of deep scientific vision and scientific farsightedness. In this chapter, the author deeply targets the science of application of genetic algorithm in fluidization engineering and petroleum refining. Chemical process engineering is in the midst of immense scientific regeneration and vision. In this chapter, the author deeply comprehends the scientific success of the application of genetic algorithm in designing petroleum-refining units. The other avenues of research endeavor in this chapter are the vast scientific knowledge advancement in the field of fluidization and chemical process engineering. Fluidization engineering is the major pillar of scientific advancement and design of fluidized catalytic cracking unit (FCCU) in a petroleum refinery. The vision and the challenge of human scientific endeavor in petroleum engineering science and chemical engineering are immense and far-reaching. In this chapter, the author deeply elucidates on the need of evolutionary computation and genetic algorithm in designing petroleum-refining units. The scientific acuity and the scientific foresight of design of FCCU are deeply and lucidly explained in this chapter. The other visionary areas of research pursuit are the applications of composites in chemical engineering science. Composite science is in the state of newer scientific rejuvenation. This chapter also unfolds the vast scientific success, the scientific ingenuity, and the insight behind composite science.

11.1 INTRODUCTION

The world of challenges in petroleum engineering science and chemical process engineering are ever-growing and surpassing vast and versatile scientific frontiers. Petroleum refining involves the vision of operation of fluidized catalytic cracking unit (FCCU). FCCU is the heart of a petroleum refinery. Chemical process engineering, petroleum-engineering science, and applied mathematics will all today lead a long and visionary way in the true realization of scientific vision and vast scientific fortitude. Depletion of fossil fuel resources and the global climate change are the burning issues of human civilization today. In this chapter, the author deeply elucidates the success of human scientific endeavor in chemical-process engineering and petroleum engineering with a deep focus toward furtherance of science and engineering. Human civilization and scientific research pursuit today stands in the midst of deep scientific introspection and scientific acuity. Applied mathematics and genetic algorithm (GA) are today the visionary mathematical tools of tomorrow. Scientific foresight, the vast scientific ingenuity, and the futuristic vision of applied mathematical tools will all be the pallbearers toward a newer visionary era in the field of GA and evolutionary computation. In this chapter, the author rigorously points out toward the vision and the challenge of science and engineering in GA applications in chemical engineering and petroleum engineering science. The utmost necessity of human civilization is energy and environmental sustainability. Energy security, provision of food, shelter, and water are the pillars of human's progress today. Scientific provenance, deep scientific ingenuity, and the challenges and the vision of engineering science will all lead a long and visionary way in the true realization of energy and environmental sustainability today. Applied mathematical tools and applied computer science are the scientific imperatives of today's visionary world of engineering science. GA and evolutionary computation are the necessities of scientific endeavor in engineering and technology today. In the similar vein, fluidization and fluidized bed cracking are the challenges of modern science. The author in this well-researched chapter pointedly focuses on the success of scientific research pursuit in fluidized bed catalytic cracking and the application areas of GA. GA and bioinspired computation are the pillars of engineering science of petroleum refining in modern day human civilization. The author deeply enumerates the vision, the challenges, and the targets of modern science with the sole purpose of advancements of science and engineering.

11.2 THE VISION OF THIS STUDY

Technology and engineering science are the cornerstones of scientific endeavor today. Petroleum refining and petroleum engineering science are the utmost necessities of human scientific progress today. The vision of chemical process engineering and petroleum engineering today are far-reaching and surpassing vast and versatile scientific frontiers. This chapter unravels the application areas of GA, evolutionary computation, and bioinspired computation in petroleum refining and also elucidates the success of the entire domain of petroleum engineering science in the quest for furtherance of global science and engineering. GA and applied mathematical tools are the needs of science and engineering today. For the purpose of optimization of robust systems, GA and bioinspired computation are the utmost needs of scientific endeavor. The world of chemical process engineering and applied mathematics are today in the similar manner witnessing vast and versatile challenges and scientific fortitude. The vision of this research endeavor is to target the scientific challenges and the futuristic vision of GA and evolutionary computation applications in chemical process engineering and petroleum engineering. Due to the depletion of fossil fuel sources, the world of petroleum engineering science and petroleum refining are highly challenged and visionary scientific endeavor are of utmost need. The efficiency of the petroleum-refining units today stands as an important challenge to the future of human civilization. The author in this chapter deeply elucidates the scientific success, the scientific profundity, and the farsightedness in the field of GA and fluidization engineering. Another area of research pursuit in this chapter is the field of composite science and its application areas in chemical engineering. Composites are the next generation smart materials. This chapter truly unfolds the scientific intricacies behind composites applications.

11.3 WHAT DO YOU MEAN BY GENETIC ALGORITHM AND MULTIOBJECTIVE OPTIMIZATION?

The science of GA and multiobjective optimization are today huge colossus with a definite vision and a vast scientific farsightedness of its own. Both evolutionary computation and the science of optimization are linked with each other by an unsevered umbilical cord. Human scientific ingenuity and deep scientific insight are today in the midst of deep distress as petroleum engineering science and petroleum refining moves from one visionary paradigm to another. Multiobjective optimization and multiobjective simulated annealing are the cornerstones of scientific research pursuit in applied mathematics and applied computer science today. In this chapter, the author pointedly focuses on the scientific astuteness, the deep scientific provenance, and the futuristic vision in the application of GA in the design of chemical and petroleum engineering systems. Today GA and multiobjective optimization are veritably linked by an unsevered umbilical cord. Human scientific farsightedness, the vast vision of optimization science, and the futuristic vision of chemical process engineering and petroleum engineering science will all lead a long and visionary way in the true realization of engineering and science today. Bioinspired computation and GA are similarly connected by an umbilical cord with much vision and scientific might.

In computer science and operation research, a GA is a metaheuristic inspired by the process of natural selection that belongs to the larger class of evolutionary computation and algorithm. GAs are vastly used to generate high-end solutions to optimization and search problems by relying on bioinspired operators such as mutation, crossover, and selection. Scientific research pursuit in computational techniques are today in the path of regeneration and vision. GAs are of immense need in designing engineering systems. In a GA, a population of candidate solutions (called individuals, creatures, or phenotypes) to an optimization problem is evolved toward a better solution. Multiobjective optimization is an area of multicriteria decision-making, that is, concerned with mathematical optimization problems involving more than one objective function to be optimized simultaneously. The world of scientific challenges and scientific ardor, the futuristic vision of optimization science, and the vast scientific ingenuity will all lead a long and visionary way toward the true understanding of GA and other evolutionary algorithms.

11.3.1 THE SCIENTIFIC DOCTRINE OF COMPOSITES AND CHEMICAL ENGINEERING SCIENCE

Science and engineering of material science are moving at a rapid pace surpassing one visionary boundary over another. Composite science and material science are in the path of newer scientific rejuvenation. Composites are the smart materials of tomorrow. Scientific ingenuity, human's immense scientific prowess, and the futuristic vision of chemical engineering science will all lead a long and visionary way in the true emancipation of technology and engineering science today. Composite is any material made of more than one material. Modern composites are made of two components: a fiber and a matrix. The fiber is most often glass but sometimes Kevlar, carbon fiber, or polyethylene. The challenges and the vision of chemical process engineering applications are vast and versatile. Today, there is tremendous necessity of smart materials such as composites. In this chapter, the author deeply unravels the success of composite applications and the vast strides of chemical engineering science.

11.4 THE VAST SCIENTIFIC DOCTRINE OF FLUIDIZATION ENGINEERING

Fluidization engineering and the operation of FCCU today are the scientific needs of energy engineering and energy sustainability today. Human's immense scientific prowess and grit are the torchbearers toward a newer era in the field of petroleum engineering and chemical process engineering. Science and technology of petroleum refining are highly advanced today. Applied areas of chemical process engineering and petroleum engineering are today in the process of scientific rejuvenation. The world today stands in the midst of deep catastrophe and vast scientific introspection as global water crisis, heavy metal drinking water contamination, and depletion of fossil fuel resources are raging the entire scientific scenario. Depletion of fossil fuel resources is an enigmatic research question in petroleum engineering science. Thus, there is a need of the application of applied mathematics, applied computer science, and novel computational methods. Evolutionary computation, GA, and bioinspired computation are the application areas in the design of petroleum engineering and chemical engineering systems. Design of FCCU in a petroleum refinery is a scientific imperative toward the furtherance of the science and technology of petroleum engineering science.

11.5 THE VISION AND THE CHALLENGE OF OPERATION OF AN FCCU

The vision and the challenge of operation of a FCCU are today immense and surpassing vast and versatile scientific boundaries. Human scientific endeavor, human scientific ingenuity, and the vast scientific profundity are all the torchbearers toward a newer era in science and engineering today. Petroleum engineering and energy engineering are the fountain heads of science and engineering endeavor today. Technology has practically no answers to the evergrowing global issue of depletion of fossil fuel resources. Human's immense scientific prowess and grit, man's scientific provenance, and the futuristic vision of petroleum engineering science will lead a long and effective way in the true realization of energy sustainability and energy security today. Fluidized catalytic cracking (FCC) is a heart of petroleum refining today. The challenges and the definite vision of FCC application are witnessing drastic changes. Application of multiobjective optimization and multiobjective-simulated annealing in the design of a petroleum engineering system are the cornerstones of vast scientific research pursuit in modeling, simulation, and optimization of FCCU today. Technological validation, scientific motivation, and the utmost needs of the human society are today the pallbearers toward a newer era in energy sustainability and energy security today. Petroleum engineering science today stands in the midst of a deep scientific catastrophe. Thus the need of the scientific ingenuity and scientific vision of modeling, simulation, and optimization of a petroleum engineering system.

11.6 CHEMICAL PROCESS ENGINEERING AND THE SCIENCE OF SUSTAINABILITY

Chemical process engineering, petroleum engineering, and other diverse areas of science and engineering are today linked with the science of sustainability. Nanotechnology is another branch of science which has immense scientific potential and is veritably linked to other branches of engineering. Nanovision is the scientific vision of tomorrow. The visionary words of Dr. Gro Harlem Brundtland, former Prime Minister of Norway, on the science of "sustainability" needs to be reenvisioned and restructured with the passage of scientific history and time. In this chapter, the author reiterates the vast scientific success, the immense scientific profundity, and the scientific potential behind GA applications in the design of chemical engineering and petroleum engineering systems. Petroleum engineering science and petroleum refining today stand in the midst of deep scientific revelation and vast and versatile scientific ingenuity. Technology and engineering has few answers to the scientific intricacies of global climate change and depletion of fossil fuel resources and thus needs to be envisioned and restructured as human civilization and scientific endeavor moves toward a newer direction. Water science, drinking water treatment, and industrial wastewater treatment today stand amidst deep astuteness and scientific forbearance. Today, the science of sustainability encompasses diverse areas of engineering and science such as chemical process engineering and environmental protection. Global water crisis and industrial wastewater issue have plunged science and engineering to a deep abyss. Thus, the science of sustainability stands in the midst of deep concern and vast scientific discernment. In the similar vein, science of sustainability needs to be readdressed and reenvisioned with the progress of human civilization. In this treatise, the author pointedly focuses on the scientific success, the vast scientific rigor, and the scientific enigma behind global concerns for sustainable development, petroleum engineering science, and petroleum refining. Technology and science needs to be redrafted if the global concerns for water shortage are not mitigated. Here comes the immense importance of chemical process engineering and the greater emancipation of science and engineering.

11.7 THE FUTURISTIC VISION OF CHEMICAL ENGINEERING

Chemical engineering today stands in the crossroads of vision and scientific fortitude. The needs of chemical engineering are immense and scientifically inspiring. The futuristic vision of chemical engineering and the world of challenges in nanotechnology are challenging the global scientific firmament today. Human civilization today stands in the midst of unending calamity as global climate change and vicious and vituperative issue of depletion of fossil fuel resources confront the human scientific research pursuit. Research splendor, the futuristic and definite vision of chemical engineering, and the utmost needs of human advancements are all the torchbearers toward a newer eon in engineering emancipation today. In the similar vein, environmental engineering domain is highly challenged in the midst of global climate change, the issue of environmental sustainability, and the burning issue of loss of ecological biodiversity. Science and technology of chemical engineering are today huge colossus with a definite and purposeful vision of its own. The future of chemical engineering today encompasses environmental science, mitigation of environmental disasters, and the vast need of modeling, simulation, and optimization. The scientific success and the vast scientific understanding need to be reenvisioned and readdressed with the progress of scientific history and time. Chemical process engineering today encompasses vast avenues of petroleum refining and petroleum engineering. Besides, the vast scientific and academic rigor of chemical process engineering also involves environmental protection. Human scientific endeavor's immense prowess, the needs of drinking water treatment and industrial wastewater treatment, and the futuristic vision of environmental engineering science will surely lead a long and visionary way in the true emancipation of chemical process engineering. Thus, the futuristic vision of chemical process engineering involves environmental protection, global energy, and environmental sustainability. In this chapter, the author rigorously points toward the needs, vision, and scientific ingenuity of global chemical engineering. The author also reiterates and pronounces the vast vision of chemical process engineering in the quest toward furtherance of science and engineering. Water purification, drinking water treatment, and industrial wastewater treatment are the imminent needs of human civilization today. This treatise unfolds and unravels the necessities of human civilization today such as the needs of energy and environmental sustainability.

11.8 RECENT SCIENTIFIC ENDEAVOR IN THE FIELD OF GENETIC ALGORITHM APPLICATIONS

GA and evolutionary computation are the fountainhead of chemical process design and the design of chemical and petroleum engineering systems. The vision and the vast challenges of science and engineering needs to be more addressed and reorganized as human civilization faces the global environmental and energy crisis. Human's vast scientific grit and prowess, the immediate needs of scientific validation, and the vast world of scientific profundity will all be the successful torchbearers toward a newer visionary eon of energy security and environmental sustainability.

Coley[1] with lucid and cogent insight deeply discussed GAs for scientists and engineers. Human scientific endeavor and human scientific ingenuity today are in the process of vast regeneration and scientific rejuvenation. GA today is a fountainhead of applied computational tools in chemical engineering and petroleum engineering today. Applied computer science, the vast vision of computational techniques, and applied mathematics are the needs of human society and human scientific endeavor today. In this book, the author discussed with vast farsightedness applications of GA, improvement of the algorithm, comparison of GA and biological terminology, mutation, selection, elitism, crossover, and initialization. Science and engineering of GA are today surpassing vast and versatile scientific frontiers.[1] The author in this book also touches upon foundations of GA, advanced operators, and the applications of GA.[1] GAs are numerical optimization algorithms inspired by both natural selection and natural genetics.[1] The method is a general one, capable of being applied to a diverse range of research endeavor. A typical GA might consist of the following: (1) a number, or population, of guesses of the solution to the problem; (2) a way of calculating how good or bad the individual solutions within the population; (3) a method for mixing fragments of the better solutions to form new, on average even better solutions; and (4) a mutation operator to avoid permanent loss of diversity within the solutions.[1] Thus, the scientific vision and the scientific foresight are revealed and investigated in this well-researched treatise. Technology of GA has advanced a lot over the years. Human scientific research pursuit in GA needs to be reenvisioned and reenvisaged with the progress of applied mathematics, applied computer science, chemical engineering, and petroleum engineering science.[1] GA today has applications in design of chemical engineering and petroleum engineering systems. Human scientific ingenuity and human scientific provenance are today at its helm as GA applications assumes immense importance in today's modern scientific endeavor.[1] In this chapter, the author vastly comprehends the need of GA in the furtherance of evolutionary computation and applied mathematics in present day human civilization and present day scientific endeavor.

Mitchell[2] with lucid and cogent insight discussed GAs and its applications in diverse areas of science and engineering. Human scientific acuity, scientific farsightedness, and the vast and wide scientific doctrine of GA and multiobjective optimization will all lead a long and visionary way in the true emancipation and the true realization of applied mathematics and applied computer science. The author deeply discussed (1) an overview of GA, (2) genetics algorithm in problem solving, (3) GAs in scientific models, (3) theoretical predictions of GA, and (4) future directions in the emancipation of GA. Science and engineering arises from the very human desire to understand and control the human civilization.[2] Over the course of human history, humans have slowly rebuilt a grand edifice of knowledge and scientific might which is applied mathematics and applied computer science. GA and evolutionary computation are two giant branches of scientific research pursuit in applied computer science and the holistic domain of applied science.[2] The advent of electronic computers without any doubt has been the most revolutionary development in the domain of science and engineering today. The ongoing revolution with much scientific profundity has increased the human ability to predict and control nature in ways that were difficult to predict one century ago. Human civilization and human scientific research pursuit thus today stands in the midst of scientific wisdom and deep scientific truth. The goals of creating artificial intelligence and artificial life can be traced back to the advent of computers eons back. The vast challenge and the vision of computational techniques and evolutionary computation in the similar vein are witnessing immense revamping and needs to be thoroughly reenvisioned and reenvisaged.[2] In the 1950s and 1960s, several computer scientists around the world independently studied and emancipated evolutionary systems with the sole objective that evolution could be used as an effective computational tool for optimization of engineering systems.[2] The vision and idea of all these systems were to evolve a population of candidate solutions to a given problem, using operators using natural genetic variation and other natural selection. Applied computer science and the world of scientific emancipation are today in the path of a newer regeneration. The advent of computer science and computational techniques are the miracles and wonders of human civilization today. Technology and engineering science today stands amidst scientific introspection and vast vision. GAs were invented by John Holland in the 1960s and were developed by Holland and his students and colleagues at the University of Michigan in the 1960s and the 1970s.[2] In contrast to the evolution strategies and evolutionary programing, Holland's goal was not to design algorithms to solve specific goals, but rather to formally study the theory of adaptation as it occurs in nature and to develop ways in which the mechanisms of natural adaptation might be incorporated into computer systems. Diverse areas of engineering and science such as chemical process engineering and petroleum engineering are today connected to GA and bioinspired computation with immense scientific might and technological fortitude. This paper opens up and unravels the scientific intricacies of GA, the deep future thoughts, and the world of difficulties in incorporating GA in robust engineering systems.[2] Technology is at its helm in today's world of scientific endeavor. GA has transformed from a scientific idea to an immensely rebuilt phenomenon in applied mathematics and applied computer science. Today is the visionary world of applied mathematics and computational techniques. Dr. John Holland's research endeavor coincided with the vast scientific revolution in computational techniques and applied computer science. This paper widely reviews the scientific intricacies in the GA applications in diverse areas of science and technology.[2]

Bodenhofer[3] in a winter school discussed the theory and applications of GA. Scientific motivation, deep scientific revelation, and scientific acuity are the pillars of this paper. Today, GA and its applications in diverse areas of science and engineering are crossing vast and versatile scientific boundaries. The author in this treatise touched upon (1) genetic operations on binary strings, (2) analysis, (3) variants, (4) GA variants for real-valued optimization problems, (5) tuning of fuzzy systems using GAs, (6) genetic programing, and (7) classifier systems.[3] Human's immense scientific grit and determination, man's vast futuristic vision and the needs of the human society will all lead a long and visionary way in the true realization of applied mathematics and optimization science. Applying mathematics to a difficult problem of the real scientific world mostly means, at first, modeling the problem mathematically, maybe with hard restrictions, idealizations, or simplifications and then solving the mathematical problem. This entire endeavor needs immense scientific introspection.[3] Applied mathematics and applied computer science in today's world involves (1) artificial neural networks, (2) fuzzy control, (3) simulated Annealing, and lastly (4) GA.[3] The world today is faced with vast and versatile scientific issues and replete with scientific intricacies as regards engineering and science. Today is the world of genetic adaptations and bioinspired computation. In this treatise, the author pointedly focuses on the needs of GA and evolutionary computation toward the furtherance of science and engineering.[3]

Bandyopadhyay et al.[4] with cogent insight deeply discussed a simulated annealing-based multiobjective optimization algorithm (MOSA). Human scientific research pursuit and human civilization are today in the path of newer scientific regeneration and vast scientific intricacies. Simulated annealing has veritably unraveled the scientific barriers and the scientific dilemma in high-powered computation and robust engineering research questions. Applied mathematics and applied computer science are today witnessing immense scientific rejuvenation.[4] This paper describes a simulated annealing based MOSA that incorporates the concept of archive to provide a set of trade-off solutions for the problem under consideration. Technological vision, the vast scientific motivation, and the futuristic vision of computational science today are the torchbearers toward a newer avenue of scientific emancipation in engineering science.[4] To investigate and determine the acceptance probability of a new solution vis-à-vis the current solution, an elaborate procedure is followed and explained that takes into account the domination status of the new solution with the current solution, as well as those in the archive.[4] Multiobjective simulated annealing today is in the path of newer scientific regeneration and scientific vision. Human scientific progress and academic rigor are the pillars of the science of optimization today. This paper gives a vast glimpse on the scientific success, the scientific regeneration, and the scientific potential behind the true realization of applied mathematics and applied computer science.

Deb[5] with vast scientific conscience in a well-researched report discussed multiobjective optimization using evolutionary algorithms. From the definition, multiobjective optimization involves optimizing a number of objectives simultaneously. Scientific conscience, the scientific and technological profundity, and the vast scientific prowess will today evolve into a newer avenue of emancipation and fortitude in decades to come in the domain of optimization. Today, the challenge and the vision of science and engineering are immense and forthright. The prowess of science and engineering are in the similar vein immense and ever-growing. The problems and research questions of optimization becomes challenging when the objectives are in conflict with each other, that is, the optimal solution, of an objective function is different from each other. Starting with the parameterized techniques in the early nineties, the so-called evolutionary multiobjective optimization (EMO) algorithms are now a well-established field of research endeavor.[5] Today, scientific research questions and research barriers have few answers to the need of energy sustainability.[5] Energy sustainability is today linked with diverse areas with engineering science such as chemical process engineering and petroleum engineering. Thus, GA and multiobjective optimization are veritably linked with design of chemical engineering and petroleum engineering systems. In this chapter, the author gives a brief description to the operating principles and outlined the current research and development initiatives in EMO problems.[5]

Back et al.[6] with deep and lucid farsightedness discussed basic algorithms and operators in evolutionary computation. In this book, the authors delineated (1) a brief introduction to evolutionary computation, (2) possible applications of evolutionary computation, (3) advantages and disadvantages of evolutionary computation over other approaches, (4) principles of evolutionary processes, (5) principles of genetics, (6) a brief history of evolutionary computation, (7) evolutionary algorithms and their standard instances, (8) representations, (9) selection, and (10) search operators.[6] Scientific and technological vision and vast scientific profundity and conscience are the needs of science and engineering today. This treatise explores and unravels the scientific success, the vision, and the imminent needs of applied mathematics and computational science. As a well-recognized field, evolutionary computation is still young. The term itself was invented in 1991 and it brought together authors around the globe who are doing research in evolutionary strategies.[6] These techniques of GAs, evolutionary strategies, and evolutionary programing have one fundamental commonality: they each involve the reproduction, random variation, competition, and selection of conflicting individuals in a population. The authors deeply comprehend these issues and intricacies. Evolution is an optimization process. The science of optimization does not imply perfection, yet evolution can discover highly precise functional solutions to particular problems posed by an organism's environment. In this book, the authors deeply comprehends the human scientific progress, the scientific genre, and the vast engineering profundity in the field of evolutionary computation and computational science.[6]

Technology and engineering science are today in the path of newer scientific evolutionary and vast regeneration. Computational science and applied computer science are the utmost needs of scientific endeavor today. Evolutionary computation, evolutionary programing, and GA are the imminent needs of design of engineering systems today. The author in this entire chapter discusses with scientific conscience and foresight the vast scientific potential and the scientific elegance of the GA applications and its relevance to fluidization engineering.

11.9 RECENT SCIENTIFIC RESEARCH PURSUIT IN THE FIELD OF FLUIDIZATION ENGINEERING

Scientific research pursuit in the field of fluidization engineering is today surpassing vast and versatile scientific frontiers. The needs of the human society, the vast technological and scientific validation, and the world of scientific truth are the veritable torchbearers toward a newer era in the field of the energy sustainability. Fluidization engineering and the operation of FCCU are the necessities of energy engineering and energy sustainability today. Global research and development initiatives in the field of energy engineering and energy sustainability are moving toward a newer era of immense vision and scientific forbearance. The efficiency and the robustness of petroleum refining and petroleum engineering science needs to be more reenvisioned and reenvisaged as human scientific progress moves from one visionary paradigm toward another. In this chapter, the author with insight and vision treads a weary path toward the scientific truth in GA applications of design of petroleum-refining units. Human's vast scientific prowess, man's immense scientific grit, and the world of scientific validation will surely usher in a new era in the field of energy sustainability, chemical engineering, and petroleum engineering science. Fluidization engineering today is in the path toward newer scientific regeneration and vast scientific revamping.

Elnashaie[7] discussed with immense insight modeling, simulation, and optimization of industrial fixed-bed catalytic reactors. The scientific challenges and the vision of technology and engineering science are immense and ever-growing in today's modern day human civilization. In this book, the authors discussed systems theory and principles for developing mathematical models of industrial fixed-bed catalytic reactors. The authors touched upon (1) chemisorption and catalysis; (2) intrinsic kinetics of gas–solid catalytic reactions; (3) practical relevance of bifurcation, instability, and chaos in catalytic reactors; (4) effect of diffusional resistances, the single pellet problem; (5) the overall reactor model; (6) physicochemical parameters for industrial steam reformers; and (7) numerical techniques for the solution of equations.[7] This book primarily deals with the mathematical modeling of industrial fixed-bed reactors which have vast importance in petrochemical plants and petroleum refineries. Human scientific success, acuity, and deep ingenuity are vastly pronounced in this book. Despite the relatively simple and passive external appearance of fixed-bed catalytic reactors, the processes taking place within the boundaries of the system and their interactions are quite complex and can give rise to rather complicated issues in design, safe operation, and optimization. The authors deeply comprehend the utmost need of engineering science and technology in the furtherance of design of chemical engineering and petroleum engineering systems.[7]

Sadeghbeigi[8] with vast scientific insight discussed and elucidated design, operation, and trouble-shooting of FCC facilities in the *Fluidized Catalytic Cracking Handbook.* Science and engineering of FCC are today in the path of newer scientific vision and regeneration. The author in this book discussed (1) FCC-feed characterization, (2) FCC catalysts, (3) chemistry of FCC reactions, (4) unit monitoring and control, (5) products and economics, (6) project management and hardware design, (7) debottlenecking and optimization design, and (8) emerging trends in FCC.[8] Scientific ingenuity and deep scientific discernment are the cornerstones of this treatise. Scientific endeavor in the field of FCC are assuming immense global importance as global concerns for depletion of fossil fuel resources are challenging the vast scientific firmament. FCC is the heart of petroleum refining in human civilization's scientific endeavor in petroleum engineering science.[8] The author in this well-researched book unravels and explores the hidden scientific truth and the vast scientific profundity in FCCU and its operation. Human scientific challenges and human scientific vision in petroleum engineering science are changing the entire petroleum engineering scenario. The need for energy sustainability has changed drastically the global energy scenario today. This book gives a vast glimpse of the wide scientific understanding and the scientific acuity in the research pursuit in the catalytic cracking operations in a petroleum refinery.

Raychaudhuri[9] with deep and cogent insight discussed the fundamentals of petroleum and petrochemical engineering. Technological ingenuity and vast scientific validation of petroleum refining are changing the engineering science scenario today. The world of petroleum engineering science and energy sustainability are witnessing immense challenges as human scientific research pursuit evolves into a newer era of scientific regeneration.[9] The author discussed (1) the crude petroleum oil; (2) petroleum products and test methods; (3) processing operations in a petroleum refinery; (4) the vast world of lubricating oil and grease; (5) petrochemicals; (6) offsite facilities, power, and utilities; (7) the relevant material and energy balances; (8) the operation of heat-exchangers and pipe-still furnaces; (9) distillation and stripping; (10) extraction; (11) reactor calculations; (12) instrumentation and control in a refinery; and (13) plant management and economics.[9] This is a comprehensive treatise and a watershed text in the field of petroleum and petrochemical engineering. Petroleum fuels, such as gasoline and diesel, are the major fuels for all transportation vehicles. Commodities manufactured from petrochemicals, for example, plastics, rubbers, and synthetic fibers, derived totally from petroleum, have become part and parcel of human life and human scientific progress. In fact, petroleum is a nonrenewable fossilized mass, the amount of which is being exhausted with everyday human consumption. Here comes the importance of reenvisioning and revamping of petroleum engineering science with the global concern for depletion of fossil fuel resources.[9] The vast importance of renewable energy paradigm, the immense scientific vision, and the needs of human scientific progress will all lead a long and effective way in the true realization of energy sustainability and energy security. The author in this paper deeply explores the vast scientific potential and the societal needs of petroleum engineering science and energy sustainability.

Petroleum refining and FCCU are today the opposite sides of the visionary coin. Human scientific regeneration needs to be more envisioned and revamped as human civilization moves forward. Technology and engineering have few answers to the global concerns for fossil fuel depletion. In this entire chapter, the author deeply elucidates the scientific vision of the application of GA in designing petroleum engineering systems and its vast scientific emancipation.

11.10 FLUIDIZATION ENGINEERING AND FCCU

Fluidization engineering and FCCU are the heart of a chemical process and a petroleum refinery. Vast scientific vision, the scientific ardor, and the immense scientific prowess of human civilization will all today lead a long and visionary way in the true emancipation and the true realization of petroleum refining and petroleum engineering science today. The truth and vision of science in engineering and technology are slowly unfolding as human scientific progress faces deep crisis and unending scientific introspection. Petroleum refining is a pivotal parameter toward the economic growth of human civilization. Today, scientific forays and research and development initiatives in diverse areas of engineering and science stand in the deep abyss of vision and deep comprehension. Thus, there is a need of applied mathematical tools and applied computer science. Here also comes the importance of application of applied mathematical tools for design of petroleum-refining units. Fluidization engineering is today moving toward a newer visionary era vision and deep scientific determination. FCCU is the heart of a petroleum refinery. The need and the scientific explanation behind design of the riser reactor in a FCCU need to be envisioned and readdressed with the passage of scientific history and time. Energy engineering and the science of energy sustainability are today unraveling the vast world of applied petroleum engineering and evolutionary computation. Evolutionary computation involves GA, multiobjective optimization, and multiobjective simulated annealing. Technological validation, the vast scientific acuity, and the immense scientific challenges will surely lead a long and visionary way to the successful realization of petroleum engineering as well as energy sustainability.

11.11 RECENT SCIENTIFIC ADVANCES IN FCCU OPERATION

Unit operations of chemical engineering are the fountainhead of human scientific endeavor in chemical process engineering today. Chemical process engineering and petroleum engineering science are in the process of newer scientific regeneration and are connected to nanotechnology by an unsevered umbilical cord. FCC today is in the process of newer scientific revamping and is in the path of newer visionary direction.

Koratiya et al.[10] with lucid insight deeply discussed modeling, simulation, and optimization of a FCC downer reactor. Downer reactor, in which gas and solids move downward cocurrently, has unique features such as it involves high-severity operation at the initial stage with the benefit of near-plug flow reactor.[10] The challenge and the vision of petroleum refining are today immense and path breaking. Literature have shown that the downer have a higher advantage than a riser. This technology is explored with vision and scientific fortitude in this paper.[10] In this paper, mathematical model for downer reactor have been lucidly developed, in which a five lump model is used to characterize the feed composition and the products, where gas oil crack to give lighter fractions and coke. Human scientific prowess, the vast scientific ingenuity, and the world of challenges will evolve toward a newer scientific understanding in FCC design in today's scientific research pursuit. Today, GA is applied in the design of petroleum and chemical engineering systems. Optimization study of FCCU downer reactor to maximize its profitability and satisfy real-life constraints are studied with deep scientific vision in this paper. Application of Non-dominated sorting GA (NSGA-II) is done and is used to solve a two-objective function optimization problem. FCC is the heart of a petroleum refinery and vast technological acuity and scientific validation are of utmost need in the true realization of energy sustainability and energy security. The challenge and the vision needs to be reenvisioned and streamlined with the passage of scientific history and time.

Khandalekar[11] discussed with a broad scientific foresight, in his Master's thesis, control and optimization of FCC process. Technological vision, the deep scientific insight and the vast world of scientific validation are the veritable pillars of this research work. The FCCU receives multiple feeds consisting of high boiling components from several other refining units and cracks these feed components into lighter components. Technology and engineering science of FCCU is highly advanced today and surpassing vast and versatile scientific frontiers. The objectives of FCCU control systems are (1) safety-reverse catalyst flow protection, (2) heat balance, (3) carbon balance, (4) conversion, (5) interactions, and (6) constraints.[11] The prime objective of this treatise is the development of a dynamic simulator for a fluidized catalytic cracker. A riser model is envisioned and includes only the coke-yield model for the reactor riser and does not predict the composition of the products. A yield model was incorporated into the present model. The author deeply comprehends the scientific needs, the scientific profundity, and the vast scientific farsightedness in the control and optimization of a FCCU.[11]

Science and engineering are today in the path of newer regeneration and an avenue toward newer scientific emancipation. Technology has today fewer answers to the global energy crisis and fossil fuel depletion. In this entire chapter, the author deeply elucidates the success of optimization science, the futuristic vision of GA, and the scientific emancipation of fluidization engineering.

11.12 ENERGY SUSTAINABILITY AND THE VISION FOR THE FUTURE

Modern scientific research pursuit and modern human civilization are today moving in the path toward newer scientific regeneration. Provision of basic human needs such as food, water, and electricity are the necessities of human civilization today. Petroleum engineering science and petroleum refining today stand in the midst of scientific vision and vast scientific provenance. The challenges and the vision of scientific research pursuit in energy engineering and science of sustainability are immense and scientifically thought provoking. Science and technology have few answers to the growing global crisis of climate change, loss of ecological biodiversity, and depletion of fossil fuel resources. Sustainable development, whether it is energy, environmental, social, and economic, is the scientific imperatives of modern civilization today. Human's immense scientific prowess, the vast scientific provenance of modern science's research pursuit, and the futuristic vision of energy engineering and environmental engineering will all lead a long and visionary way in the true emancipation of energy and environmental sustainability today. Energy security today stands in the midst of deep distress and vast scientific comprehension in today's world of modern science.[12–14] Depletion of fossil fuel resources is a major impediment to scientific endeavor in energy engineering today. In such a crucial juncture of scientific history and time, mankind's scientific prowess in applied mathematics, applied computer science, and the world of challenges in petroleum engineering assume immense importance. GA, evolutionary computation, and bioinspired computation are the needs of human scientific endeavor today. The author in this chapter repeatedly pronounces the vast vision of sustainability science, the vast and visionary world of energy engineering, and the necessities of petroleum engineering science. Evolutionary computation and the world of petroleum engineering are the two opposite sides of the visionary coin. This chapter reflects and elucidates with deep scientific insight the needs of energy and environmental sustainability to human society with the sole aim of furtherance of science and engineering.[12–14]

11.13 MODERN SCIENCE AND THE DIFFICULTIES AND CHALLENGES

Modern science today stands in the midst of vast scientific fortitude and vision. Technology and engineering science need to be revamped and reenvisioned as human confronts the global burning issues of climate change and depletion of fossil fuel resources. The state of environment is veritably dismal. The difficulties and the challenges are scientifically inspiring and at the same time evergrowing with the passage of scientific history and time. Chemical process engineering, environmental engineering, and petroleum engineering science are highly challenged as human civilization moves toward a newer visionary future. Environmental and energy sustainability in the similar manner stands in the midst of introspection and deep comprehension. In this chapter, the author repeatedly pronounces the scientific success, the vast scientific acuity, and the scientific profundity behind sustainable development. The world of petroleum engineering science and chemical process engineering are veritably linked with sustainable development today. Holistic sustainable development is the utmost need of the hour as human civilization moves forward toward a newer scientific regeneration. Global climate change and the enigmatic problem of industrial wastewater treatment and drinking water treatment are the burning issues of human today. The world today is facing immense challenges as regards scientific emancipation in petroleum engineering and chemical engineering. Technology and engineering science has today few answers to the growing crisis of climate change and fossil fuel resources depletion. In this chapter, the author pointedly focuses on the scientific success, the scientific ingenuity, and the grave concerns of human scientific endeavor in petroleum refining.[12–14]

11.14 CONCLUSION AND FUTURE SCIENTIFIC PERSPECTIVES

Today, the challenges of modern civilization are immense and thought provoking. Scientific vision and vast scientific foresight are the utmost necessities of science and technology today. The author rigorously points toward the scientific necessity of GA and its application in petroleum refining. Petroleum engineering today stands in the midst of deep scientific ingenuity and scientific fortitude. Future scientific perspectives in the field of chemical engineering and applied mathematical tools such as GA are far-reaching and need to be reenvisioned with the passage of scientific history and time. Energy security and energy sustainability today stands in the midst of deep scientific introspection and vast vision. The world stands in the midst of deep scientific crisis as the science of energy and environmental sustainability stands highly challenged and deeply strained. Petroleum refining and petroleum engineering science needs to be revamped and streamlined as depletion of fossil fuel resources challenged the vast scientific firmament. In this chapter, the author reiterates the success of mathematical tools applications in the design of petroleum-refining units. GA and evolutionary computation are the fountainhead of optimization techniques today. Multiobjective optimization and Multiobjective simulated annealing are linked to GA and bioinspired computation by an unsevered umbilical cord. Technology and engineering science today have few answers to the global issue of depletion of fossil fuel resources and global climate change. Here comes the need of energy and environmental sustainability. The entire upshot of this chapter surpasses scientific imagination and vast scientific profundity as science and engineering moves forward. This chapter should be a veritable eye-opener toward the scientific needs of applied mathematics and applied computer science in the quest toward scientific emancipation of energy and environmental sustainability. This entire body of research work pointedly focuses on the human scientific ingenuity and the deep scientific foresight in the application of GA and evolutionary computation in diverse areas of science and engineering. The challenge and the vision need to be reenvisioned and reenvisaged as human moves toward sustainable development. The veritable upshot of this chapter covers the scientific ingenuity and the deep scientific cognizance in the field of evolutionary computation and its applications in design of chemical and petroleum engineering systems. Human's immense scientific prowess, man's vast scientific vision, and the world of scientific challenges will all lead a long and visionary way in the true emancipation and the true realization of successful sustainability. The future research trends in the field of composite science and polymer science are immense. This chapter throws vast insight and depicts profoundly the need of composites in engineering applications of tomorrow.

ACKNOWLEDGMENT

The author deeply acknowledges with immense respect the contributions of his late father, Shri Subimal Palit, an eminent textile engineer from India who taught the author the rudiments of chemical engineering.

KEYWORDS

- **petroleum**
- **genetic**
- **fluidization**
- **vision**
- **composites**
- **cracking**

REFERENCES

1. Coley, D. A. *An Introduction to Genetic Algorithms for Scientists and Engineers*; World Scientific Publishing Co. Pte. Ltd.: Singapore, 1999.
2. Mitchell, M. *An Introduction to Genetic Algorithms*; The MIT Press: USA, 1999.
3. Bodenhofer, U. *Genetic Algorithms: Theory and Applications (Lecture Notes)*, 3rd ed.; Winter, Johannes Kepler University: Linz, Austria 2003/2004.
4. Bandyopadhyay, S.; Saha, S.; Maulik, U.; Deb, K. A Simulated Annealing-based Multiobjective Optimization Algorithm: AMOSA. *IEEE Trans. Evol. Comput.* **2008,** *12* (3), 269–282.
5. Deb, K. Multi-objective Optimization Using Evolutionary Algorithms: An Introduction. *KANGAL Report Number 2011003*; IIT Kanpur Genetic Algorithm Laboratory: Kanpur, India, 2011.
6. Back, T.; Fogel, D. B.; Michalewicz, Z. *Evolutionary Computation-1: Basic Algorithms and Operators*; Institute of Physics Publishing: Philadelphia, PA, 2000.
7. Elnashaie, S. S. E. H.; Elshishini, S. S. *Modeling, Simulation and Optimization of Industrial Fixed Bed Reactors*; Gordon and Breach Science Publishers, S.A.: Great Britain, 1993.
8. Sadeghbeigi, R. *Fluid Catalytic Cracking Handbook: Design, Operation and Troubleshooting of FCC Facilities*; Gulf Publishing Company: Houston, TX, 2000.
9. Chaudhuri, U. R. *Fundamentals of Petroleum and Petrochemical Engineering*; CRC Press; Taylor and Francis Group: USA, 2011.

10. Koratiya, K. R.; Kumar, S.; Sinha, S. Modeling, Simulation and Optimization of FCC Downer Reactor. *Petrol. Coal* **2010,** *52* (3), 183–192.
11. Khandalekar, P. Control and Optimization of Fluidized Catalytic Cracking Process. Master of Science Thesis, Texas Tech University, USA, 1993.
12. Ramteke, M.; Gupta, S. K. Kinetic Modeling and Reactor Simulation and Optimization of Industrial Important Polymerization Processes: A Perspective. *Int. J. Chem. React. Eng.* **2011,** *9*, 1-54.
13. Sawaragi, Y.; Nakayama, H.; Tanino, T. Theory of Multi-objective Optimization. *Mathematics in Science and Engineering*; Academic Press: New York, USA 1985.
14. Rangaiah, G. P. Multi-objective Optimization: Techniques and Applications in Chemical Engineering. In *Advances in Process Systems Engineering*; Rangaiah, G. P., Ed.; World Scientific Publishing Co. Pte. Ltd.: Singapore, 2009; Vol. 1.

WEB REFERENCES

www.pslc.ws/macrog/composit.htm.
https://netcomposites.com/guide-tools/guide/introduction/polymer-composites/.
https://en.wikipedia.org/wiki/Composite_material.
https://www.intechopen.com/...polymers.../introduction-of-fibre-reinforced-polymers.
eng.thesaurus.rusnano.com/wiki/article1003.
stmjournals.com/Journal-of-Polymer-and-Composites.html.
https://www.hindawi.com/journals/ijps/si/569490/cfp/.
nptel.ac.in/courses/113105028/32.
https://en.wikipedia.org/wiki/Evolutionary_computation
https://www.mitpressjournals.org/loi/evco.
https://www.techopedia.com/definition/19218/evolutionary-computation.
https://onlinecourses.nptel.ac.in/noc18_ch10.
https://www.amazon.in/Fluidization-Engineering-Chemical-D-Kunii/.../0409902330.
https://mfix.netl.doe.gov/...fluidization.../Fundamentals%20of%20Fluidization%20Par.
https://en.wikipedia.org/wiki/Fluidized_bed_reactor.

CHAPTER 12

COMPUTATIONAL FLUID DYNAMICS TECHNIQUES AND ITS APPLICATIONS IN CHEMICAL ENGINEERING, PETROLEUM ENGINEERING, AND THE VAST DOMAIN OF COMPOSITE SCIENCE

SUKANCHAN PALIT[1,2,*]

[1]*Department of Chemical Engineering, University of Petroleum and Energy Studies, Energy Acres, P.O. Bidholi via Premnagar, Dehradun 248007, Uttarakhand, India*

[2]*43, Judges Bagan, P.O. Haridevpur, Kolkata 700082, India*

[*]*E-mail: sukanchan68@gmail.com, sukanchan92@gmail.com*

ABSTRACT

Human civilization and human scientific endeavor are today witnessing immense challenges and drastic changes. Science and technology of computational fluid dynamics (CFD) are in the midst of deep scientific vision and introspection. This chapter gives a wider view of the scientific success and the deep scientific potential behind fluid dynamics and CFD with special emphasis on design of chemical engineering and petroleum engineering systems. Human civilization's scientific prowess, the scientific vision, and the technological profundity will all lead a long and visionary way in the true emancipation and true realization of fluid dynamics and applied mathematical tools today. Applied mathematics and computational techniques today stand in the midst of vision and deep comprehension as human moves forward. In the present scientific scenario, CFD are of immense importance in chemical engineering and petroleum engineering. Fluid dynamics, most precisely CFD, has diverse applications in engineering science particularly chemical engineering and petroleum engineering. Multiphase flow is an area of immense research endeavor today. Fluid flow principles are in the heart of the vast domain of chemical engineering and mechanical engineering. CFD is the branch of fluid mechanics providing a cost-effective means of simulating real flows by the numerical solution of the governing equations. Technological vision and the scientific aim of CFD are enumerated in details in this well-researched treatise. The other goal and mission of this study are the applications of CFD in composite manufacturing processes. Technological validation and vast scientific motivation in the field of CFD applications in composite manufacture are the immediate necessities of research pursuit today. The author unfolds the scientific intricacies in the field of composites and polymer science.

12.1 INTRODUCTION

Computational fluid dynamics (CFD), chemical engineering, and petroleum engineering are the areas of immense scientific advancements today. Fluid dynamics have diverse applications in almost every branches of engineering today. Scientific vision and immense scientific understanding are the hallmarks of scientific endeavor in fluid dynamics and CFD in particular. Technological vision and deep motivation are reframing the academic rigor in CFD. Today, the vast areas of chemical engineering and unit operations of chemical engineering in particular are moving toward newer scientific regeneration and scientific profundity. Energy and environmental sustainability are the concerns and at the same time true vision of human civilization today. Slow depletion of fossil fuel resources and loss of ecological biodiversity are of immense concerns to human civilization today. Thus the need for the application of computational tools in the furtherance of chemical engineering and petroleum engineering science today. CFD are changing the face of instinctive scientific research pursuit and ushering in a new era in chemical engineering and petroleum engineering. The author in this treatise deeply delineates the recent scientific endeavor, the scientific potential, and the vast scientific sagacity in CFD applications in chemical engineering and petroleum engineering science. Composite science and material science are entering a newer phase in the field of chemical engineering and chemical process technology. In this chapter, the author vastly unravels the deep scientific truth, the scientific fortitude, and the scientific ingenuity in the field of CFD applications in composite and polymer manufacturing processes.

12.2 THE VISION OF THIS STUDY

CFD and its vast and varied applications are witnessing new vision and an era of newer innovations. This area of scientific research pursuit is highly advanced today. Fluid dynamics and unit operations of chemical engineering are surpassing visionary scientific frontiers. This study describes with deep and cogent insight the vast world of CFD with detailed explanation of its applications and engineering science. Chemical engineering and petroleum engineering science are two branches of engineering science which are witnessing vast and drastic challenges. Energy and environmental sustainability today stand in the midst of immense introspection and vision. Energy sustainability and petroleum engineering science are today linked by an unsevered umbilical cord. The author in this well-researched treatise pointedly focuses on the vast applications of CFD and the ever-growing grave concern on global energy sustainability. The vision of Dr. Gro Harlem Brundtland, the former Prime Minister of Norway on the concept and science of sustainability, need to be redefined and readdressed with each step of scientific rigor. Human civilization and human scientific endeavor today needs to be reenvisioned as science and engineering moves toward a newer era. The major vision of this chapter is to target the world of challenges in CFD applications in chemical and petroleum engineering. The other vision of the study also encompasses composite science and CFD applications. This is a newer avenue of research pursuit today. The author with deep scientific conscience unfolds the need of CFD in manufacturing process of composites. Manufacturing engineering and polymer science are two wide branches of research pursuit today. This well-researched treatise unfolds and unravels the vast scientific vision in the future of composite science today.

12.3 THE SCOPE OF THIS STUDY

CFD and the whole world of computational techniques are today in the stage of newer development and newer innovations. These areas of research endeavor are veritably crossing scientific boundaries. Human civilization's immense scientific prowess, the intricacies of computational sciences, and the vast futuristic vision will all today lead a long and visionary way in the true application and realization of fluid dynamics and applied mathematics. The scope of research pursuit in CFD is wide, varied, and versatile. Scientific and technological profundity are the utmost need of the hour in research pursuit in engineering science and CFD applications today. Aerospace engineering, chemical engineering, and petroleum engineering are few applications in the diverse applications of CFD in science and engineering. Multiphase flow, fluid flow, and chemical reaction engineering has immense applications of CFD. Today is the world of multiphase flow in design of chemical engineering and petroleum engineering systems. This scope of fluid dynamics and CFD in particular are today opening up new innovations and new scientific understanding in decades to come. Aerospace engineering has vast applications of CFD. Composite science is another large avenue of science and engineering today. This chapter opens vast avenues of scientific intricacies in the field of composite science and polymer science. CFD has also applications in composite science and its manufacture. CFD is a versatile area of scientific research pursuit today. It has vast and versatile applications in polymer science, material science, and composite science also. Thus, the scope of the study also involves the vast and versatile areas of composite manufacture and the application of CFD in the manufacturing process.

12.4 LITERATURE REVIEW

Scientific vision and vast scientific discernment are in the state of immense barriers and hurdles. Today research pursuit in the field of applied mathematics, computer science, and CFD are witnessing newer challenges and innovations. In chemical process engineering applications, fluid flow, heat transfer, and chemical kinetics are immense needs in the path toward furtherance of science and engineering. Multiphase flow or two-phase flow are important areas of study in chemical process design and design of chemical reactors. Here comes the utmost need of CFD and other mathematical tools. The author in this treatise rigorously points toward the vast scientific prowess, the scientific sagacity, and the deep scientific revelations in applied mathematics applications in chemical engineering.

Application of CFD in the scientific rigor of chemical process engineering is of immense importance in the avenues of scientific profundity today. CFD is the thorough analysis of systems involving fluid flow, heat transfer, and associated phenomena such as chemical reactions by means of computer-based simulations. Human scientific endeavor and human civilization are today witnessing immense challenges. These scientific challenges can be overcome by the vast applications of applied science, applied mathematics, and computer science in engineering science. In this section, the author pointedly focuses on the significant achievements of CFD applications to human scientific endeavor. Versteeg and Malalasekara[5] redefined and reemphasized CFD in lucid details. CFD is a powerful tool and spans a wide range of industrial and nonindustrial application domains. Some of the examples are (1) aerodynamics of aircrafts and vehicles, (2) hydrodynamics of ships, (3) power-plant combustion, (4) turbomachinery, (5) electrical and electronic engineering, (6) chemical process engineering, (7) external and internal environment of buildings, (8) marine engineering, (9) environmental engineering, (10) hydrology and oceanography, (11) meteorology, and (12) biomedical engineering.[5] Scientific research pursuit and technological profundity in CFD are witnessing immense upheavals. The intricacies of application of CFD and applied mathematics to diverse domains of engineering are delineated in details in this treatise. The ultimate aim of developments in the CFD field is to provide a capability comparable with CAE (computer-aided engineering) tools such as stress analysis codes.[5] The vast availability of affordable high-performance computing hardware and the introduction of user-friendly interfaces has led to a vigorous upsurge of scientific interest, and CFD has entered into the wider industrial community in the 1960s.[5] The challenge and the vision of science and engineering after 1960s has undergone remarkable revamping and science and technology surged ahead in a new era. Versteeg and Malalasekara[5] discussed in deep details the overall structure of a CFD code and discussed the role of veritable building blocks. CFD codes are structured around the numerical algorithms that can tackle fluid flow problems.[5] In their treatment of the physics of fluid flows, the authors have added a summary of the basic ideas encompassing large-eddy simulation and direct numerical simulation. Over the last decade of the twentieth century, a number of new discretization techniques and solution approaches have come to the forefront in commercial CFD codes.[5] Science and technology are highly challenged and the vision widened as CFD undergoes drastic changes.[5]

Applied mathematics and CFD are today in a state of immense scientific regeneration. Application of CFD results in vigorous industrial research and design crucially hinges on confidence in its outcomes. The authors discussed the domain of uncertainty in CFD results. The challenge of science and engineering are immense and far-reaching. So this treatise targets on the industrial research area of CFD.[5]

12.4.1 WHAT DO YOU MEAN BY COMPOSITE SCIENCE?

A composite material is a material made from two or more constituent materials with significantly different chemical and physical properties that when combined give a different and unique property. Scientific success, the vast scientific stewardship, and the scientific ingenuity behind composite science and polymer science needs to be reenvisioned and reframed with the progress of science, history, and the visionary timeframe. The new composite material may be preferred for many reasons: common examples include materials which are stronger, lighter, and less expensive when compared to conventional materials. Today, CFD have applications in every manufacturing processes including composites. This chapter opens up newer scientific thoughts, newer ingenuity, and deep scientific truth in CFD applications in composite manufacturing.

12.4.2 COMPOSITE SCIENCE AND THE ENGINEERING VISION FOR THE FUTURE

The world of material science and polymer science are moving from one visionary paradigm toward another. Technology and engineering science are highly challenged as regards application of computational tools in composite manufacturing process and the vast world of polymer science. Mankind's immense scientific grit, man's vast scientific prowess, and the technological ingenuity will all today lead a long and visionary way in the true emancipation of polymer science and nanotechnology. In the similar manner, composite science and its manufacturing process are in the process of newer scientific rejuvenation. The engineering vision of composite science is today vast and versatile. In this chapter, the author with immense scientific truth and scientific fortitude unravels the scientific ingenuity behind composite applications and CFD technology. Applied mathematical tools are of immense scientific necessity today. This chapter opens up new windows of scientific innovation and instinct in the field of composites in years to come.

12.5 SCIENTIFIC DOCTRINE AND DEFINITION OF CFD

Scientific doctrine and definition of CFD are vigorously changing the face of applied mathematics and computational science. The world of challenges and the scientific vision in the field of research pursuit in CFD need to be revamped and restructured with the passage of scientific history and time. Mathematical tools and computational techniques are the utmost need of science and engineering today. Sustainability vision whether it is energy or environment needs to be readdressed and reenvisaged as science and engineering move forward toward a newer era. Today, fluid mechanics and CFD applications stand in the midst of deep vision and scientific introspection. Multiphase flow is a crucial scientific endeavor of immense interest. The immense scientific prowess of human, the immediate scientific needs of modern society, and the vast vision and profundity of engineering science will lead a long and visionary way in the true emancipation and the true realization of applied science and applied engineering today. Scientific doctrine in the field of chemical process engineering, petroleum engineering science, environmental engineering, and other diverse branches in engineering are in the path of immense vision and regeneration. The challenge of CFD today is in its application in multiphase flow and design of engineering systems such as chemical engineering systems and petroleum refining units. Today, scientific doctrine in engineering science is surpassing vast and versatile frontiers. Mathematical tools such as CFD need to be scientifically validated with the passage of scientific history and time. Human's immense technological prowess, the need for computer science in design of engineering systems, and the need for scientific validation will all lead a long and effective way in the true emancipation of CFD tools today. CFD or fluid dynamics as a whole are in the path of newer vision with the progress in space technology and nuclear research. Both space science and nuclear technology needs the application of computational technique such as CFD.

12.6 APPLICATIONS OF FLUID DYNAMICS IN ENGINEERING SCIENCE

Fluid dynamics in engineering science has diverse scientific applications and is opening up new avenues of scientific innovation and scientific instinct in the coming decades. The major areas of scientific endeavor are mechanical engineering, chemical process engineering, and aerospace engineering. Today CFD also has significant applications in biomedical engineering and medical science also. Human scientific vision today stands in the midst of deep scientific crisis and unimaginable disaster. Scientific forbearance and scientific optimism in environmental engineering and petroleum engineering science are in a state of immense disaster. Technology has few answers to the unimaginable crisis in environmental protection. This area now encompasses the vast domain of CFD. The scientific and academic rigors in CFD applications are changing the face of research pursuit today. Multiphase flow today is in a state of immense scientific regeneration with vast and varied applications. Scientific ingenuity, deep scientific progeny, and human's immense scientific regeneration will all lead a long and effective way in the true realization of applied mathematics and computational techniques today. Petroleum engineering science and chemical process technology are the areas of engineering science which need to be envisioned in every respects with the application of CFD. Modern science and its immense scientific profundity and ingenuity are today in the path of regeneration. Fluid dynamics and chemical process engineering are two opposite sides of the visionary scientific coin. Fluid dynamics and multiphase flow are in the forefront of chemical engineering research and development initiative today. The vast world of multiphase flow still remains unexplored. Engineering science and technology need to be envisioned with the passage of scientific history and time as regards application of CFD to modern science. CFD will veritably open up new doors of innovation and instinct in days to come.

12.6.1 APPLICATIONS OF CFD IN COMPOSITE TECHNOLOGY

The challenge and the vision of composite technology are immense and far-reaching. Scientific ingenuity, scientific truth, and scientific profundity are the pillars of scientific validation and vast technological motivation today. A composite material is a material made of two or more constituent materials with significantly different physical and chemical properties, that, when combined, produce a material with characteristics different from the individual components. The individual components remain separate and distinct within the finished product. The new product may be preferred for many reasons: common examples include materials which are stronger, lighter, or less expensive when compared to conventional materials. CFD have tremendous applications in every branch of manufacturing technology today. This chapter unfolds the scientific intricacies, the vast technological prowess, and the scientific insight in the field of composite manufacturing and CFD. CFD has been used to simulate the flow during vacuum infusion, sheet molding compound, the autoclave process, and twin-screw extrusion. The scientific vision behind CFD applications in composite manufacturing is today surpassing one visionary frontier over another. In this entire treatise, the author pointedly focuses on the need of CFD in engineering science and the manufacturing technology of composites, polymers, and smart materials.

12.7 VISIONARY SCIENTIFIC ENDEAVOR IN FLUID DYNAMICS

Fluid dynamics and CFD are the cornerstones of the science of mechanical engineering today. Chemical process engineering and its engineering profundity needs to be drastically reenvisioned and reenvisaged as science and technology moves forward. Today scientific endeavor stands in the midst of vision and forbearance. The vision of the scientific revolutionaries and proponents of fluid dynamics today needs to be reemphasized and are the need of the hour. Scientific vision and scientific forbearance in the field of fluid dynamics needs to be redefined as scientific and academic rigor in chemical process engineering and petroleum engineering surges forward. Mechanical engineering and aerospace engineering are also cornerstones of research endeavor in CFD today. Technology of CFD needs to be revisited and reenvisaged as science and engineering of fluid flow and multiphase reactors enters into a newer scientific paradigm. In this section, the author pointedly focuses on the vast scientific endeavor and the scientific vision in the application of CFD in different avenues of engineering science. Technology and engineering science are today surpassing vast visionary boundaries and crossing one visionary paradigm over another. This entire chapter is a vast eye-opener to the scientific intricacies and the scientific struggles in the application of CFD.

12.8 SIGNIFICANT SCIENTIFIC RESEARCH PURSUIT IN CFD

Scientific research pursuits in CFD are witnessing immense and dramatic challenges. History of mechanical engineering science, the immense scientific prowess, and the technological profundity behind multiphase flow are all the pallbearers toward a newer redefining of CFD today. The methodology of CFD and the immense scientific endeavor behind this vision will all lead a long and visionary way in the true emancipation of multiphase flow today. Technology and engineering science are today in the path of newer scientific rejuvenation globally. Today is the scientific world of nuclear engineering and space research. Mathematical techniques and computational science are the needs for the furtherance of engineering science. In this chapter, the author stresses on the diverse applications of CFD in chemical engineering and petroleum engineering. Science is prudent, visionary, and evergrowing in present day human civilization. Human scientific endeavor today stands in the midst of deep challenges and unending scientific introspection. This entire treatise gives a vast glimpse on the scientific success and deep scientific profundity in the application of CFD and advanced fluid dynamics in the design of petroleum engineering and chemical engineering systems. The author in this section delineates the significant scientific research pursuit in the field of CFD applications in different areas of engineering and science.

Sayma[1] lucidly in a watershed text dealt the domain of CFD. CFD is the branch of fluid dynamics providing a cost-effective means of simulating real flows by the numerical solution of differential equations. Technological motivation and scientific validation are the cornerstones of this book. The governing equations for Newtonian fluid dynamics, namely the Navier–Stokes equations, have been in use in the scientific world for the last 150 years.[1] Technology has advanced at a rapid pace since then. However, the development of reduced forms of these equations is still an active area of research endeavor, in particular, the turbulent closure problem of the Reynolds-averaged Navier–Stokes equation.[1] For non-Newtonian fluid dynamics and its visionary domain, chemically reacting flows and two-phase flows, the theoretical development is at less advanced stage.[1] The world of challenges in engineering science, the immense scientific barriers of application of computational methods, and the immense scientific prowess of fluid dynamics in particular will all lead a long and visionary way in the true realization of mathematical tools such as CFD today. Technology and engineering science are today in the path of newer scientific overhauling and deep regeneration. In this book, the author reiterates the vast scientific success of CFD in its application in diverse areas of engineering science. Computational techniques replace the governing partial differential equations with system of algebraic equations that are much easier to solve using computers. Human scientific endeavor, the vast research acumen, and the farsightedness of applied mathematics are the torchbearers toward a newer visionary era in the field of CFD. The scientific status of CFD applications in different areas of engineering science is vast and versatile. This book is a vast eye-opener on the immense scientific adjudication and the deep scientific discerning in the field of applied mathematics and computer science. The steady improvement of computing power, since the 1950s, thus has led to the emergence of CFD domain.[1] Deep scientific imagination and vast scientific discernment are the pillars of this well-researched treatise.

Anderson[2] deeply discussed with cogent insight CFD and its basics with applications. Technological advancements and vast scientific validation are the pivots of this research foray. Technology and engineering science of fluid dynamics and applied mathematics are today crossing vast and versatile scientific boundaries. Today, applied mathematics and computer science stand in the midst of scientific vision and scientific ingenuity.[2] Technology today has few answers to the application areas of CFD. In this book, the author pointedly focuses on the philosophy of CFD, the governing equations of fluid dynamics, their derivations, a discussion of their physical meaning, and a presentation of forms particularly suitable to CFD. Human scientific fortitude, deep scientific profundity, and the vast world of scientific hurdles are the necessities of innovation and vision today. Anderson[2] deeply reiterates the human scientific success of the application of CFD in design of engineering systems and the vast technological profundity in the understanding of CFD. In the other areas of technological innovations, the author delineates are mathematical behavior of partial differential equations and its impact on CFD. Some application areas the author touches upon are numerical solutions of quasi-one-dimensional nozzle flows, numerical solution of a two-dimensional supersonic flow, and Prandtl–Meyer expansion wave and incompressible Couette flow.[2] The author successfully winds up the entire treatise with some advanced topics of modern CFD and finally the future of CFD.[2]

Shaw[3] discussed with vast farsightedness the use of CFD. CFD today is in the path of newer scientific regeneration and deep scientific vision. This book poignantly depicts fluids in motion and numerical solutions of partial differential equations. Technological and scientific validation, the world of scientific challenges, and the vast scientific vision of computational techniques will lead a long and visionary way in the true emancipation of CFD. CFD is in the path of newer scientific rejuvenation. In this research endeavor, the author lucidly dealt with computer-based analysis procedures and tools in CFD. Vast scientific motivation and deep technological validation and innovation are the hallmarks of this book.[3] The author also pointedly focuses on flow problems in engineering, how to build a mesh, setting the fluid flow parameters, obtaining a solution and analyzing the result. The vision and the challenges of science and engineering are immense in present day human civilization. This book is a veritable eye-opener toward the immense scientific potential and the scientific ingenuity in CFD applications in engineering and science. The author ends this treatise with case studies and the modeling of flows with additional complexity. Human scientific progress in computational techniques today stands in the midst of deep scientific introspection and vast vision.[3] CFD is one of the major computational tools and it needs to be envisioned and restructured with the passage of scientific history and time. The author in this book deeply treads a difficult path in the vast scientific emancipation of CFD.[3]

Patankar[4] discussed with deep and cogent insight numerical heat transfer and fluid flow. Human's immense scientific ingenuity, man's immense vision, and the world of challenges in computational techniques will all lead a long and visionary way in the true realization of computer science and applied mathematics today. This is a watershed text in the field of CFD. Human scientific regeneration and deep scientific rejuvenation are the hallmarks of this well-researched book.[4] The author deliberates on mathematical description of physical phenomena, discretization methods in CFD, heat conduction, convection and diffusion, calculation of the flow field, and some illustrations and discussions of CFD applications.[4] The book is concerned with heat and mass transfer, fluid flow, chemical reaction, and other related processes that occur in engineering design, in the natural environment and in living organisms in the environment. Today, CFD is part and parcel of engineering equipment design, mechanical engineering design, chemical process engineering, and aerospace engineering. Nearly, all methods of power production involve fluid flow and heat transfer as essential processes. The same processes govern the heating and air-conditioning of buildings.[4] Major areas of chemical and metallurgical industries use components such as furnaces, heat exchangers, condensers, and reactors, where thermofluids have vast applications. Aircrafts and rockets depend on their functioning to fluid flow, heat transfer, and chemical reaction.[4] In the design of electrical machinery and electronic circuits, heat transfer is often the limiting factor. Here comes the immense importance of CFD, numerical heat transfer, and fluid flow. Science and engineering of fluid dynamics are entering into a newer visionary era of scientific rejuvenation and vast scientific regeneration. This book is primarily aimed at developing a general method of prediction for heat and mass transfer, fluid flow, and related processes. A redefining characteristics of the numerical methods developed in this book is that they are deeply related to physical considerations rather than mathematical manipulations. Advantages of theoretical calculation over experimental investigations are (1) low cost, (2) speed, (3) ability to simulate realistic conditions, and (4) ability to simulate ideal conditions.[4] Technology and engineering science are today in the path of vast scientific rejuvenation and deep scientific comprehension. The science and engineering of CFD will today veritably open new knowledge dimensions and new windows of innovation and scientific instinct in decades to come.[4]

Versteeg and Malalasekara[5] deeply delineated with vast scientific farsightedness an introduction to CFD. This well-researched textbook touches upon conservation laws of fluid flow and boundary conditions, turbulence and its modeling, the finite volume method for diffusion problems, the finite volume method for convection–diffusion problems, solution algorithms for pressure–velocity coupling in steady flows, the finite volume method for unsteady flows, CFD modeling of combustion, and numerical calculation of radiative heat transfer.[5] Human scientific ingenuity, the vast technological profundity, and the vast needs scientific needs of computational techniques will veritably lead a long and effective way in the true realization and true emancipation of applied mathematics and CFD today. Scientific research pursuits in the field of CFD are crossing vast and versatile boundaries. This book is a watershed text in the field of CFD and fluid dynamics as a whole. Technological vision, the scientific prowess, and the needs of computational science are the torchbearers of scientific vision in CFD.[5]

Karthik[6] discussed with lucid and cogent insight turbulence models and their applications. Scientific research pursuits in CFD today are opening new knowledge dimensions in the field of engineering science and technology.[6] CFD can be applied in diverse areas of science and engineering such as aerospace engineering, chemical process technology, and petroleum engineering. Human and human scientific endeavor today stands in the midst of deep scientific introspection and scientific fortitude. The vast scientific prowess of engineering science, the scientific grit of mankind, and the utmost vision of computer science will all lead a long and visionary way in the true emancipation of CFD today.[6] In this presentation, the author deeply delineates turbulence models introduction, Boussinesq hypothesis, eddy viscosity concept, zero-equation model, one-equation model, two-equation model, algebraic stress model, Reynolds stress model, and the vast and varied applications areas.[6] CFD today is in the path of newer scientific regeneration and deep scientific vision. A turbulence model is a procedure to close the system of mean flow equations. The technology of CFD is highly advanced today and surpassing cross-boundary research domains. Human scientific ingenuity, the needs of petroleum engineering and chemical process engineering, and the vast scientific vision are the veritable torchbearers toward a newer era in the field of science and engineering today. For most engineering applications, it is unnecessary to resolve the details of the turbulent calculations.[6] Turbulent models allow the calculation of the mean flow without first calculating the full time-dependent flow field.[6] For a turbulence to be successful it (1) must have wide applicability, (2) be accurate, (3) simple, and (4) economical to run.[6] Human technological profundity and vast scientific ingenuity are at its helm as science and engineering surge forward. In a similar manner, CFD is today ushering in a newer era in applied mathematics and computational science. The challenge and the vision of teaching CFD to undergraduate and postgraduate students are immense and groundbreaking. The technological challenges and the profundity of engineering science are today opening up new scientific instincts and newer vision.[6]

Zubanov et al.[7] discussed with deep and cogent farsightedness the technique for simulation of transient combustion processes in the rocket engine operating with gaseous fuel “Hydrogen and Oxygen.”[7] The article delineates the method for simulation of transient combustion processes in the rocket engine. Science and engineering are two huge colossuses with a deep and pragmatic vision of its own. Application of CFD today is at its helm as technology and engineering science tread forward toward a scientific era.[7] The engine operates on gaseous propellant: oxygen and hydrogen. Combustion simulation was performed using the ANSYS CFX software.[7] Technological profundity and deep scientific and academic rigor are the cornerstones of this article. Three reaction mechanisms for the stationary mode were deeply investigated and described in minute details.

Reaction mechanisms were taken from several sources and deeply verified. The method for converting ozone properties from the Shomate equation to the NASA-polynomial format was described in minute details. Investigations and scrutiny of unsteady processes of rocket engines including the process of ignition were proposed as the area of application of the described simulation technique.[7] The science and engineering of CFD is surpassing visionary frontiers and today drastic and dramatic challenges. In this article, the author deeply discusses one particular application of CFD in combustion science and opens up new knowledge dimensions in the field of scientific validation and technological vision in decades to come.[7] Chemical reaction engineering and reaction kinetics are at the heart of a chemical process and investigations concerning this arena are highly recommended in the pursuit of technology and engineering science. Human scientific vision, the vast scientific prowess of engineering science, and the immense application areas of applied mathematics will all lead a long and visionary way in the true realization of computational techniques and computer science today.[7]

Pinho[8] deeply delineated with cogent insight and vast vision numerical studies of propane–air mixture combustion in a burner element. This is a watershed text in the field of application of CFD. Engineering science and technology are today in the path of immense vision and deep ingenuity. This study considers numerical simulations of the combustion of propane with air, in a burner element due to high temperature and velocity gradients in the combustion chamber.[8] Engineering science of CFD is highly advanced today and needs to be vehemently readdressed as science and engineering surges forward toward a newer scientific vision. The technological vision of CFD needs to be readdressed as science and engineering moves forward toward a newer visionary era. When early commercial CFD packages became available more than 20 years ago, simulating the complex physics inside combustion chambers was already one of the visionary and target applications,[8] thus ushered in a new era in the field of applied physics, applied mathematics, and computational science. Human scientific ingenuity and deep scientific vision are the hallmarks toward a newer era of computer science.[8] Today, with increasing maturity of CFD technique and computational power, one needs to focus in numerical simulation of combustion in the area of nonequilibrium chemistry and multiphase flow. Numerical analysis of multiphase flow today is moving in the path of newer scientific regeneration and vast scientific and technological vision.[8]

Donini[9] discussed with immense foresight and deep thought advanced turbulent combustion modeling for gas turbine application. Technological advancements, the needs of computational tools, and the success of scientific endeavor are the forerunners toward a newer era in the field of CFD. Application of CFD in engineering science and its immense vision are the cornerstones of this research endeavor.[9] The numerical modeling of reactive flows for industrial applications has gained immense importance and has a continued growth of interest in the last few decades. This especially applies to the design of applications for which experimental investigations are extremely complex and expensive, such as gas turbines burners.[9] Numerical simulation is foreseen to provide a tremendous increase in gas turbine combustors design efficiency and quality over the next future. Scientific fortitude, scientific vision, and vast scientific profundity are the torchbearers toward a newer scientific genre in the field of CFD applications.[9] Multiphase flow and design of chemical engineering and petroleum engineering systems are today linked by a visionary umbilical cord.[9] The status of research pursuit in CFD is crossing vast and versatile scientific frontiers.[9]

Chung[10] discussed with vast foresight CFD and its immense vision. This book is intended for the beginner as well as the practitioner in CFD. It includes two major computational methods, mainly finite difference methods and finite element methods as veritably applied to the numerical solution of fluid dynamics and heat transfer problems.[10] An equal importance on both the methods is equally attempted. Modern science today is in the path of newer scientific revelation and vast scientific discernment.[10] The advantages of this watershed text are that the two computational methods are consolidated in the same treatise. Scientific research pursuits in today's human civilization are traversing vast and visionary boundaries. CFD today stands in the crucial juncture of introspection and vision. In this book, the author reiterates the vast scientific success of CFD applications and its wide vision. The author overviews briefly one-dimensional computations by finite difference methods, computations by finite element methods, and computations by finite volume methods. The author (Chung[10]) also touches upon incompressible viscous flows via finite difference methods and compressible flows via finite difference methods. Again, the author delineates incompressible and compressible flows by finite element methods. Technology of CFD is far-reaching and visionary as regards its applications to diverse areas of science and engineering.[10] The book ends its investigations with deep insight to applications such as applications to turbulence, applications to chemically reactive flows and combustion, applications to acoustics, applications to combined mode radiative heat transfer, applications to multiphase flow, applications to electromagnetic flows, and finally applications to relativistic astrophysical flows.[10] Human scientific ingenuity in the field of CFD applications is scaling visionary heights today. Multiphase flow and scientific research pursuit in CFD applications are the torchbearers toward a greater realization of applied mathematics and computer science today. This book is a phenomenal piece of reference work in the domain of fluid mechanics and particularly CFD and will surpass scientific frontiers of immense vision in decades to come.[10]

12.8.1 SIGNIFICANT RESEARCH ENDEAVOR IN CFD APPLICATIONS IN CHEMICAL PROCESS ENGINEERING

Human scientific instinct and scientific research pursuit are today in the path of scientific revelation and deep vision. Fluid dynamics and its immense academic rigor are in the avenues of scientific regeneration and vast vision. Today, chemical process engineering and applied chemistry are in the threshold of a newer scientific genre of immense scientific comprehension. Technology and engineering science of CFD applications and computational techniques applications needs to be envisioned and addressed with the passage of scientific history and time. One of the primary applications of CFD is its vast scientific understanding in the field of chemical reactor engineering. Harris[11] deeply comprehended with cogent insight CFD for chemical reactor engineering. Science of chemical reactor engineering and chemical reaction kinetics are surpassing vast and versatile scientific frontiers. Unit operations in chemical engineering and unit processes are the challenges and the vision of science and engineering today. CFD involves the numerical solution of conservation equations for mass, momentum, and energy in a vast flow geometry of interest, together with additional sets of equations envisioning the problem at hand.[11] Recent progress in the predictions of flow in baffled stirred tank reactors is deeply reviewed. Improved turbulence modeling and high-performance computing have a decisive role to play in the future applications of CFD in chemical reaction engineering. Examples cited in this paper are non-Newtonian and nonisothermal flow in an extruder.[11] There are other examples of CFD applications such as gaseous turbulent flow in a reactor with competing parallel and consecutive reactions.[11] Kaushal and Sharma[12] discussed with immense farsightedness the concept of CFD and its applications in food-processing equipment design. Technology and engineering science of food processing and CFD applications are surpassing vast and versatile scientific boundaries. This paper reviews the applications of CFD in food processing including cleaning of storage tanks, design of dryers, sterilizers, freezers, mixers etc.[12] CFD is a visionary technique in optimization of hygiene of closed process equipment. A veritable combination of wall shear stress, fluid exchange, and turbulence conditions can definitely predict areas which are not properly cleaned. Thus the need of CFD tools and its diverse applications. Today chemical process engineering is a huge colossus with a definite and targeted vision of its own. Computational tools are the utmost needs of research pursuit in chemical process technology today. Scientific prowess, deep scientific profundity, and the utmost need of modern science in human society will all lead a veritable way in the true realization of applied computer science and computational techniques.[12]

12.8.2 SIGNIFICANT RESEARCH PURSUIT IN CFD APPLICATIONS IN PETROLEUM ENGINEERING

Technology and engineering science are moving at a drastic pace crossing one visionary boundary over another. In the similar manner, petroleum engineering and petroleum refining needs to be envisioned and revamped over the global fuel crisis. Depletion of fossil fuel resources is a veritable bane to human civilization and human scientific pursuit. Thus the need for innovation and vision. Abdi et al.[13] with deep and cogent insight delineated applications of CFD in natural gas processing and transportation. In this chapter, two examples of CFD applications in natural gas processing and transportation are presented with minute details. A commercial software package (FLUENT) was used in these detailed investigation.[13] The aim and objective of the studies are briefly discussed, the methodology outlined, and boundary conditions deeply delineated in minute details. The high demand for natural gas has encouraged the energy industry toward the further discovery of remote offshore reservoirs. Scientific world today stands in the midst of vision and deep scientific provenance.[13] This chapter is a veritable eye-opener toward a greater visionary avenue in petroleum exploration. Common design challenges and the vast vision in all gas processing methods for offshore applications are the compactness and vast reliability of process equipment.[13] Petroleum engineering science is highly challenged today with the evergrowing concern of depletion of fossil fuel resources. Supersonic nozzles have been introduced as an alternative to treat natural gas for offshore applications and to meet the offshore requirements. Application of CFD technique has been veritably demonstrated to predict the behavior of high-pressure natural gas flowing through supersonic nozzles.[13] The pillars of CFD and its applications are demonstrated with deep insight and vast vision in this chapter. Yusof[14] with vast insight and immense farsightedness delineated in a well-researched thesis the study of multiphase flow for petroleum production using CFD. Scientific discernment, scientific revelation, and vision are the pivotal areas of this thesis. Petroleum is a highly valued natural resource where uses are never ending and the demand will be always higher than supply. Today, there should be an enhanced way to recover the oil efficiently.[14] Technology of petroleum exploration needs to be envisioned and readdressed as science and engineering surges forward. The flow of petroleum or gas through a vertical pipe from oil reservoir to platform system is significant to cater to the enhanced oil recovery. This technology needs to be envisioned as engineering science surges forward.[14] This treatise is based on this principle and is more focused on the study of multiphase flow pattern of petroleum production plus the turbulent coefficient effect on the bubbly flow in vertical pipe using CFD software.[14]

Today, the challenge and vision of petroleum engineering, petroleum exploration, and petroleum refining are immense and groundbreaking. With the growing concerns of fossil fuel depletion and the global energy crisis, humans stand between deep scientific comprehension and vision. Thus the need of innovation and scientific ingenuity. In this chapter, the author poignantly focuses on the scientific cognizance, the need of energy technology and energy sustainability, and the vast world of applied computer science and computational techniques.

12.9 AEROSPACE ENGINEERING AND THE SCIENTIFIC PROGRESS IN CFD

Human scientific progress in aerospace engineering is moving from one visionary paradigm toward another. Technology and engineering science of CFD are highly challenged and are in the path of newer scientific regeneration. Science today is a huge colossus with a deep vision and a definite purpose in its medley of cross-boundary research. Interdisciplinary research in CFD is witnessing immense scientific rejuvenation and is opening new avenues of scientific understanding and scientific vision. The entire domain of aerospace engineering is highly robust and groundbreaking today. In this section, the author poignantly depicts the vast importance and scientific imperatives in the application of CFD in aerospace engineering. Technological profundity in other branches of engineering science such as chemical process technology, petroleum engineering, and environmental protection are the necessities of scientific endeavor today. The world of fluid dynamics and hydrodynamics are in the threshold of a newer scientific rejuvenation. Modern science, aerospace engineering, and chemical process engineering need to be revamped and overhauled as human civilization surges forward. Science and technology today stands in between deep scientific revelation and profundity. The challenges of CFD applications are immense and groundbreaking. The author in this entire chapter reiterates the vast scientific success and the scientific potential behind CFD applications and its wide vision.

12.10 ADVANCED FLUID DYNAMICS AND THE DEEP SCIENTIFIC INGENUITY

Advanced fluid dynamics encompasses CFD and also has application interfaces with computational science and applied mathematics. Scientific ingenuity and deep scientific motivation are the cornerstones of cross-boundary research in chemical process engineering and advanced fluid dynamics. Technology and engineering science of fluid dynamics are highly advanced today. Technological prowess, scientific determination, and deep scientific understanding in advanced fluid dynamics will lead a long and visionary way in the true realization and true discernment of CFD today. Fluid dynamics and CFD domain are today highly challenging areas of science and engineering. CFD applications are highly relevant in many branches of engineering science such as aerospace engineering. Today is the age of nuclear science and space technology. Vast scientific ingenuity, the technological prowess, and the scientific needs of human society are all the torchbearers toward a newer visionary era in technology and engineering science. Fluid dynamics applications, its deep scientific ingenuity, and a vast scientific introspection will all lead an effective way in the true realization of scientific research pursuit in aerospace engineering and other areas of engineering science and technology.

12.11 MODERN SOCIETY AND COMPUTATIONAL SCIENCE

Modern society and computational science are today veritably linked by an unsevered umbilical cord of scientific grit and deep scientific determination. Human and human civilization today stands in the midst of introspection and deep vision. Technology has today few answers toward the global environmental crisis and global climate change. Depletion of fossil fuel resources has urged scientists and engineers to move toward newer goals of innovation and challenge. Modern society today stands in the midst of deep scientific catastrophe and introspection. Technology and engineering science in today's world are advancing at a rapid pace with the scientific forays in nuclear science and space technology. Computer science and computational techniques are the cornerstones of scientific and technological advancements today. Applied mathematics and the vast domain of mathematical tools and techniques are changing veritably the scientific firmament. Modern society and scientific research pursuit today needs to be restructured and overhauled as engineering science and technology ushers in a new era of vision and ingenuity. Today, science and engineering are in the path of a new visionary era in nuclear science and space technology. On the other side of the visionary coin is the grave global concern of depletion of fossil fuel resources, climate change, and energy security. Petroleum engineering science and chemical process technology are the two vast avenues of application of computational tools in design of engineering systems. Human scientific ingenuity thus needs to be reorganized and reenvisioned. Global concerns for energy sustainability and energy security are changing the entire mindset of scientists and engineers. Modern society and human scientific endeavor today are in the midst of deep scientific fortitude and scientific ingenuity. Computational science, applied mathematics, and engineering science today are the necessities of human scientific progress and the wide avenues of human civilization. Science and engineering of CFD today stands in the midst of deep scientific comprehension and vast scientific ingenuity. Modern science in a similar manner is in the process of scientific regeneration. Technology and engineering science today stands highly challenged as human surges forward. CFD today in the similar manner stands in the midst of scientific intricacies and vision. In this chapter, the author reiterates the scientific success, the vast scientific potential, and the immense scientific imagination behind CFD applications in engineering and science.

12.12 SUSTAINABILITY AND THE PROGRESS IN SCIENCE AND TECHNOLOGY

Energy and environmental sustainability are the cornerstones of scientific research pursuit in present day human civilization. Sustainable development today stands in the midst of immense technological profundity and scientific introspection. Human scientific progress today depends on forays in research endeavor in energy and environmental sustainability. Energy security, the scientific paradigm of energy and environmental sustainability today, will lead a long and visionary way in the true realization of global sustainable development. Renewable and nonrenewable energy domains are the vision and challenges of science and engineering today. Solar energy, wind energy, and energy from biomass are changing the scientific fabric of energy engineering today. In this treatise, the author reiterates the contribution of applied mathematics and computational science in the design of engineering systems. The visionary words of Dr. Gro Harlem Brundtland, former Prime Minister of Norway, on sustainability need to be envisioned and readdressed with the passage of scientific history and the global visionary timeframe. Global research and development initiatives in energy security, energy, and environmental sustainability are in a state of immense concern and deep catastrophe. Technology and engineering science have lack of proper answers to the ever-growing concerns of depletion of fossil fuel resources and global climate change. Here comes the immediate need of mathematical tools and computational science. The visionary avenues of research and development initiatives in energy sustainability and human's immense scientific prowess will all lead a long and effective way in the true emancipation of energy security today. Today human scientific progress is in a state of immense scientific challenges and a deep state of technological stress. Scientific profundity, deep scientific ingenuity, and vision are the necessities of research endeavor today. Thus, sustainable development and energy and environmental sustainability will open up vast windows of innovation and instinct in decades to come.

12.13 FUTURE RECOMMENDATIONS AND FUTURE THOUGHTS

The vision of human civilization and the rigor behind scientific endeavor are today far-reaching and immensely ever-growing. Applied science, applied mathematics, and engineering science are changing the face of human scientific endeavor today. Scientific rejuvenation in research pursuit in fluid dynamics is entering into a new era of immense scientific potential. Today, the immense scientific prowess behind CFD applications is veritably challenging the deep scientific landscape and might. Future recommendations in the study of CFD are changing and targeted toward more avenues in multiphase flow. Today, sustainable development, whether it is energy or environment, is the cornerstone of research pursuit. Technological vision, the vast scientific profundity, and the concerns for energy and environment are all leading a long and visionary way in the true emancipation of mechanical engineering, petroleum engineering, and chemical engineering science. Future thoughts and future avenues of research endeavor are the areas of immense need today in the path of science and life. Newer vision and newer innovation are the challenges of CFD application today. Aerospace engineering and rocket propulsion technology are the two vibrant areas of CFD application today. These areas of science need to be reframed and reenvisioned as civilization surges ahead. Today, technology has few answers to the intricacies of CFD. Thus, there is a need for a greater scientific understanding and scientific discernment in fluid dynamics and mechanical engineering science.

Scientific validation and technological revamping are the utmost need of the hour today. Future recommendations and future avenues of research should be targeted toward more innovative research endeavor in the field of CFD applications in diverse areas of engineering and science. Fluid dynamics applications are today revolutionizing the scientific landscape and ushering in a new era in mechanical engineering and chemical engineering science.

12.14 FUTURE TRENDS IN CFD RESEARCH

CFD and fluid dynamics are today changing the scientific paradigm of human scientific endeavor. Fluid flow, heat transfer, and chemical reaction engineering today encompasses the vast and versatile domain of CFD. The future of human civilization and human scientific endeavor lies in the hands of engineers and scientists. Technological vision and scientific profundity need to be reemphasized and reenvisaged with the progress of human civilization. CFD research has profound applications in mechanical engineering, chemical process engineering, and aerospace engineering. These areas of research pursuit need to be realigned and revisited with the progress of scientific history and visionary timeframe. Today is the age of nuclear science and space technology. Every nation on earth is surging forward with vision and might toward a newer emancipation of science and technology. Human vision and scientific forbearance and candor are changing the face of scientific research pursuit in nuclear science and space technology. In such a crucial juxtaposition of science and engineering, human vision in applied mathematics and computational techniques need to be overhauled. Here comes the importance of CFD and mechanical and aerospace engineering applications. CFD research is equally important in manufacturing process and is highly relevant in composite manufacturing process. CFD has vast and versatile applications in chemical engineering, petroleum engineering, and diverse branches of engineering science. Future research and development initiatives should be targeted toward more CFD applications in composite and polymer science.

12.15 CONCLUSION, SUMMARY, AND FUTURE PERSPECTIVES

Science of applied mathematics and chemical process engineering are today advancing at a rapid pace. CFD and mechanical engineering today stand in the midst of deep scientific introspection and immense vision. Technology today is a huge colossus with a definite vision and a definite willpower of its own. There are vast and versatile answers to the intricacies of fluid dynamics and CFD today. Chemical process engineering and mechanical engineering are two branches of engineering which are bonded with each other like an unsevered umbilical cord. In this treatise, the author deeply comprehends the success of mathematical tools and computational techniques in its applications in applied science and engineering. CFD today stands in the verge of newer scientific regeneration. Future perspectives and future research trends are targeted toward newer innovations and newer visionary areas in scientific forays. Scientific validation of computational techniques is the utmost need of the hour today. Today, the vision of science and technology are evergrowing, promising, and surpassing scientific boundaries. This treatise gives a vast glimpse of the huge arena of CFD with the sole vision of furtherance of science and engineering. Today, both nuclear science and space technology stand in the midst of technological vision and scientific profundity. CFD tools are veritably applied in every branch of engineering science and encompass nuclear science and space technology as well. The author in this treatise pointedly focuses on the scientific vision, the vast scientific stature, and the future avenues of research in the field of CFD and chemical engineering science. Human scientific vision and deep scientific grit are the cornerstones of research pursuit in CFD and engineering science today. This treatise rigorously points toward the immense scientific potential and the success of scientific validation in CFD applications. Human scientific endeavor and applications of computational techniques will all lead a long and visionary way in the true emancipation of scientific progress today. The world of challenges in the field of composite materials and polymer science are vast and versatile. In this chapter, the author deeply comprehends the scientific intricacies in the application of CFD in composite manufacturing processes. This scientific challenge, the scientific travails, and the scientific profundity are deeply elucidated in this well-researched treatise.

Human civilization and human scientific endeavor today stands in the midst of deep scientific vision and immense scientific forbearance. Technological advancements and scientific achievements in engineering science are witnessing immense challenges and extensive revamping. In this chapter, the author repeatedly urges the scientific community the need for mathematical tools for the furtherance of science and engineering. Applied mathematics and applied computer science will witness a new beginning as science and engineering moves from one visionary paradigm toward another.

KEYWORDS

- **petroleum**
- **chemical**
- **vision**
- **computational**
- **fluid**
- **composites**
- **dynamics**

REFERENCES

1. Sayma, A. *Computational Fluid Dynamics*; Abdulnaser Sayma & Ventus Publishing ApS, 2009. ISBN-978-87-7681-430-4.
2. Anderson, J. D. *Computational Fluid Dynamics: The Basics with Applications*; McGraw Hill Inc.: USA, 1995.
3. Shaw, C. T. *Using Computational Fluid Dynamics*; Prentice Hall: USA, 1992.
4. Patankar, S. V. *Numerical Heat Transfer and Heat Flow*; Hemisphere Publishing Corporation: USA, 1980.
5. Versteeg, H. K.; Malalasekara, W. *An Introduction to Computational Fluid Dynamics: The Finite Volume Method*; Pearson Education Limited: UK, 1995.
6. Karthik, T. S. D. In *Turbulence Models and Their Applications*; 10th Indo-German Winter Academy, Department of Mechanical Engineering, Indian Institute of Technology: Chennai, India, Dec 11-17, 2011 (Guide: Prof. Franz Durst).
7. Zubanov, V. M.; Stepanov, D. V.; Shabliy, L. S. In *The Technique for Simulation of Transient Combustion Processes in the Rocket Engine Operating with Gaseous Fuel "Hydrogen and Oxygen,"* International Conference on Information Technologies in Business and Industry; IOP Publishing Ltd.: UK, Sept 21-26, 2016.
8. Pinho, C. E. L., Delgado, J. M. P. Q.; Pilao, R.; Conde, J.; Pinho, C. *Numerical Study of Propane-air Mixture Combustion in a Burner Element, Defect and Diffusion Forum;* Trans Tech Publications: Switzerland, 2008; Vol. 273–276, pp 144–149.
9. Donini, A. Advanced Turbulent Combustion Modeling for Gas Turbine Application. PhD Dissertation, Eindhoven University of Technology, Netherlands, 2014.
10. Chung, T. J. *Computational Fluid Dynamics*; Cambridge University Press: Cambridge, UK, 2002.
11. Harris, C. K.; Roekaerts, D.; Rosendal, F. J. J.; Buitendijk, F. G. J.; Daskopoulos, Ph.; Vreenegoor, A. J. N.; Wang, H. Computational Fluid Dynamics for Chemical Reactor Engineering. *Chem. Eng. Sci.* **1996,** *51* (10), 1569–1594, June 1–6, 2008, Hong Kong.
12. Kaushal, P.; Sharma, H. K. Concept of Computational Fluid Dynamics (CFD) and Its Applications in Food Processing Equipment Design. *J. Food Process. Technol.* **2012,** *3* (1), 2–7.
13. Abdi, A. M.; Jassim, E.; Haghighi, M.; Muzychka, Y. Applications of CFD in Natural Gas Processing and Transportation (Chapter). In *Computational Fluid Dynamics;* Hyoung, W. O. H., Ed.; InTech Publishers: Croatia, 2010; p 420.
14. Yusof, N. B. H. The Study of Multiphase Flow for Petroleum Production Using Computational Fluid Dynamics (CFD). Bachelor of Chemical Engineering (Gas Technology) Thesis, Faculty of Chemical and Natural Resources Engineering, University Malaysia, Pahang, Malaysia, 2013.

WEB REFERENCES

https://en.wikipedia.org/wiki/Composite_material.
www.polymerjournals.com/PolymersPolymerComposites.asp.
https://www.intechopen.com/...polymers.../introduction-of-fibre-reinforced-polymers-.
onlinelibrary.wiley.com › Materials Science › Composites.

https://www.slideshare.net/RichardPradeep/polymers-and-polymer-composites.
nptel.ac.in/courses/113105028/32.
stmjournals.com/Journal-of-Polymer-and-Composites.html.
https://www.hindawi.com/journals/ijps/si/569490/cfp/.
https://www.tandfonline.com/toc/gcfd20/current
www.inderscience.com/jhome.php?jcode=PCFD.
www.mcgill.ca/mecheng/facultystaff/staff/wagdihabash/journal.
www.engpaper.com/cfd-2016.htm.
www.scirp.org/journal/ojfd/.
https://www.hindawi.com/journals/ijce/2013/917373/.
www.scimagojr.com/journalsearch.php?q=14393&tip=sid.
https://www.cfd-online.com/Links/finddocs.html.

CHAPTER 13

EVOLUTIONARY COMPUTATION, MULTIOBJECTIVE OPTIMIZATION, SIMULATED ANNEALING, AND COMPOSITES AND POLYMER SCIENCE: A SHORT AND FAR-REACHING REVIEW

SUKANCHAN PALIT[1,2,*]

[1]*Department of Chemical Engineering, University of Petroleum and Energy Studies, Energy Acres, P.O. Bidholi via Premnagar, Dehradun 248007, Uttarakhand, India*

[2]*43, Judges Bagan, P.O. Haridevpur, Kolkata 700082, India*

[*]*E-mail: sukanchan68@gmail.com; sukanchan92@gmail.com*

ABSTRACT

The world of science and technology is today moving from one visionary paradigm toward another. The challenges and the vision of science and technology are immense and groundbreaking. Applied science and applied mathematics are the hallmarks of scientific research today. Applied mathematical tools such as evolutionary computation are veritably changing the scientific landscape of optimization science. In this chapter, the author rigorously points toward the scientific vision behind evolutionary computation, multi-objective optimization, and multi-objective-simulated annealing. Multi-objective-simulated annealing is a latent yet a far-reaching area of scientific research pursuit today. The author deeply discusses the barriers of science in applications of multi-objective optimization and multi-objective-simulated annealing. Applied mathematics, computing, and evolutionary computation are today in a state of immense scientific rejuvenation as human civilization moves forward. Nontraditional optimization techniques are today changing the vast scientific scenario of endeavor and vision. Optimization science and diverse applications of optimization in engineering and science are changing the face of scientific research pursuit. Human scientific vision today is in a state of immense scientific disaster with the growing concerns of environmental disasters, loss of ecological biodiversity, and depletion of fossil fuel resources. Technology is highly challenged nowadays. In this chapter, the author delineated in lucid details the vast scientific challenges, the scientific aura, and the deep scientific profundity behind applied mathematics and computational techniques. Human scientific rejuvenation in computational techniques application is changing the face of scientific rigor today. This treatise pointedly focuses on the vast vision and the vast scientific regeneration in the field of evolutionary computation. The author in this treatise also opens up a new chapter in the field of composites and polymer science and the application of computational techniques such as optimization and simulated annealing. This technology of application of simulated annealing in manufacturing process is new yet highly advanced. This well-researched treatise opens up the scientific intricacies behind the world of composites and polymer science.

13.1 INTRODUCTION

Science and engineering today are moving forward at a rapid pace in the present day human civilization. Evolutionary computation and multi-objective optimization are veritably changing the scientific landscape of applied mathematics and computer science. As human civilization and human scientific endeavor surges ahead with new vision and newer innovations, computational techniques, and computer science assumes immense importance. Science today is a huge colossus with a definite vision and definite willpower of its own. In this treatise, the author rigorously points out toward the scientific success, the scientific potential, and the deep scientific profundity behind evolutionary computation applications in engineering and science. Today, chemical process engineering, petroleum engineering, and environmental engineering stand in the midst of disaster as well as deep comprehension. This chapter pointedly focuses on the intricacies and the barriers in mathematical applications of evolutionary computation, multi-objective optimization, and multi-objective-simulated annealing. The vision and challenge of computer science are enormous and far-reaching today. This treatise gleans and reveals the vast technological profundity of computational techniques and evolutionary computation with the sole aim of furtherance of science and engineering. Scientific research pursuit in the field of application of optimization and simulated annealing in manufacturing processes of composites and polymer science are dealt in lucid details in this chapter.

13.2 THE AIM AND OBJECTIVE OF THIS STUDY

The challenge and aim of engineering science are immense and far-reaching today. Human civilization and human scientific endeavor today stands in the midst of deep introspection and vision. The sole purpose of this study goes beyond scientific imagination and targets the intricacies of evolutionary computation. The world of optimization science is witnessing immense scientific upheavals and scientific adjudication. The main vision and the objective of this study surpasses scientific imagination and scientific vision. The target of this scientific endeavor is to bring to the forefront the scientific intricacies of multi-objective optimization, multi-objective-simulated annealing, and other avenues of evolutionary computation and differential evolution. Human scientific endeavor and human progress today stand in the midst of vision and deep cognizance. The authors in this treatise widely proclaim the immense scientific intricacies and the scientific forbearance in the application of genetic algorithm (GA) in solving optimization problems. Multi-objective-simulated annealing and its immense scientific endeavor are the other cornerstones of this well-researched treatise. Human scientific and technological vision and deep scientific discerning in applied science and applied mathematics are in a state of new rejuvenation. This treatise opens up newer avenues and newer innovations in the field of evolutionary computation and differential evolution.

Engineering science and applied science are today in the process of major restructuring and upheaval. Computer science and information technology are today in the need of immense restructuring as civilization faces immense scientific travails. The major focal point of this well-researched treatise is to deeply emancipate the success of application of evolutionary computation and optimization in designing petroleum engineering and chemical engineering systems. Today, human scientific ardor and vision in design of engineering systems are highly challenged. Technological reframing as regards application of evolutionary computation are the hallmarks of science and technology today. This chapter deeply pronounces the deep scientific ingenuity and the vast scientific needs of evolutionary computation and optimization in the success of science. The other vast areas of this treatise are the domain of composite and polymer science and its vast manufacturing processes.

13.3 EVOLUTIONARY COMPUTATION AND FUTURE OF SCIENCE AND TECHNOLOGY

Evolutionary computation and its immense scientific avenues today stand in the midst of vision and scientific fortitude. The intricacies and hurdles of research pursuit in evolutionary computation are witnessing immense regeneration and upheavals. Science and technology of mathematical techniques are moving at a rapid pace surpassing wide and vast visionary frontiers. Evolutionary computation, multi-objective optimization, and multi-objective-simulated annealing are the hallmarks of this well-researched treatise. The future of science and technology of applied mathematics is bright and groundbreaking. Energy engineering and computer science are ushering in a new era of scientific regeneration and deep scientific rejuvenation. Sustainability, whether it is energy or environment, is the utmost need of human society today. Sustainable development and the futuristic vision of science and engineering will all lead a long and visionary way in the true emancipation of applied science, applied mathematics, and computational science. Human scientific vision and scientific and academic rigor are highly challenged today as civilization moves forward. In this treatise, the author with immense vision and farsightedness reiterates the success of evolutionary computation application in the future human scientific progress of civilization. Technological barriers, scientific steadfastness, and vast scientific travails are the necessities of human civilization and scientific endeavor today. The future of science and technology today lies in the vast realm of computer science and applied mathematics. Mankind's immense scientific grit, the definite vision, and human scientific endurance will all lead a long and visionary way in the true emancipation of engineering science and technology today. Science is today deeply ironical and vastly visionary as nuclear age, space technology, and information technology surge ahead surpassing one scientific paradigm over another. Computational techniques and applied computer science are the needs of human scientific research pursuit today.

Evolutionary computation, nontraditional computational techniques, and bio-inspired computation are the future revolutionary directions in science and technology today. This treatise deals deeply with immense scientific profundity the veritable wonder of computer science, computational techniques, and information technology in the road toward furtherance of science and engineering in the present century.

In today's present day human civilization, the success of technology and engineering science today veritably depends upon the scientific urgency, the vast scientific provenance and the visionary realm of computer science and diverse avenues of engineering. In this treatise, the author with deep scientific fervor delineates the vast area of validation of petroleum engineering and chemical process engineering with the veritable need of computational techniques in designing engineering systems.

13.4 WHAT DO YOU MEAN BY MULTI-OBJECTIVE OPTIMIZATION?

Applied mathematics and computational science are today in the path of a newer scientific rejuvenation and a newer scientific doctrine. Human scientific progress and the application areas of engineering science are today linked by an unsevered umbilical cord. Applied mathematics, evolutionary computation, and computational science have diverse areas of applications in engineering and science. Multi-objective optimization and multi-objective-simulated annealing are the two visionary scientific frontiers. Human civilization and deep scientific research pursuit in applied mathematics are changing the scientific firmament of vision and introspection today. In this treatise, the author with deep and cogent insight, enumerates the immense scientific potential, the scientific forbearance, and the vast vision behind both multi-objective optimization and multi-objective-simulated annealing.

Multi-objective optimization is an area of multicriteria decision-making that is concerned with mathematical optimization problems involving more than one objective function to be simultaneously optimized.

13.5 WHAT DO YOU MEAN BY MULTI-OBJECTIVE-SIMULATED ANNEALING?

Multi-objective-simulated annealing is one of the visionary areas of scientific research pursuit in applied mathematics and engineering science today. Multi-objective-simulated annealing is a branch of mathematical endeavor which encompasses the tenets of metallurgical engineering and materials science. Vision of science, the challenges of computational techniques, and the futuristic vision of applied mathematics will all lead a long and visionary way in the true emancipation and the true realization of computer science and mathematical tools such as multi-objective-simulated annealing. Human civilization and human scientific research pursuit today stands in the midst of deep scientific revelation and vast scientific vision. The scientific prowess of mathematical techniques, the vast scientific determination, and the futuristic vision of evolutionary computation will surely lead a long and visionary way in the true emancipation and the true realization of engineering science and applied mathematics. Science today has a definite and purposeful vision which is crossing vast boundaries. Technology and engineering science are huge colossus with a definite vision and definite will power of its own.

Simulated annealing is a highly probabilistic technique for approximating the global optimum of a given function. Importantly, it is a metaheuristic to approximate global optimization in a large search space. In this chapter, the technological profundity of multi-objective-simulated annealing is delineated in minute details.

13.5.1 WHAT DO YOU MEAN BY COMPOSITE SCIENCE AND POLYMER SCIENCE?

A composite material is a material which is formed from two or other materials of different physical and chemical properties, which when combined, forms a material having properties distinct from the other materials. The new material may be preferred for many reasons: common examples include material which are stronger, lighter, and less expensive when compared to traditional materials. Typical engineered composite materials include (1) reinforced concrete and masonry, (2) composite wood such as plywood, (3) reinforced plastics, (4) ceramic matrix composites, (5) metal-matrix composites and (6) advanced composites. A polymer is a large molecule or macromolecule composed of many repeated subunits. Because of their broad range of properties, both synthetic and natural polymers play essential roles in everyday life. The challenge and the vision of composite manufacturing processes are immense and versatile. In this chapter, the author with vast scientific conscience unravels the scientific intricacies in the field of computational tools in composite manufacturing processes.

13.6 SCIENTIFIC DOCTRINE AND SCIENTIFIC COGNIZANCE BEHIND MULTI-OBJECTIVE OPTIMIZATION

Technological advancements in the 21st century are surpassing vast and versatile scientific frontiers. The challenge and the vision of optimization science need to be reenvisioned and revamped with the passage of scientific history and timeframe. Multi-objective optimization and multi-objective-simulated annealing are today the frontiers of applied mathematics. Human civilization and human scientific endeavor today stands in the midst of vision and scientific fortitude. Applied mathematics and computer science are in the path of newer scientific rejuvenation and deep scientific determination. Technology and science of optimization are huge colossus with a definite vision and deep scientific profundity of its own. The scientific doctrine and the deep scientific cognizance of multi-objective optimization today need to be reshaped and reenvisaged. The technological and scientific profundity needs to be addressed and envisioned as regards application of energy and environmental sustainability to human society today. Research and development initiatives in applied mathematics and computer science are today ushering in a newer era of vision and deep cognizance. Research and development forays into multi-objective optimization and multi-objective-simulated annealing are surpassing vast and versatile scientific frontiers. Technology has few answers to the scientific intricacies and the deep scientific vision behind applied mathematics and the science of optimization. Design of petroleum engineering and chemical engineering systems today veritably needs the application of multi-objective optimization and multi-objective-simulated annealing. Over the last few decades, optimization has garnered immense importance and deep scientific vision. In this treatise, the author repeatedly stresses on the vast scientific success, the immense scientific imagination and the utmost need of optimization science in the furtherance of science and engineering. Today, the world of science stands in the midst of deep scientific comprehension and unending scientific vision. Scientific doctrine and vast scientific cognizance in the field of multi-objective optimization are today surpassing scientific frontiers and are today involved in cross-boundary research. Scientific cognizance in the field of optimization and multi-objective optimization are changing the vast scientific firmament. Petroleum engineering, chemical process engineering, and energy technology are today the needs of human civilization. Energy engineering primarily renewable energy technology is the fountainhead of scientific justification and scientific judgment in present day human civilization. In these areas also, multi-objective optimization applications in design of engineering systems assumes vast importance. Solar energy, biomass energy, and energy from sea waves are the today's avenues of scientific research which needs to be vastly envisioned and reenvisaged with the passage of human history and visionary timeframe.

13.7 SCIENTIFIC VISION BEHIND MULTI-OBJECTIVE-SIMULATED ANNEALING

Multi-objective-simulated annealing is a branch of scientific endeavor which is replete with vision and scientific forbearance. Human civilization's immense scientific acuity, scientific prowess, and the futuristic vision will all lead a long and visionary way in the true emancipation of optimization science today. Multi-objective-simulated annealing is an avenue of science which is a branch of multi-objective optimization. Applied mathematics, applied science, and computer science are today in the path of newer innovation and newer scientific rejuvenation. Human scientific progress in present day civilization is in the path of glory and scientific revival. The world of challenges and the scientific intricacies in evolutionary computation applications today are in the road to newer scientific and academic rigor. Multi-objective-simulated annealing is a comparatively new branch of scientific endeavor. The scientific needs of the human society, the vast scientific motivation, and the ultimate futuristic vision of mathematical tools will veritably lead a long and visionary way in the true validation of evolutionary computation applications.

13.8 TECHNOLOGICAL VISION, SCIENTIFIC PROFUNDITY, AND SCIENTIFIC GRIT BEHIND MULTI-OBJECTIVE-SIMULATED ANNEALING

Technological and scientific visions in the research pursuit in multi-objective-simulated annealing are in the path of regeneration and deep rejuvenation. The scientific grit and the futuristic vision of optimization science are the pallbearers toward a greater visionary era in the field of applied mathematics. Scientific profundity and deep scientific vision are the hallmarks of research pursuit. In this treatise, the author rigorously points out toward the human scientific progress in applied mathematics with a deep vision toward furtherance of science and engineering. Mathematical modeling is the need of the hour in the furtherance of science and engineering of chemical process engineering and petroleum engineering science. The technology of mathematical modeling is today surpassing vast and versatile scientific frontiers. In this treatise, the author rigorously points out toward the intricacies of the science of mathematical modeling with greater concentration on multi-objective optimization and multi-objective-simulated annealing. Today, the world of science and technology is moving toward one visionary paradigm toward another. The engineering world of human civilization and scientific endeavor needs to be reframed and reenvisioned as basic engineering sciences and applied sciences witnesses immense scientific upheavals. Technological vision, scientific profundity, and vast scientific grit behind multi-objective-simulated annealing are ushering in a new era in computer science, applied mathematics, and engineering sciences. Simulated annealing and bio-inspired optimization are the veritable challenges of today. Engineering sciences such as the avenues of chemical process engineering and petroleum engineering are the veritable needs of human society today. The world today stands in the midst of deep scientific introspection and immense scientific travails and barriers. As space technology and nuclear engineering usher in a new era in the world of science and technology, petroleum engineering and chemical process engineering need to be reenvisioned and reenvisaged with the passage of scientific history and time.

13.9 THE SUCCESS OF SCIENCE OF SUSTAINABILITY

Sustainability in any form needs to be reenvisioned and restructured with the passage of scientific history and time. Energy sustainability and environmental sustainability are the areas of grave concern in present day human civilization. Energy engineering, electrical engineering science, and chemical process engineering are in the path of newer scientific regeneration and deep rejuvenation. Climate change, loss of ecological biodiversity, and the immediate needs of the human society are the pallbearers toward a newer visionary era in the field of petroleum engineering science and chemical process engineering. Loss of fossil fuel resources is a danger to human civilization and is a cause of grave scientific concern. Sustainable development and infrastructure development are the immediate necessities of human scientific endeavor in present day human civilization. Technology of petroleum engineering science is highly advanced today and crossing vast and versatile scientific frontiers. The success of sustainability science, the futuristic vision of energy engineering and environmental engineering science, and the utmost needs of human society will lead a long and visionary way in the true realization and a deep emancipation of science and technology in present day human civilization. Sustainability as defined by Dr. Gro Harlem Brundtland, former Prime Minister of Norway, today needs to be restructured and realigned as human scientific endeavor surges ahead. Environmental and energy sustainability are the needs of human society today. Human scientific endeavor's immense prowess, the vision, and the challenge of engineering science and the energy needs of human society will today lead a long and visionary way in the true realization of science of sustainability. The science of sustainability today is in the process of newer scientific regeneration and deep scientific rejuvenation. Human's immense scientific vision and prowess, the vast technological foresight, and the need of energy sustainability to human society will lead a long and visionary way in the true realization of energy engineering and energy technology. Energy engineering and technology can be classified into renewable energy and nonrenewable energy avenues. Sustainability whether it is energy or environmental can only be successfully realized if technology and engineering science can be envisioned and restructured with the passage of scientific history and time. Applied mathematics and computer science are moving toward a newer generation of scientific vision and deep scientific introspection.

13.10 ENERGY SUSTAINABILITY AND NEED OF MATHEMATICAL TOOLS

Mathematical techniques in design of chemical and petroleum engineering systems are changing the face of scientific firmament today. Human civilization and human scientific research pursuit today stands in the midst of scientific provenance and vision. Energy and environmental sustainability are the cornerstones of human civilization today. Sustainable development primarily energy and environmental sustainability today are in the state of immense scientific provenance and restructuring. Mathematical tools, the needs for scientific sustainability, and the future vision of energy engineering, petroleum engineering, and chemical process will lead a long and visionary way in the true realization of energy sustainability. Today is the age of nuclear science and space technology. Evolutionary computing today is a wonder of science. The civilization's immense scientific prowess and the wonders of computational science are the forerunners toward a greater visionary era in engineering science and technology. Human scientific progress and human scientific genre are the challenges and vision of computer science and engineering. Energy sustainability and the application of mathematical tools in diverse areas of technology and engineering science are the two opposite sides of the visionary coin today. Petroleum refining and petroleum engineering science are two areas of scientific research pursuit where optimization and bio-inspired optimization have vast applications. Design of chemical engineering and petroleum engineering systems also can be done with the help of mathematical tools such as multi-objective optimization and multi-objective-simulated annealing.

13.11 RECENT SCIENTIFIC ENDEAVOR IN APPLICATION OF MULTI-OBJECTIVE OPTIMIZATION

Science and technology of multi-objective optimization are surpassing vast and versatile scientific frontiers. Technology and engineering science are highly challenged today as human civilization moves forward. Multi-objective optimization and applied mathematics are today in the path of newer scientific regeneration as human scientific endeavor and human scientific civilization faces immense challenges. Research pursuit in applied mathematics and chemical process engineering, the vast scientific prowess of human, and the veritable needs of human society will all lead a long and visionary way in the true realization of science and technology today. Multi-objective optimization can be applied in the design of chemical engineering and petroleum engineering systems. Multi-objective-simulated annealing is not a new concept yet it is visionary. In this treatise, the author rigorously points out toward the vast scientific success and the technological profundity in the application of optimization science in the fields of chemical process engineering and petroleum engineering science.

Elnashaie and Elshishini[1] deeply discussed with cogent insight modeling, simulation, and optimization of industrial fixed-bed reactors.

Technological advancements, the scientific ingenuity of optimization science, and the needs of computational tools will lead a long and visionary way in the true emancipation of design of chemical engineering and petroleum engineering systems. Design of a fluidized as well as fixed-bed reactor involves the application of mathematical tools such as multi-objective optimization and bio-inspired computation or GA.[1] Human scientific endeavor today is in the path of newer regeneration and newer innovation. This treatise deals primarily with the mathematical modeling of industrial fixed-bed reactors which is extremely important in petroleum refinery and petrochemical industries. The vision and the challenge of multi-objective optimization science are today ushering in a new era in the field of science and engineering. Some of the important areas of application of modeling in this treatise are the ammonia production line which consists of about seven fixed-bed gas–solid catalytic reactors. Human scientific vision, deep scientific profundity, and scientific provenance are of immense necessity in the path toward scientific research pursuit in chemical process engineering and petroleum engineering science today. The list of fixed-bed catalytic reactors used in the fertilizer industries and petroleum industries is endless, and despite some of their valid disadvantages and the common and theoretically justified criticisms directed toward these types of contacting equipment in favor of more sophisticated configurations such as bubbling fluidized beds, circulating fluidized beds, and so on; they remain by far the most dominant configuration in petrochemical and petroleum refining industries.[1] Today the world of energy sustainability and energy security are in the path of immense scientific barriers and vast scientific introspection. This entire book gives a vast glimpse of the success of mathematical modeling in the design of petroleum engineering and chemical engineering systems.[1]

Khandalekar[2] deeply depicted with vast scientific farsightedness control and optimization of fluidized catalytic cracking (FCC) processes. The vast scientific ingenuity and deep profundity of the engineering science of FCC processes are depicted poignantly in this thesis. Today is the world of deep scientific vision and scientific rejuvenation. Technological profundity are at its helm as science and engineering surges forward surpassing visionary frontiers.[2] The fluidized catalytic cracking unit (FCCU) receives multiple feeds consisting of high boiling components from several other refining units and cracks these streams into lighter and more valuable components. Human civilization and human scientific endeavor today stand in the midst of scientific revelation and deep scientific discernment. In this treatise, the author discusses the control and optimization strategies in the operation of a FCCU. After further processing in the initial stages of FCCU, the FCCU product streams are blended with streams from other refinery units to produce a number of products, for example, various grades of gasoline.[2] The vast vision of control and optimization of FCCUs are of deep and utmost necessity as engineering science of petroleum refining moves forward. Technological validation and scientific profundity in the field of design of a FCCU are the hallmarks of this thesis. Due to its large throughput and ability to produce gasoline which is more valuable than the feed, economic operation of FCCU is of vital importance in the overall running of a petroleum refinery. Generally, the fluid catalytic cracking (FCC) process can be divided into three major sections: (1) reactor riser, (2) regenerator, and (3) main fractionators. In a reactor riser, the feed is injected into a hot stream of regenerated catalyst.[2] The reaction occurs over a short contact time in a riser before the catalyst and reacted products are separated in the riser.[2] The vision and the challenge of scientific endeavor and deep scientific validation are today ushering in a new era in the field of FCCU operation. Science today is replete with immense prudence and vast prowess as human civilization moves forward. The catalyst residence time in the riser is of 10 s. The short contact time minimizes the excessive deactivation of catalysts and gives better yields of more enriched product such as gasoline.[2] Human scientific ingenuity and the vast vision of science are the torchbearers toward newer innovations and newer concepts in the field of petroleum refinery operation. Fluidized catalytic cracker is one of the most difficult processes in a petroleum refinery to control. It has multivariable, strongly interacting and highly nonlinear control problem. In addition, multiple hard equipment constraints and operating constraints are present. Technological challenges and scientific travails are the necessities of the greater emancipation of control and optimization of FCCU.[2] With its internal feedback loop created by the circulating catalyst and its veritable complex dynamic responses, the FCC is a truly robust and fascinating process to analyze and control. The challenge and the vision is ever-growing as science and engineering surges forward. The nonlinearity of the responses needs to be taken into account as the FCCU operation ushers in new concepts and new future thoughts. Economic objectives on manipulated variables, such as the need to maximize the charge rate or cracking severity, or both, compete with operating constraints and further complicate the control issue. The objectives of FCCU control systems are

1. safety reverse catalyst flow protection;
2. heat balance—maintain reactor and regenerator temperature;
3. carbon balance—catalyst regeneration to improve yields;
4. conversion—achieve most profitable level depending on economic constraints;
5. interactions—overcome process variable interactions to achieve effective control[2]; and
6. constraints—operate up to the physical limits restricting further profitability.[2]

In this thesis, the author deeply delineates the scientific success, the scientific vision, and the vast scientific profundity in the FCCU operation. Validation of science and engineering of petroleum refining are today moving in the path of newer vision and newer scientific regeneration. This thesis is a veritable eye-opener toward the vast unknown world of control and optimization. This challenge of FCCU operation gives a vast glimpse in the evergrowing issue of process control and optimization.

Occelli[3] discussed with deep and cogent insight advances in FCC along with testing, characterization, and environmental regulations. To refiners, the challenges and the vision are an ever-growing process. After overcoming oil supply limitations from Middle East politics and the obstacles of fuel reformulation and the rising crude prices, the industry is now facing is facing an increasing number of mandates by governmental bodies worldwide at a time when there is a decline in demand of transportation fuels based on traditional fossil feedstocks.[3] The immense scientific prowess of human research pursuit, the vast scientific profundity, and the needs of science in human progress will all lead a long way in the true emancipation and the true realization of petroleum engineering science today. FCC stands as a major scientific endeavor in the path toward true emancipation in petroleum engineering science today. Depletion of fossil fuel resources is a global issue of immense concern. Evolutionary computation, multi-objective optimization, and multi-objective-simulated annealing needs to be reenvisioned and reenvisaged along with petroleum engineering science and petroleum refining as global research and development initiatives surges forward.[3]

Sadeghbeigi[4] deeply discussed with immense lucidity FCC process and provides practical and proven recommendations to improve the performance and reliability of FCC operations. FCC process is today veritably linked with deep scientific vision and unending scientific ingenuity.

This entire book gives a wider reflection on the process description of FCCU. Human's immense scientific prowess, scientific judgment, and the technological needs of human society will today lead a long and visionary way in the true realization of energy security and energy sustainability. Petroleum engineering science and energy sustainability are two opposite sides of the visionary coin.[4] FCC process continues to play an important role in an integrated refinery as the primary conversion process. For many refiners, the cat cracker is the veritable key to profitability in that the successful operation of this unit determines whether or not the refiner can remain competitive in today's global market. The vision and the challenge of science, the deep scientific profundity of petroleum refinery, and the success of engineering will all today lead an effective way in the true emancipation of petroleum refining. In this book, the author discusses with cogent insight FCC feed characterization, FCC catalysts, chemistry of FCC reactions, unit monitoring and control, products and economics, project management and hardware design, trouble shooting, optimization, and lastly emerging trends in FCC.[4] Optimization science today is in the path of immense scientific regeneration. Today, multi-objective optimization and multi-objective-simulated annealing are linked with the design of petroleum and chemical engineering systems. This book gives a vast glimpse of the science of FCC process and optimization and veritably opens up new windows of innovation and scientific instinct.

Ray Chaudhuri[5] discussed with immense farsightedness the fundamentals of petroleum and petrochemical engineering. Scientific revelation and vast scientific vision are the hallmarks of this book. The author in this extensive research work dealt with composition and properties of crude oil, petroleum products and test methods, processing operations in petroleum refinery, the composition of lubricants and grease, the wide vision of petrochemicals, offsite facilities, power and utilities, heat-exchanger and pipe-still furnaces, distillation and stripping, extraction, elements of pipeline transfer facilities, instrumentation and control in a refinery, and plant management and economics. Today, mathematical tools are immensely used in design and operation of a petroleum-refining unit. These include multi-objective optimization and GA. Today, application of multi-objective-simulated annealing in operation and design of a petroleum-refining unit such as FCCU is a new innovation and a newer scientific endeavor. This entire book is a veritable eye-opener toward a newer scientific regeneration in the field of petroleum refining and mathematical techniques such as optimization and simulated annealing.[5]

The present day human civilization today stands in the midst of deep introspection and vision. Human scientific challenges today are immense as mankind faces the issue of depletion of fossil fuel resources and the wide world of energy security and energy sustainability. In this entire chapter, the author reiterates the scientific vision and the scientific divination in the application of applied mathematics in design of chemical engineering and petroleum engineering systems.

13.12 SIGNIFICANT RESEARCH ENDEAVOR IN MULTI-OBJECTIVE-SIMULATED ANNEALING

Multi-objective-simulated annealing is a branch of multi-objective optimization. Human society and human are today in the path of newer rejuvenation and newer scientific vision. Science and technology in present day human civilization stands in the midst of deep catastrophe and vast scientific comprehension. Scientific forays and scientific research pursuit in the field of multi-objective-simulated annealing are today surpassing cross-boundary research. This is an area of metallurgical engineering. Today, materials science and metallurgical engineering are the areas of frontier science and frontier engineering science. Multi-objective-simulated annealing can be applied to the design of chemical engineering and petroleum engineering systems also. This is a newer and visionary avenue of scientific research pursuit today. Human scientific vision, deep scientific profundity, and the vast aisles of scientific ingenuity will all lead a long and visionary way in the true realization of metallurgical engineering, chemical process engineering, and energy technology.

Smith[6] in his well-researched doctoral thesis delineated simulated annealing techniques for multi-objective optimization. The vision and the challenge of the science of optimization are today surpassing vast and versatile scientific frontiers. Computational optimization is today ushering in a new era in the field of applied mathematics and computational techniques. Simulated annealing is a single objective optimization technique, which is probably convergent, making it an important technique for extension to multi-objective optimization. Simulated annealing technique today is in the path of newer restructuring and newer envisioning.[6] Previous proposals for extending simulated annealing to the multi-objective case have mostly taken the form of a traditional single-objective-simulated annealer, optimizing a composite (often summed) function of the objectives. The vast vision of engineering science of optimization and GA are today veritably ushering in a new knowledge dimension in diverse areas of science and engineering. Scientific ingenuity and deep scientific provenance are the areas of utmost need in simulated annealing techniques and its application in today's scientific research pursuit. The author in the first part of the thesis deals with introducing an alternate method for multi-objective-simulated annealing, vastly dealing with the dominance relation which operates without assigning preference information to the objectives.[6] A widely detailed nongeneric improvements to this robust algorithm are presently with vision and ingenuity. This new method is shown to exhibit rapid convergence to the desired set, depending upon the properties of the problem. This new technique is widely applied to the commercial optimization of CDMA mobile telecommunication networks and is shown to perform well upon this problem.[6] The second stage of this thesis contains an investigation into the effects upon convergence of a range of optimizer properties. In the starting of this thesis, the motivation for investigating simulated annealing methods for multi-objective optimization is discussed in details. Many problems are optimization problems, whereby the configuration of a system must be determined which will veritably maximize the performance of the system. In this treatise, the author pointedly focuses on the newer innovations, the scientific adjudication, and the deep scientific vision behind simulated annealing applications in diverse areas of engineering and science. Simulated annealing and other novel areas of multi-objective optimization are changing the vast scientific firmament of endeavor and vision today.[6]

Bandopadhyay[7] deeply comprehended and discussed in detail a simulated annealing-based multi-objective optimization-based algorithm. This well-researched treatise describes a simulated annealing-based multi-objective optimization algorithm that incorporates the concept of archive to provide a set of tradeoff solutions for the problem under consideration.[7] Scientific vision and deep technological ingenuity are the veritable hallmarks of this paper. To determine the acceptance probability of a new solution, an elaborate procedure is followed that takes into account the domination status of the new solution with the current solution, as well as those in archive. A measure of the amount of domination between two solutions is also used for the purpose. A complexity analysis of the proposed algorithm is extensively provided. Simulated annealing is today in a state of immense scientific comprehension and deep vision. An extensive and farsighted study of the proposed algorithm with two other existing and well-known multi-objective evolutionary algorithms demonstrates and pronounces the effectiveness of the former with respect to five existing performance measures, and several test problems of varying degrees of mathematical difficulty.[7]

Rennard[8] deeply discussed in a comprehensive treatise nature-inspired computing for economics and management. Bio-inspired computing and nature-inspired computing are today in the avenues of newer scientific regeneration and vast scientific vision.[8] Almost all evolutionary economics models worked out in past few decades are dynamical ones and are focused on far-from-equilibrium analysis. Technological validation and deep scientific motivation are the pillars of this well-researched treatise. This treatise deeply discussed the evolutionary model of industrial dynamics.[8] The challenge and the vision of science need to be readdressed and reframed as human scientific endeavor and the path of human civilization surges forward. This evolutionary model describes the behavior of a number of competing firms producing functionally equivalent products. The author in this well-researched treatise deeply and lucidly comprehends the vastly relevant research domain of innovation and industry evolution. The most important weakness of the orthodox economics seems to be the lack of innovation and vision in their models and concentration on equilibrium analysis.[8] Innovation and vision can be considered as the pivot of modern economic evolution. Scientific ingenuity, vast scientific vision, and the wide scientific cognizance of evolutionary computation will all lead a long and visionary way in the true realization of energy security, energy sustainability, and the vast world of renewable energy technology. The author in this treatise presents only a small sample of the simulation results just to show how similarities of the proposed model's behavior relate to real industrial innovative processes. The targets and the vision of science of nature-inspired computing are today opening up new avenues of scientific thoughts, scientific imagination, and vast scientific profundity in decades to come.[8]

Chakraborti[9] discussed with lucid and cogent insight the unique contributions of multi-objective evolutionary and GAs in material research. This is a watershed treatise, which envelops and envisions the vision and the challenge of evolutionary computation applications in diverse areas of engineering and science.[9] The current state of the art of materials and metallurgical engineering research using multi-objective genetic and evolutionary algorithms are presented with critical analysis. The basic concepts of multi-objective optimization and Pareto optimality are delineated in simple terms and the advantages of an evolutionary approach are vastly emphasized.[9] Current materials research endeavor in this area is summarized, focusing the significant achievements till date and the specific needs for improvement. Emancipation of engineering science and technology are the veritable pillars of this well-researched treatise. Numerous gradient-based methods are available for calculating the pareto-optimality.[9] They however require calculating one optimal solution at a time. The targets and the vision of material science and metallurgical engineering are ever-growing and groundbreaking in present day human civilization. The author in this treatise with cogent insight ponders upon the scientific success, the scientific vision, and the scientific farsightedness in multi-objective optimization in materials research and metallurgical engineering forays.[9]

Ziane[10] deeply discussed with cogent insight simulated annealing optimization for multi-objective economic dispatch solution. This paper delineates a multi-objective-simulated annealing optimization to solve a dynamic generation dispatch problem.[10] Technological profundity, the scientific rigor, and academic excellence are the hallmarks of simulated annealing techniques and GA tools today. Science of optimization is ushering in a new era in engineering science. The authors in this paper with vast scientific vision discussed the importance of application of simulated annealing in the field of economics.[10]

Li and Landa-Silva[11] deeply discussed with scientific ingenuity evolutionary multi-objective-simulated annealing with adaptive and competitive search directions.[11] The vision and the challenge of simulated annealing are today opening up new directions of scientific thoughts and scientific doctrine in decades to come. In this paper, the authors propose a population-based implementation of simulated annealing to tackle multi-objective optimization problems, in particular those of combinatorial in nature. The proposed algorithm is called evolutionary multi-objective-simulated annealing algorithm, which veritably combines local and evolutionary approach by incorporating two distinctive features.[11] Human scientific vision, mankind's immense scientific prowess, and the world of scientific challenges will all lead a long and visionary way in the true emancipation and true realization of optimization science, applied mathematics, and computer science. The first feature is to tune the weight vectors of scalarizing functions (i.e., search directions) for selection during local search using a two-phase strategy.[11] The second feature is the competition between members of current population with similar weight vectors.[11] Technological advancements, deep scientific motivation, and the needs of applied mathematics toward human progress will go a long and effective way in the emancipation of engineering science and technology in present day human civilization.[11]

Alrefaei[12] deeply discussed with cogent insight simulated annealing for multi-objective stochastic optimization. Science and engineering of optimization are today entering into a newer visionary era with a drastic pace. In many areas of optimization, technology has few answers. This paper addresses the multi-objective stochastic optimization problem that arises in many real-world applications especially in the supply chain management and optimization. Abiding by this objective, a simulated annealing algorithm is presented and used for solving this problem. The algorithm uses the hill-climbing criterion to escape from local minimality trap.[12] Engineering science and technology are today surpassing one visionary scientific paradigm over another. This paper also introduces a new Pareto set for stochastic optimization problems and vastly demonstrates the application of simulated annealing on this Pareto set. This is a watershed text in the field of optimization and will surely lead a long and effective way in the true emancipation of applied mathematics and computational science.[12]

Safaei[13] in another well-researched paper delineates lucidly multi-objective-simulated annealing for a maintenance workforce scheduling problem as a case study. Scientific vision and deep scientific profundity are the hallmark of this paper. A multi-objective-simulated annealing algorithm is discussed in this paper to solve a real maintenance workforce scheduling problem aimed at simultaneously minimizing the workforce cost and maximizing the equipment availability.[13] Human scientific endeavor in optimization and applied mathematics are today in a visionary era of scientific regeneration. Simulated annealing and other nontraditional optimization techniques are challenging the scientific firmament today. This paper is a vast eye-opener to the success of science of optimization.

13.13 OPTIMIZATION, CHEMICAL PROCESS ENGINEERING, AND PETROLEUM ENGINEERING SCIENCE

The science and technology of optimization are today facing immense scientific challenges as human civilization and human scientific endeavor surges forward toward a newer visionary era. The scientific revelation and deep provenance of both chemical process engineering and petroleum engineering are today ushering in a new era in science today. Technology has diverse answers toward the success of multi-objective optimization and GA in the field of design of chemical engineering and petroleum engineering. Design of petroleum-refining units today stands in the midst of vision and deep scientific divination. Petroleum refining and petroleum engineering science are of utmost need in the path toward scientific emancipation today. Depletion of fossil fuel resources, environmental engineering catastrophes and loss of ecological biodiversity have urged the scientific domain today to gear forward toward newer vision and newer innovations. Thus, here comes the importance of applied mathematics, computational techniques, and mathematical tools. Multi-objective optimization, evolutionary computation, and GA will all lead a long and visionary way in the true emancipation of petroleum engineering science and chemical process engineering. Human scientific progeny, deep divination of science, and the scientific ingenuity behind application of computational techniques are the torchbearers toward a newer era in science and engineering.

13.14 GA, BIO-INSPIRED COMPUTATION AND THE FUTURE OF ENGINEERING SCIENCE

GA and bio-inspired computation today are in the path of newer scientific regeneration. Mimicking biology is a newer avenue of scientific research pursuit in computational science and applied mathematics. The vision and the challenge of bio-inspired computation are changing the face of human scientific endeavor and mathematical ingenuity. The future of engineering science and technology today lies in the hands of basic and applied science such as applied mathematics. Human scientific vision and technological motivation in GA and nontraditional mathematical techniques are the utmost need of the hour in the path of scientific research pursuit today. Global burning challenges such as depletion of fossil fuel resources and climate change are changing the face of environmental engineering and petroleum engineering. Design of a petroleum-refining unit such as FCCU and other unit operations in a petroleum refinery needs the contribution of computational techniques such as optimization, GA, and bio-mimicked optimization. In this chapter, the author reiterates the tremendous scientific potential and the deep technological ingenuity behind multi-objective optimization and its applications in chemical engineering and petroleum engineering. The author also stresses the immense importance of bio-inspired optimization in the design of engineered systems. Science and engineering are today huge colossus with a definite and a vast vision of its own. The future of computational science and engineering science lies at the hands of engineers and scientists. GA will witness a new beginning if there is an effective contribution of bio-inspired optimization in the scientific success of design of petroleum and chemical engineering systems.

Design of a petroleum-refining unit such as FCCU is a visionary scientific endeavor today. Unit operations in chemical engineering play a pivotal role in the furtherance of science and engineering of chemical process technology today. Petroleum engineering and chemical process engineering are two opposite sides of the visionary coin of scientific endeavor today. The challenge and vision today lie in the application of applied mathematics and computational techniques in the design of unit operations in chemical engineering and petroleum engineering operations. Petroleum refinery is a deep visionary and scientific imperative in the furtherance of global issues in energy security and energy sustainability. Environmental and energy sustainability are the needs of human today. Technology has few answers to the intricate questions of energy and environmental sustainability. Science and engineering are today in the midst of immense stress and deep catastrophes with the growing concerns of energy sustainability. Here comes the necessity of petroleum refinery and energy technology both renewable as well as renewable. Future of engineering science stands in the midst of immense vision and scientific divination. Human's immense scientific prowess thus is highly challenged as science surges forward. This entire treatise is a veritable eye-opener toward the scientific importance of energy engineering and computational science.

13.15 SIGNIFICANT SCIENTIFIC RESEARCH PURSUIT IN APPLICATIONS OF GA

GA and its applications today stand in the midst of deep scientific introspection and vast vision of engineering science. GA is a branch of bio-inspired optimization. Technological ingenuity and scientific profundity are the cornerstones of the visionary future of GA today. Today, GA is witnessing difficult challenges as regards its applications in design of engineering systems. Science today is vastly prudent and path breaking as regards application of mathematical tools in diverse areas of engineering and technology. In this section, the author with deep insight delineates the scientific success, the scientific challenges, and the vast scientific travails

GA today faces in the furtherance of avenues of engineering science. Nontraditional GA tools are the success of science today. This section also gives vast glimpse to the nontraditional areas of GA.

Kasat[14] deeply discussed with immense lucidity multi-objective optimization of industrial FCCUs in a petroleum refinery using elitist nondominated sorting GA. Deep scientific ingenuity and technological profundity are the hallmarks of this treatise. Today is the world of computational techniques and applied computer science. Nontraditional GA are changing the face of scientific research pursuit in design of petroleum-refining units. This study provides insights into the optimal operation of the FCCU. A five-lump model is used to characterize the feed and products. The model is tuned using industrial data.[14] Today, scientific vision in GA applications, the vast domain of petroleum refining and the utmost needs of energy technology will all lead a long and visionary way in the true emancipation of global energy sustainability and petroleum engineering science. The world of challenges in petroleum refining is opening up new windows of innovation and vision. In this research pursuit, the elitist nondominated sorting GA (NSGA-II) is used to solve a three objective function optimization problem. The hallmark of this paper is immensely visionary and groundbreaking and brings forward toward the scientific landscape a new era of scientific regeneration in computational techniques.[14]

Kasat and Gupta[15] opened a new chapter in the field of multi-objective optimization and GA. They lucidly delineated multi-objective optimization of an industrial FCCU using GA with the jumping genes operator.[15] Human scientific vision and vast scientific divination are in the path of immense rejuvenation. The multi-objective optimization of industrial operations using GA and its variants, often requires immense computational science. Thus, there is a need of GA and bio-inspired optimization. Because of the large computational time, any technique to speed up the computation is desirable. Scientific ingenuity is the utmost need due to this reason.[15] The binary coded elitist NSGA-II is adapted, and the new code, NSGA-II-JG is used to obtain solutions for the multi-objective optimization of an industrial FCCU.[15] Technological advancements and scientific forays are the torchbearers toward a newer visionary era and a newer innovation in GA. This research work is a watershed endeavor in the field of GA and veritably opens up a new window of scientific instinct and scientific regeneration.[15]

13.16 THE WORLD OF CHALLENGES AND THE VISION BEHIND BIO-INSPIRED COMPUTATION

Human scientific ingenuity and scientific research pursuit in bio-inspired computation are breaking vast scientific frontiers. The vision and the challenges are immense and deeply inspiring as technology and engineering surges forward. Bio-mimicking of optimization science and applications of nontraditional GAs are changing the face of scientific endeavor and deep scientific profundity today. In the entire chapter, the author rigorously points toward the human scientific vision in the design of engineering systems primarily chemical engineering and petroleum engineering systems. Environmental engineering and environmental protection is another area of GA applications today. The science of computational science mainly bio-inspired computation is the necessities of scientific research pursuit today. This chapter vastly comprehends the areas of applied mathematics, computer science, and engineering science in a visionary attempt toward furtherance of science and technology. Scientific provenance and deep scientific divination in optimization applications are delineated and discussed in minute details in this chapter. Human civilization and human scientific endeavor today stands in the midst of deep scientific introspection and catastrophe with the depletion of fossil fuel resources, climate change, and loss of ecological biodiversity. The world of scientific challenges and barriers lies in the hands of engineers and scientists. Thus, there is a need of applied mathematics and basic and applied science. Bio-inspired computation is a branch of deep scientific endeavor of applied mathematics and computer science. Today, the scientific world is technology driven. Computer science and applied mathematics are challenging the entire scientific landscape.

13.17 MULTI-OBJECTIVE-SIMULATED ANNEALING AND THE FUTURE OF COMPUTATIONAL SCIENCE

Computational science and computational techniques are the new innovations and the newer visionary avenues of research pursuit today. Human scientific vision needs to be reframed as regards application of optimization, multi-objective-simulated annealing, and computational techniques toward the emancipation of diverse areas of engineering science today. Science and technology are today in the path of newer scientific regeneration and today are huge colossus with a vast vision of its own. Today, the world is moving toward a newer direction in space technology and nuclear science. Energy and environmental sustainability should be the cornerstone of every research pursuit in present day human civilization. Computer science and computational techniques are today in the path of newer scientific regeneration. Human scientific vision, scientific challenges, and scientific travails are highly tested in present day human civilization. The science of optimization and its vast scientific cognizance are the parameters of utmost importance in the visionary path of petroleum engineering, chemical process engineering, and other diverse areas of engineering science. Design of chemical and petroleum engineering systems such as design of petroleum-refining units are of immense importance in the successful emancipation of energy sustainability today.

13.17.1 COMPOSITE SCIENCE, POLYMER SCIENCE AND THE SCIENTIFIC VISION OF TOMORROW

Composites and polymers are the smart and innovative materials of tomorrow. Human's immense scientific prowess, man's scientific grit, and determination and the futuristic vision of computational techniques will all lead a long and visionary way in the true emancipation of manufacturing processes of composites and polymers. The status and the vision of composites manufacturing are vast and groundbreaking. Application of optimization in composite manufacturing processes is challenging the vast scientific firmament. Composite science and polymer science will surely be the forerunners toward a newer era in computational techniques and material science.

13.18 FUTURE RECOMMENDATIONS OF THE STUDY AND FUTURE FLOW OF THOUGHTS

Evolutionary computation today is in the path of newer scientific regeneration. Human scientific vision, scientific, and academic rigor today stand in the midst of deep scientific introspection and scientific resurrection. Science of GA and multi-objective-simulated annealing are surpassing visionary scientific boundaries. Applied mathematics, chemical process engineering, and petroleum engineering are in the threshold of a new scientific genre, scientific motivation, and deep scientific paradigm. Science and engineering are today huge colossus with a definite and purposeful vision of its own. Loss of ecological biodiversity, climate change, loss of fossil fuel resources, and frequent environmental catastrophes are changing the root cause of scientific firmament. Future research in evolutionary computation is groundbreaking and needs to be restructured and revamped as science and engineering surges forward. Applied mathematics and computational techniques are ushering in a new era in science and engineering. The challenge and vision of computational science and applied mathematics today are vast and versatile. In this treatise, the author pointedly focuses on the scientific success, the vast vision, and the intricate challenges in the path toward scientific emancipation in evolutionary computation. Technology and engineering science has few answers to the application areas of evolutionary computation. This treatise is a vast overview of the vision behind evolutionary computation, multi-objective optimization, and multi-objective-simulated annealing. The crux of this treatise goes beyond scientific imagination and scientific discernment and thus ushers in a new era in applied mathematics and computer science. The science of optimization and GA are today in the path of newer scientific regeneration. Future of applications of optimization and applied mathematics in chemical process engineering, petroleum engineering, and other diverse areas of science and engineering are vast, varied, and groundbreaking.

13.19 FUTURE RESEARCH TRENDS IN EVOLUTIONARY COMPUTATION

Evolutionary computation today stands in the midst of immense scientific introspection and unending scientific imagination. Technology and engineering science are in the crossroads of new scientific history and a newer scientific divination. Computational science and applied mathematics are the challenges of the future. Scientific and academic rigor in the field of evolutionary computation and optimization are witnessing immense restructuring. Today, human scientific vision and human scientific progress are in a state of immense distress and catastrophe. Modern science and modern technology are today in the path of newer scientific reenvisioning. The vast challenge and vision of evolutionary computation, the imperatives of science and technology and the advancements of human society will today lead a long and visionary way in the true emancipation of applied science and applied mathematics. Future research paradigm in evolutionary computation and bio-inspired optimization should be targeted toward design of chemical and petroleum engineering systems. FCC reactor and fixed-bed catalytic cracking reactor design today wholly depend on the science of optimization and evolutionary computation. Simulated annealing and multi-objective optimization need to be reframed and restructured as human scientific endeavor moves toward a newer visionary paradigm. Evolutionary computation and the world of bio-inspired optimization are challenging the vast scientific fabric of vision, might, and forbearance today. The future of evolutionary computation is vastly visionary and has no bounds. This treatise unfolds the scientific success, the scientific and technological ingenuity and the immense scientific revelation behind evolutionary computation and multi-objective-simulated annealing.

Technology and engineering science are today in the process newer scientific regeneration and newer vision. Renewable energy technologies are the hallmarks of scientific endeavor and scientific profundity today. Future of energy engineering and fossil fuel technologies are targeted toward newer vision and newer innovations. In this chapter, the author depicts lucidly the vast scientific success, the scientific farsightedness and the vision behind applied mathematics applications in design of engineering systems. Thus, the world of challenges will surely usher in a newer eon in the field of computational tools such as simulated annealing.

13.20 CONCLUSION

The progress of human civilization and the scientific endeavor today stand in the midst of deep scientific introspection and scientific ingenuity. Energy technology, petroleum engineering, and chemical process engineering are the new avenues of scientific vision in present day human civilization. Computational tools, applied mathematics, and the world of engineering science are highly challenged as the human surges forward toward a newer era of vision and forbearance. The science of optimization and evolutionary computation today stand in the midst of deep scientific comprehension and unending vision. Today, the present day human civilization is an age of internet revolution, space technology, and nuclear engineering science. This entire treatise deeply delineates the success of applied mathematics and computational techniques toward greater emancipation of optimization science and diverse areas of technology. Evolutionary computation, multi-objective optimization, and simulated annealing are today in the path of immense scientific travails and vast scientific introspection. The challenges of human scientific research pursuit such as loss of ecological biodiversity and depletion of fossil fuel resources are changing the wide scientific firmament. The author in this treatise drastically challenges the deep-rooted scientific endeavor in multi-objective optimization and simulated annealing. Today, mathematicians and engineers are playing decisive roles in the true emancipation of optimization and bio-inspired optimization. GA and bio-inspired optimization are two decisive areas of scientific pursuit and deep scientific emancipation. The author with deep and cogent insight unravels the scientific success and the vast technological profundity in the application of bio-inspired optimization in the areas of engineering science and technology. The scientific innovation, the scientific instinct, and the deep scientific progeny will surely go a long and visionary way in the true realization and the true rebuilding of optimization science and GA. Provision of basic human needs such as water, energy, and electricity stand as scientific imperatives toward a greater emancipation of engineering science and technology. Engineering science, petroleum engineering, and chemical process engineering are the visionary avenues of research pursuit today. This treatise widely opens up new windows of innovation and scientific instinct in decades to come. The vast world of composite science and polymer science are elucidated in deep details in this chapter. The vision of composite manufacturing process is immense and groundbreaking. This chapter will surely be an eye-opener in the application of mathematical tools in engineering science.

KEYWORDS

- **optimization**
- **multi-objective**
- **simulated**
- **annealing**
- **composites**
- **vision**

REFERENCES

1. Elnashaie, S. S.; Elshishini, E. H. *Modeling, Simulation and Optimization of Industrial Fixed Bed Catalytic Reactors*; Gordon and Breach Publishers: Langhorn, PA, 1993.
2. Khandalekar, P. D. Control and Optimization of Fluidized Catalytic Cracking Process. Master of Science Thesis, Texas Technological University, USA, 1993.
3. Occelli, M. L. *Advances in Fluid Catalytic Cracking-Testing, Characterization and Environmental Regulations*; CRC Press; Taylor and Francis Group: USA, 2010.
4. Sadeghbeigi, R. *Fluid Catalytic Cracking Handbook: Design, Operation and Troubleshooting of FCC Facilities*, 2nd ed.; Gulf Professional Publishing: USA, 2000.
5. Chaudhuri, U. R. *Fundamentals of Petroleum and Petrochemical Engineering*; CRC Press; Taylor and Francis Group: USA, 2011.
6. Smith, K. I. A Study of Simulated Annealing Techniques for Multi-objective Optimization. Doctoral Thesis, University of Exeter, Exeter, UK, 2006.
7. Bandopadhyay, S.; Saha, S.; Maulik, U.; Deb, K. A Simulated Annealing Based Multiobjective Optimization Algorithm: AMOSA. *IEEE Trans. Evolution. Comput.* **2008,** *12* (3), 269–283.
8. Rennard, J.-P. *Handbook of Research on Nature-inspired Computing for Economics and Management*; Idea Group Reference: Hershey, PA, 2007.
9. Chakraborti, N. Critical Assessment. 3: The Unique Contributions of Multi-objective Evolutionary and Genetic Algorithms in Materials Research. *Mater. Sci. Technol.* **2014,** *30* (11), 1259–1262.
10. Ziane, I.; Benhamida, F.; Graa, A. Simulated Annealing Optimization for Multi-objective Economic Dispatch Solution. *Leonardo J. Sci.* **2014,** *13* (25), 43–56.
11. Li, H.; Landa-Silva, D. Evolutionary Multi-objective Simulated Annealing with Adaptive and Competitive Search Direction. In *IEEE Congress on Evolutionary Computation (CEC 2008)*; pp 3310–3317.
12. Alrefaei, M.; Diabat, A.; Alawneh, A.; Al-Aomar, R.; Faisal, M. N. Simulated Annealing for Multiobjective Stochastic Optimization. *Int. J. Sci. Appl. Inf. Technol.* **2013,** *2* (2), 18–21.
13. Safaei, N.; Banjevic, D.; Jardine, A. K. S. Multiobjective Simulated Annealing for a Maintenance Workforce Scheduling Problem: A Case Study. In *Simulated Annealing*; Cher Ming Tan, C., Ed.; I-Tech Education and Publishing: Vienna, Austria, 2008.
14. Kasat, R. B.; Kunzru, D.; Saraf, D. N.; Gupta, S. K. Multiobjective Optimization of Industrial FCC Units Using Elitist Nondominated Sorting Genetic Algorithm. *Ind. Eng. Chem. Res.* **2002,** *41*, 4765–4776.
15. Kasat, R. B.; Gupta, S. K. Multi-objective Optimization of an Industrial Fluidized Bed Catalytic Cracking Unit (FCCU) Using Genetic Algorithm (GA) with Jumping Genes Operator. *Comput. Chem. Eng.* **2003,** *27*, 1785–1800.

WEB REFERENCES

https://en.wikipedia.org/wiki/Polymer.
https://www.britannica.com/science/polymer.
www.mdpi.com/journal/polymers.

pslc.ws/macrog/kidsmac/basics.htm.
https://en.wikipedia.org/wiki/Composite_material.
compositeslab.com/composites-101/what-are-composites/.
https://www.sciencedirect.com/journal/composites.
www.pslc.ws/macrog/composit.htm.
www.rsc.org/Education/Teachers/Resources/Inspirational/resources/4.3.1.pdf.
www.acmanet.org/composites/what-are-composites.
https://compositesuk.co.uk/composite-materials/introduction.
https://en.wikipedia.org/wiki/Mathematical_optimization.
www.dictionary.com/browse/optimization.
tutorial.math.lamar.edu/Classes/CalcI/Optimization.aspx.
https://en.wikipedia.org/wiki/Simulated_annealing.
https://www.mathworks.com › Documentation Home › Simulated Annealing.
mathworld.wolfram.com › Applied Mathematics › Optimization.

SECTION IV
Biobased Composites

CHAPTER 14

PASSIVE AERATED COMPOSTING OF LEAVES AND PREDIGESTED OFFICE PAPERS

A. Y. ZAHRIM*, S. SARIAH, R. MARIANI, I. AZREEN, Y. ZULKIFLEE, and A. S. FAZLIN

Chemical Engineering Programme, Faculty of Engineering, Universiti Malaysia Sabah, Jalan UMS, 88400 Kota Kinabalu, Sabah, Malaysia

Corresponding author. E-mail: zahrim@ums.edu.my

ABSTRACT

Composting studies were conducted in lab scale for 40 days. The substrates of the compost are anaerobically treated palm oil mill effluent, leaves, and paper. The compost process was conducted in cuboid reactor with an effective volume of 0.4 m^3. During composting process, the highest temperature of 43°C was achieved at day 4. The physiochemical analyses were conducted to determine the stability and maturity of the compost. Phytotoxicity experiment was also conducted by using cabbage seed (*Brassica oleracea*) to analyze the maturity of the compost. As a result, the parameters such as moisture content, total organic carbon, and C/N were significantly decreased. The overall mass reduction was 66% and the organic matter degradation was 72.13%. The germination index of the compost was increasing at the end of the period. It shows that the compost were phytotoxic free. Overall, nutrients in the compost product of this study were in the ideal range but some of the parameters are not in the range to achieve the maturity of the compost.

14.1 INTRODUCTION

Palm oil industry is the fourth largest revenue-generating sector in Malaysia.[16] In 2016, oil palm-planted area reached 5.74 million hectares, with 1.7% increment reported as compared to the previous year. Sabah is the largest oil palm-planted state, with 1.55 million hectares or 27% of the total oil palm-planted area, followed by Sarawak with 1.51 million hectares or 26%, and Peninsular Malaysia (with 11 states) accounted for 2.68 million hectares or 47% of the total planted area.[25]

The palm oil industries significantly contribute to the economic growth and escalate the standard of living among the South-East Asian countries, especially in Malaysia. Increase of the crude palm oil production from 13.2% to 17.32 Mt had positively influenced the exports price in the major market by 7.3% to RM 64.58 billion from 60.17 billion in 2015.[25] In other perspective, palm oil mill processes industry also yields a significant pollution load. Fortunately, a large portion of the pollution load can be discharged by the generation of palm oil mill effluent (POME) alongside.[39] It was reported that 1 ton of processed fresh fruit bunch (FFB) will generate different types of wastes including the empty fruit bunch (23%), mesocarp fiber (12%), shell (5%), and POME (60%).[2]

POME is one of the final by-products of the palm oil production in a mill. It is described as liquid effluent discharges from the palm oil mill which appears as a thick brownish liquid at a temperature ranging between 80 and 90°C and a pH between 4 and 5.[53] It comprises a mixture of water, oil, total solids, and total suspended solids which are about 28% of FFB.[2] Besides that, it has a high content of suspended and dissolved organic matter (OM) and has a chemical oxygen demand of about 15,000–100,000 mg/L, averaged at 51,000 mg/L.[35] About 36% of total POME is from the combination of the wastewater discharged from sterilizer condensate, 60% of total POME is from the clarification wastewater, and the remaining 4% of POME is from the hydrocyclone wastewater.[54]

POME is the most difficult and expensive waste to manage by the mill operators. This is due to large volumes in tons that are generated at a time. The easiest and cheapest method for disposal, raw POME or partially treated POME is being settled into nearby river or land. Nevertheless, excessive amount of untreated POME deplete the water's oxygen and suffocate aquatic life.[35] This situation creates an environmental issue for the palm oil mill industry due to its overwhelming polluting characteristics. Many small and big rivers have been devastated by such disposal method as people living downstream are usually affected. POME gives adverse impact to the environment because it has a high chemical and biochemical oxygen demand and mineral content such as nitrogen and phosphorous which can cause severe pollution to the environment.[3] The dark brown color of POME also gives a bad muddy water perception especially among the villagers in a rural area.

Composting of POME is an attractive approach in converting waste into organic fertilizers. Composting can be defined as an aerobic microbiological process that converts the organic substances of wastes into stabilized humus and less complex compounds.[50] Cocomposting is different from composting as cocomposting is the simultaneous composting of two or more types of waste material.[8,29] From cocomposting, the potential has an added benefit of enhancing end-compost quality.[28] In addition, composting or cocomposting of wastes will provide conversion of wastes into a valuable product that will serve as a soil conditioner or fertilizer.[1] There were many contributing factors such as carbon to nitrogen ratio, moisture content, pH, and aeration rate that affected the composting process.[50]

Some of the challenging tasks for solid waste disposal are disposing the palm leaves and the paper product especially in metropolitan areas in most developing countries.[24] It will give adverse impact to the environment if this waste is not treated well. Several composting studies related to leaves and paper have been reported to best process the waste using recycling method.[9,31,44,50] Composting method using anaerobically digested POME (AnPOME) is also well known in a number of studies.[6,16,49]

In this study, the AnPOME is utilized as a hydrolyzing agent for leaves–paper mixture. The performance of the process was evaluated based on temperature, moisture content, OM losses, total organic carbon (TOC), mass reduction, pH level, electrical conductivity, zeta potential, and phytotoxicity.

14.2 MATERIALS AND METHODS

14.2.1 MATERIALS

Shredded paper was collected around the offices at Universiti Malaysia, Sabah. Meanwhile, POME was collected from the anaerobic pond number two in Beaufort Palm Oil Mill. The mill is located at Beaufort, Kota Kinabalu, Sabah. The fresh leaves were collected from Taman Indah Permai, Kota Kinabalu residential area. The initial weight of the compost was 249.9 kg. Table 14.1 shows the weight of each material used for composting process.

TABLE 14.1 Weight of Compost Material.

Compost material	Weight (kg)
Paper	40.0
Leaves	25.0
POME	181.4
Compost starter	3.5
Total	249.9

POME, palm oil mill effluent.

14.2.2 COMPOSTING

Prior to composting process, the paper was digested with POME for 3 days. Next, the digested paper was mixed with fresh mango leaves, followed by adding 3.5 kg of compost starter to enhance biodegradation process.[50] Then, mixture of composting materials was poured into a reactor (Fig. 14.1).

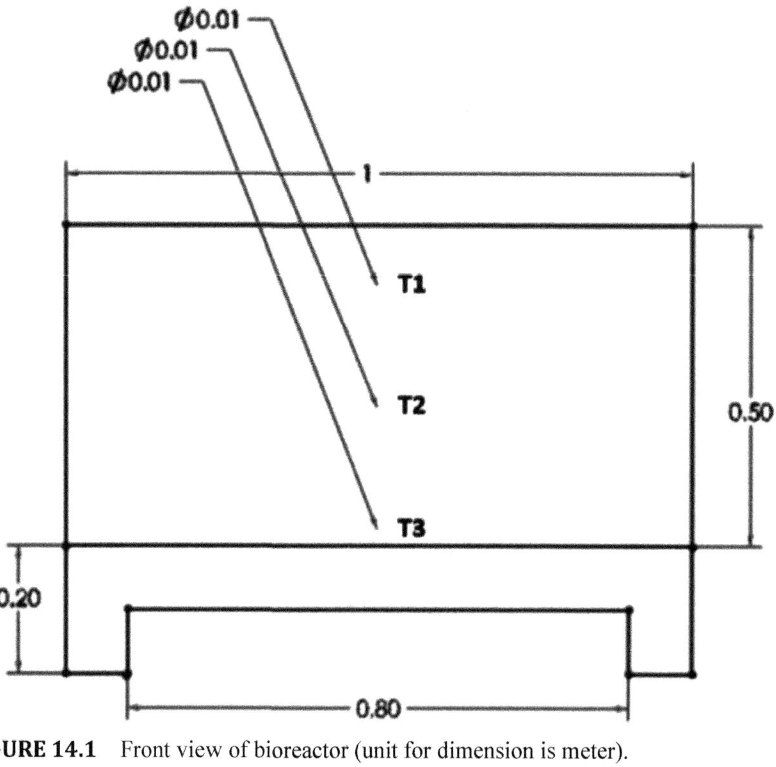

FIGURE 14.1 Front view of bioreactor (unit for dimension is meter).

A laboratory scale composting reactor (effective volume = 0.4 m^3) was built using wood. The base was built using cement to provide stability to the reactor. The reactor specification is 1 m × 1 m × 0.5 m (length × width × height). There were holes at the base of the reactor for leachate discharging as shown in Figure 14.2.

FIGURE 14.2 Base of the reactor.

The composting study was conducted at Environmental Lab, Block B, Faculty of Engineering, Universiti Malaysia, Sabah. The composting process was conducted for a period of 40 days. The sampling was taken every 10 days for physicochemical analysis.

14.2.3 PHYSIOCHEMICAL ANALYSIS

Temperature of the compost was measured daily over the period of 40 days, by using digital thermometer (Prima Long Thermometer). Temperature reading was taken at three different positions which are upper (T_1), core (T_2), and bottom (T_3) (Fig. 14.1). The ambient temperature was also recorded.

The sampling was conducted every 10 days of the composting. About 100 g of samples were collected from the center of the composting reactor. The analysis of the fresh samples was performed immediately after taking them out of the reactor. The leaves and mixture of paper and POME were weighed separately. Leaves were cut into small pieces and mixed together with the mixture of paper and POME.

Moisture content analysis was performed by drying 10 g of samples in oven at 105°C for 24 h.[47] The dried sample was then weighed and the moisture content (%) was calculated by using the formula given in the following equation:

$$\text{Moisture content (\%)} = \frac{W_{\text{crucible+sample (before drying)}} - W_{\text{crucible+sample (after drying)}}}{W_{\text{sample}}} \times 100 \quad (14.1)$$

OM and ash concentrations were determined by dry combustion at 550°C for 4 h in furnace (Thermolyne 46100 model).[47] The OM was determined as volatile solid. The percentage of ash of the sample is calculated by using the following equation:

$$\text{Ash (\%)} = \frac{W_{\text{crucible+sample(after burning)}} - W_{\text{crucible}}}{W_{\text{sample}}} \times 100\% \quad (14.2)$$

The percentage of TOC and percentage of OM were determined by eqs 14.3 and 14.4[55]:

$$\text{TOC (\%)} = \frac{100 - \text{Ash (\%)}}{1.8} = \frac{\text{\% Organic matter}}{1.8} \quad (14.3)$$

$$\text{OM loss (\%)} = 100 - 100\frac{\left[X_1\left(100 - X_n\right)\right]}{\left[X_n\left(100 - X_1\right)\right]} \quad (14.4)$$

where X_1 and X_n were the initial ash content and the ash content on each corresponding day.

A quantity of 10 g sample was mixed with 100 mL of distilled water and then filtered to obtain water-soluble extract. The water-soluble extract was used for the pH, conductivity, zeta-potential analysis, and phytotoxicity test.[41,50] A multiparameter meters (Hanna Instrument [Model: HI 9811-5]) was used to determine the pH and conductivity of the solution.

14.2.4 ZETA POTENTIAL

The zeta potential of the compost was determined on days 0, 10, 20, 30, 40, and 50 by analyzing water–extract solution. The solution was tested by using a Malvern-Zetasizer Nano Series model ZS machine to determine the zeta potential of the compost.

14.2.5 PHYTOTOXICITY ASSAY

Seed germination technique based on cabbage seeds (*Brassica oleracea*) was applied to determine the phytotoxicity of the compost. The cabbage seeds were soaked in water for 72 h to let the seeds sprouting. Only the sprouted seeds were used in the experiment.

Water-soluble extract (15 mL) and 15 cabbage seeds were placed on petri-dish that were previously covered with filter paper. Control sample was also run simultaneously by using 5 mL of distilled water instead of compost extract. The petri dishes were placed in the dark space at room temperature for 72 h. The number of germinated seeds was calculated and the growth of roots was monitored. The percentage of relative seed germination, relative root growth, and germination index (GI) were calculated using the following equations:

$$\text{RSG}\,(\%) = \frac{\text{number of seeds germinated in sample extract}}{\text{number of seeds germinated in control}} \times 100 \tag{14.5}$$

$$\text{RRG}\,(\%) = \frac{\text{root length in sample extract}}{\text{root length in control}} \times 100 \tag{14.6}$$

$$\text{GI}\,(\%) = \frac{\text{RSG} \times \text{RRG}}{100} \tag{14.7}$$

14.2.6 NUTRIENT ANALYSIS OF THE COMPOST

The nutrient content of the compost was analyzed by Sime Darby Research Sdn. Bhd located at Balung Tawau, Sabah, Malaysia.

14.2.7 STATISTICAL ANALYSIS

The average value and standard deviation of the data were calculated using Microsoft Excel. The standard error was computed and errors bars were determined for the data.

14.3 RESULTS AND DISCUSSION

14.3.1 TEMPERATURE PROFILE

The temperature variation of the compost (T_1, T_2, and T_3) in comparison with ambient temperature over the period of 40 days composting are shown in Figure 14.3. The data show a typical temperature profiles of the compost as reported previously in a number of other composting studies.[26,27,33,36,50] The temperature profiles can be represented as the stages of composting, microbial activity and is used to describe the conditions suitable for the proliferation of different microbial groups, that is, meso- and thermophiles.[21]

In this study, the compost temperatures varied between 26.6°C and 43.0°C, whereas the ambient temperature ranged from 26.7°C to 29.8°C. Basically, there are four phases that exist in the dynamics of composting, namely, temperature evolution, mesophilic, thermophilic, cooling, and maturation.[33] An increase in temperature from 26.6°C to 33.0–33.7°C during the first day indicated that the system was in the first stage. A further increase in temperature was recorded on the following day which resulted in the liberation of heat on the sides of the composter indicating the presence of biological activity. The compost temperature (T_1) reached the highest temperature of 43.0°C on day 4. This was possibly due to less ventilation in the upper part of the reactor and thus more heat might have been trapped that led to higher temperature. The energy produced in the system is due to microbial decomposition of OM is the main cause of increase in temperature.[46] On the other hand, the maximum compost temperature (T_3) at the bottom part of the reactor did not exceed 40°C. This part mainly undergoes mesophilic composting, which maintain a composting temperature less than 40°C. Since this compost (T_3) reached thermophilic temperature, it can effectively destroy pathogens and weed seed, and stable humus-like substance can be formed by converting the biodegradable solid organic into it.[52] After thermophilic phase, temperatures declined slowly to mesophilic temperature and indicated that the microbial activity has become weaker.

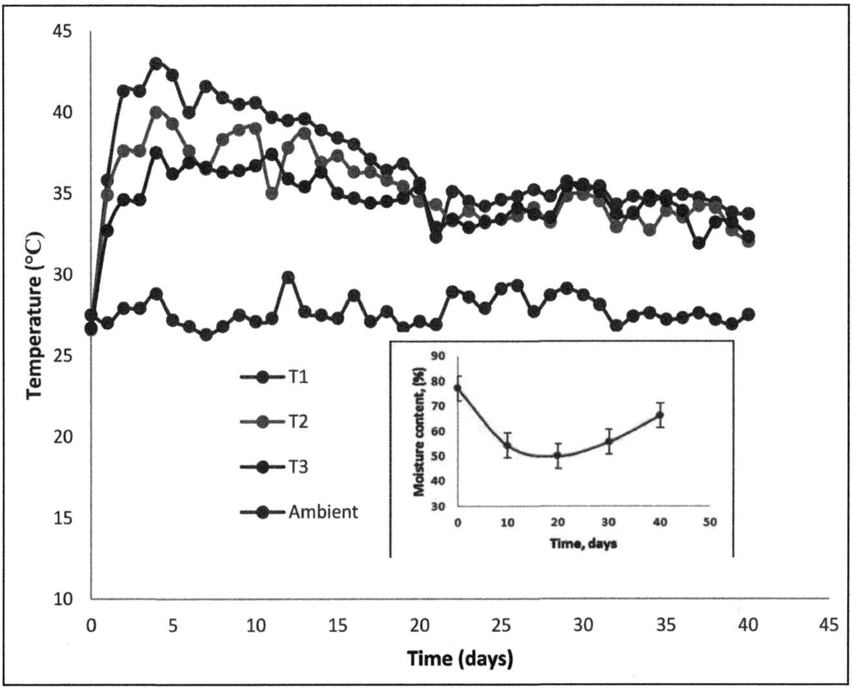

FIGURE 14.3 (See color insert.) Temperature trend during the first 40 days of the composting process. Small figure shows the moisture content profile in the compost mixture (standard errors (SE) in vertical bars).

14.3.2 MOISTURE CONTENT

Moisture content is necessary to provide a medium for the transport of dissolved nutrients required for the metabolic and physiological activities of microorganisms.[13] Physical and chemical properties of the waste material change with moisture content which acts as a transporting medium of nutrients for microbial activity.[23] The variation in moisture content values is shown in Figure 14.3. The trend of the moisture content observed in this work is similar to the findings reported by Zahrim et al.[50] Rawoteea et al.[33] and Villegas and Huiliñir.[43] According to Lin,[20] the minimal requirement for obtaining adequate microbial activities is 50%. The initial moisture content of the paper–leaves AnPOME was 77.22% and the value decreased gradually to 50% on the 20th day due to evaporation caused by thermophilic fermentation. One indication that OM decomposed was that the moisture content continuously decreased during composting.[17] This is the observation obtained during the first 20 days of the composting process. From day 20 to 40, the moisture content of the compost increased by 16% due to the lower heat generation. Consequently, the condensed vapor that was attached to the back of the reactor's lid fell back into the mixture.[42] From the results, final compost moisture content obtained was 66.15%. Since the compost moisture content did not fell below 50%, water was not added into the reactor to maintain 50–60% moisture content. It should be noted that adequate environment must be maintained to enhance microbial fermentation.

14.3.3 ORGANIC MATTER DEGRADATION DURING THE COMPOSTING PROCESS

The degradation of OM was calculated through the contents of ash (Table 14.2) to evaluate the cocomposting performance. This analysis is a valuable indication for a successful composting process.[51] OM mineralization was investigated by determining the losses of OM with the time course during cocomposting. It was found that the losses of OM increased rapidly with time and these OM losses were detected at the end of composting process (72.13%). The OM loss in this study was comparable to the results reported by Zhang et al.[51] which had a maximum OM loss of 61.66%. Other than that, Petric and Mustafić[30] reported values of 37–50% in composting of wheat straw with poultry manure. Moreover, the OM loss in composting of olive-mill pomace and poultry manure with tomato harvest stalks also reported to be in the range of 21.7–46.1%.[37]

TABLE 14.2 Ash Content in Composts During Compost Processes (Dry Weight).

Composting time (days)	0	10	20	30	40
Average ash content (%)	15.28	27.60	18.16	28.0	38.85

The rate of decomposition in the composting process can also be determined through measurement of TOC. The TOC is used as the energy sources for microorganisms during composting process and the degradation of TOC could be used to illustrate the level of compost maturity.[51] From Figure 14.4, it can be seen that the TOC (%) decreases slightly throughout the composting process. The decrease in TOC is related to the microbial respiration or amount of carbon dioxide released which depends on the degree of mineralization through microbial activities.[17,18] Thus, a larger decrease in TOC would correspond to a larger degree of decomposition by the microorganisms. In this study, a significant TOC loss was observed during the first 10 days of composting process. This is possibly due to the increased degradation of easily degradable fractions, whereas the steady decrease from day 20 onward may be influenced by hardly degradable fractions.[30]

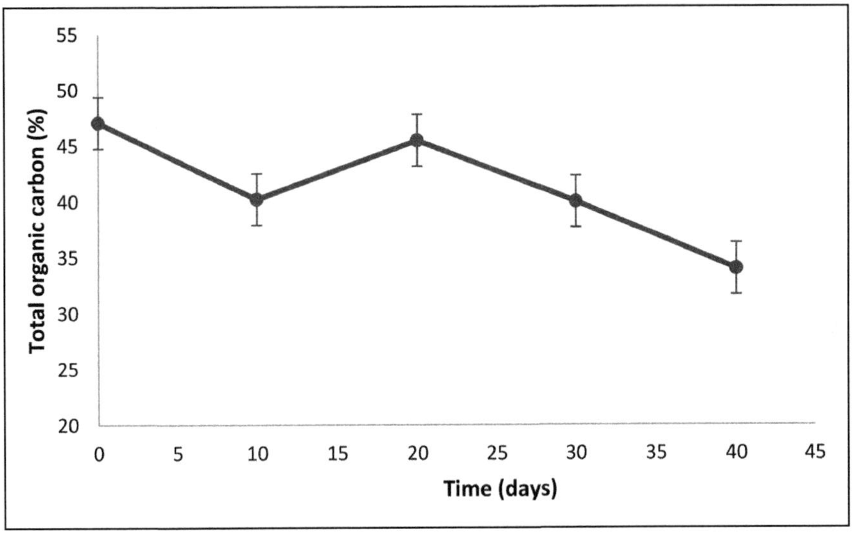

FIGURE 14.4 Profile of total organic carbon during composting process (SE in vertical bars).

14.3.4 *MASS REDUCTION*

Reactor composting was efficient in treating the total amount of 249.9 kg of paper–leaves AnPOME in 40 days, resulted in a 85-kg compost. The mass reduction (66%) after 40 days is very significant and the value is in the range of other studies (Table 14.3). The reason for these losses could be due to evaporation of water and microbial activity (C and N gas emissions).[36]

TABLE 14.3 Mass Reduction and OM Losses in Several Composting Process.

Substrate	Composting time (days)	Composting system	Highest temperature (°C)	Mass reduction (%)	OM losses (%)	References
Paper (16%), leaves (10%), anPOME (73%), and compost starter (1%)	40	Passive aerated (0.4 m^3)	43	66	72.13	This study
Paper (31%), grass-clipping (46%) and anPOME (23%)	40	Passive aerated (0.05 m^3)	31	18	ND	[50]
Vegetable (70%), meat (10%), and paper (20%)	140	Passive aerated (0.2 m^3)	58	7.1	65.13	[5]
Olive leaves (8%) and olive humid husks (92%)	90	Passive aerated (15 m^3)	22	ND	6.4	[4]
Food scrap (88%) and dry leaves (12%)	154	Passive aerated (0.2 m^3)	45.6	73.27	ND	[15]
Leaves	52	Forced aerated (0.002 m^3)	60	ND	46	[14]
Poultry manure (75%) with chestnuts leaves and burrs (25%)	103	Passive aerated (0.1088 m^3)	60	ND	6.27	[11]
Dairy cattle manure (71%) + wallboard paper (29%)	28	Passive aerated (0.431 m^3)	70–75	11.45	ND	[34]

anPOME, anaerobically digested POME; ND, not determined; OM, organic matter.

14.3.5 pH AND ELECTRIC CONDUCTIVITY

Figure 14.5 shows the pH profile and conductivity of the sample during the composting process. Based on results, the value of pH was decreasing until 6.9 at day 30. The decline in pH value at the early stage was due to establishment of anaerobic condition that causes the formation of organic acids.[7] At the end of the composting process, the pH value increased to alkaline range. This was due to the consumption of protons during the decomposition of volatile fatty acids, the generation of carbon dioxide, and the mineralization of organic nitrogen.[7]

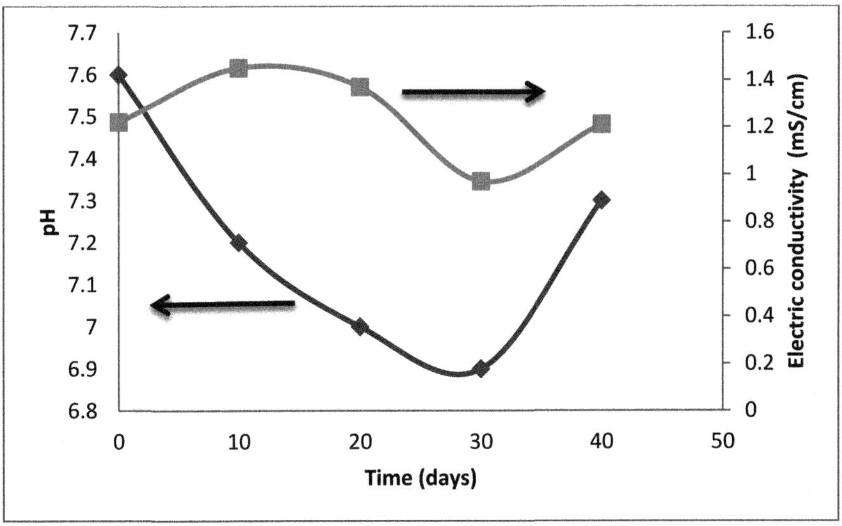

FIGURE 14.5 pH profile and electric conductivity during the composting process.

Electric conductivity reflects the salinity and suitability of the compost product to be used in agricultural industry. The electric conductivity of this study was at the range of 1.45–0.97 mS/cm. The decreasing trend of electric conductivity was due to the increasing concentration of nutrients, such as nitrate and nitrite.[7] However, the electric conductivity was at an acceptable level in terms of safe applications for plant growth for which the limit was 2.5 mS/cm.[12]

14.3.6 ZETA POTENTIAL

The zeta potential is the potential difference between the dispersion medium and the stationary layer of fluid attached to the dispersed particles.[50] It indicates the surface characteristics and electric potential variation of the residue surface during decomposition period.[19,45] Figure 14.6 shows the zeta-potential profile over the composting period. In this study, negative values of surface charge (zeta potential) were obtained in the whole composting process. As the composting progresses, the zeta-potential value of the compost decrease gradually from −8.855 mV to −20.4 mV in the first 40 days. The lowest zeta-potential value of −25.6 mV was observed on the last day of composting. The zeta potential of compost would decrease since POME is negatively charged from the beginning and tends to be more negative in the following treatments, that is, in aerobic ponds, due to the degradation of OM.[48]

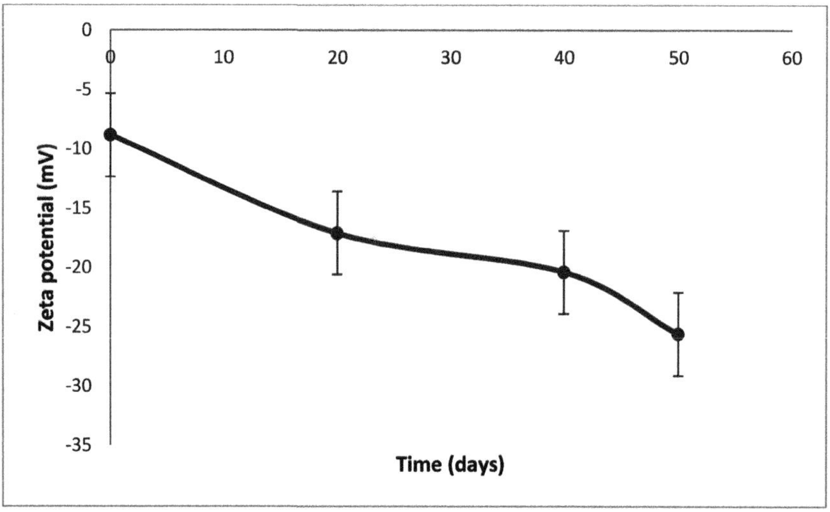

FIGURE 14.6 Zeta-potential profile during composting processes (SE in vertical bars).

14.3.7 PHYTOTOXICITY

Figure 14.7 shows increasing trend of GI profile during the first 20 days of composting period. The reason for that is the decline of the electric conductivity. Excessive salinity affects the phytotoxicity of the compost directly, but depending on the types of seeds used in the germination test. Different types of plant having different salinity tolerance. Salinity effects can be negligible when the electric conductivity readings are below 2 mS/cm.[40]

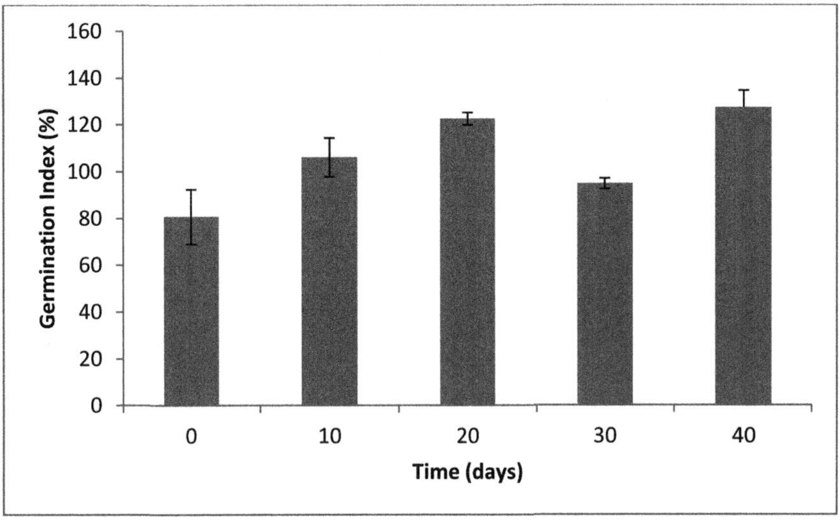

FIGURE 14.7 Profile of germination index during composting (SE in vertical bar).

However, the GI was reduced to 94.7% at day 30. The factor that could be the cause of this behavior was the pH value. The pH value at day 30 was 6.9 which was slightly acidic. The acidic condition showed the accumulation of organic acids as a result of biodegradation of soluble carbon. The organic acids were a toxic compound and have the adverse effect on the seed germination.[10,40]

Finally, the highest GI of 127.26 was achieved at day 40. Throughout the composting period, the GI was above 80% which indicated that it was phytotoxin free and had complete maturity.[32,40]

There are many reasons that effected the phytotoxicity of the compost including concentration of volatile organic acids, concentration of NH_4–N, oxygen depletion or presence of heavy metal, and the molecular weight of organic compound.[12,32] Most of these factors influence seed germination simultaneously and it is difficult to assess which parameter determines the greatest influence.[40] Besides, the GI should be interpreted with caution because the value was affected by the type of seed used and the application extraction rate.[32]

14.3.8 COMPOSTING NUTRIENT

Table 14.4 shows the initial and final nutrients content of compost produced in this work, in comparison with previous research. It can be seen that the moisture content and pH of the compost are within the range of other findings which is 50–70% for moisture content and 6–9 for pH. The decreasing value of C/N shows that the biodegradation of the OM occurred during the composting process. Nutrients in compost can improve soil condition and consequently help to reduce the dependency of inorganic fertilizers. Generally, the nutrient content including nitrogen, potassium, and sodium was increasing at the end of the composting period.

TABLE 14.4 Initial and Final Nutrient Content of the Compost Product.

Composition	Initial	Final	Recommended value for final ideal substrate (references)
Moisture content (%)	77	66	40–55 [49]
pH	7.6	7.3	5.5–6.5 [47]
Conductivity (mS/cm)	1.22	1.21	<3.0 [38]
C/N	44	29	15–20 [7]
TOC (%)	47	33	–
N (%)	1.06	1.12	–
P (%)	0.07	0.16	>0.5 (as P_2O_5) [47]
K (%)	0.21	0.72	>1.5 (as K_2O) [47]
Mg (%)	0.1	0.24	>70 mg/L [22]
Ca (%)	ND	5.08	>200 mg/L [22]
B (%)	ND	0.112	0.05–0.5 mg/L [22]

ND, not determined; TOC, total organic carbon.

14.4 CONCLUSION

In this study, a composting process was successfully conducted over 40 days using a mixture of paper, leaves, and AnPOME as the substrate. Up to 66% of mass reduction of the compost was obtained. During composting process, the highest compost temperature achieved was 43°C on day 4 of composting. The moisture contents decrease from 77% to 66%, and the pH value of the compost produced is 7.3. Both TOC and C/N decreased toward the end of composting period. The negative zeta-potential values decreased from −8.855 to −20.4 mV due to organic degradation. The results show that phytotoxicity of the compost decreases as the GI increases. Interestingly, the compost with GI of 127.26% was successfully produced.

KEYWORDS

- **composting**
- **palm oil mill effluent**
- **paper**
- **leaves**
- **phytotoxicity**
- **germination index**

REFERENCES

1. Abu Qdais, H.; Al-Widyan, M. Evaluating Composting and Co-Composting Kinetics of Various Agro-Industrial Wastes. *Int. J. Recycl. Organ. Waste Agric.* **2016,** *5*, 273–280.
2. Ahmad, A.; Buang, A.; Bhat, A. H. Renewable and Sustainable Bioenergy Production from Microalgal Co-Cultivation with Palm Oil Mill Effluent (POME): A Review. *Renew. Sustain. Energy Rev.* **2016,** *65*, 214–234.
3. Ahmed, Y.; Yaakob, Z.; Akhtar, P.; Sopian, K. Production of Biogas and Performance Evaluation of Existing Treatment Processes in Palm Oil Mill Effluent (POME). *Renew. Sustain. Energy Rev.* **2015,** *42*, 1260–1278.
4. Alfano, G.; Belli, C.; Lustrato, G.; Ranalli, G. Pile Composting of Two-phase Centrifuged Olive Husk Residues: Technical Solutions and Quality of Cured Compost. *Bioresour. Technol.* **2008,** *99*, 4694–4701.
5. Arrigoni, J. P.; Paladino, G.; Laos, F. Feasibility and Performance Evaluation of Different Low-Tech Composter Prototypes. *Int. J. Environ. Protect.* **2015,** *5*, 1.
6. Baharuddin, A. S.; Wakisaka, M.; Shirai, Y.; Abd-Aziz, S.; Abdul Rahman, N. A.; Hassan, M. A. Co-composting of Empty Fruit Bunches and Partially Treated Palm Oil Mill Effluents in Pilot Scale. *Int. J. Agric. Res.* **2009,** *4*, 69–78.
7. Bazrafshan, E.; Zarei, A.; Kord Mostafapour, F.; Poormollae, N.; Mahmoodi, S.; Zazouli, M. A. Maturity and Stability Evaluation of Composted Municipal Solid Wastes. *Health Scope* **2016,** *5*, e33202.

8. Das, M.; Uppal, H. S.; Singh, R.; Beri, S.; Mohan, K. S.; Gupta, V. C.; Adholeya, A. Co-composting of Physic Nut (*Jatropha curcas*) Deoiled Cake with Rice Straw and Different Animal Dung. *Bioresour. Technol.* **2011,** *102*, 6541–6546.
9. Francou, C.; Linères, M.; Derenne, S.; Villio-Poitrenaud, M. L.; Houot, S. Influence of Green Waste, Biowaste and Paper–Cardboard Initial Ratios on Organic Matter Transformations During Composting. *Bioresour. Technol.* **2008,** *99*, 8926–8934.
10. Gómez-Brandón, M.; Lazcano, C.; Domínguez, J. The Evaluation of Stability and Maturity during the Composting of Cattle Manure. *Chemosphere* **2008,** *70*, 436–444.
11. Guerra-Rodríguez, E.; Diaz-Raviña, M.; Vázquez, M. Co-Composting of Chestnut Burr and Leaf Litter with Solid Poultry Manure. *Bioresour. Technol.* **2001,** *78*, 107–109.
12. Himanen, M.; Hänninen, K. Composting of Bio-Waste, Aerobic and Anaerobic Sludges: Effect of Feedstock on the Process and Quality of Compost. *Bioresour. Technol.* **2011,** *102*, 2842–2852.
13. Ishak, N. F.; Ahmad, A. L.; Ismail, S. Feasibility of Anaerobic Co-Composting Empty Fruit Bunch with Activated Sludge from Palm Oil Mill Wastes for Soil Conditioner. *J. Phys. Sci.* **2014,** *25*, 77.
14. Michel Jr., F. C.; Adinarayana Reddy, C.; Forney, L. J. Yard Waste Composting: Studies Using Different Mixes of Leaves and Grassin a Laboratory Scale System. *Compost. Sci. Utiliz.* **1993,** *1*, 85–96.
15. Karnchanawong, S.; Nissaikla, S. Effects of Microbial Inoculation on Composting of Household Organic Waste Using Passive Aeration Bin. *Int. J. Recycl. Organ. Waste Agric.* **2014,** *3*, 113–119.
16. Krishnan, Y.; Bong, C. P. C.; Azman, N. F.; Zakaria, Z.; Othman, N. A.; Abdullah, N.; Ho, C. S.; Lee, C. T.; Hansen, S. B.; Hara, H. Co-composting of Palm Empty Fruit Bunch and Palm Oil Mill Effluent: Microbial Diversity and Potential Mitigation of Greenhouse Gas Emission. *J. Clean. Prod.* **2016,** *146*, 94–100.
17. Kulcu, R.; Yaldiz, O. Determination of Aeration Rate and Kinetics of Composting Some Agricultural Wastes. *Bioresour. Technol.* **2004,** *93*, 49–57.
18. Kulikowska, D. Kinetics of Organic Matter Removal and Humification Progress During Sewage Sludge Composting. *Waste Manag.* **2016,** *49*, 196–203.
19. Li, J.; Lu, J.; Li, X.; Ren, T.; Cong, R.; Zhou, L. Dynamics of Potassium Release and Adsorption on Rice Straw Residue. *PLoS One* **2014,** *9*, e90440.
20. Lin, C. A Negative-pressure Aeration System for Composting Food Wastes. *Bioresour. Technol.* **2008,** *99*, 7651–7656.
21. Liu, K.; Price, G. Evaluation of Three Composting Systems for the Management of Spent Coffee Grounds. *Bioresour. Technol.* **2011,** *102*, 7966–7974.
22. López-López, N.; López-Fabal, A. Scientia Horticulturae Compost Based Ecological Growing Media According EU Eco-label Requirements. *Sci. Hortic.* **2016,** *212*, 1–10.
23. Manu, M.; Kumar, R.; Garg, A. Performance Assessment of Improved Composting System for Food Waste with Varying Aeration and Use of Microbial Inoculum. *Bioresour. Technol.* **2017,** *234*, 167–177.
24. Moh, Y. C.; Manaf, L. A. Overview of Household Solid Waste Recycling Policy Status and Challenges in Malaysia. *Resour. Conserv. Recycl.* **2014,** *82*, 50–61.
25. MPOB. *Overview of the Malaysian Oil Palm Industry*, 2016 ed.; 2016.

26. Onursal, E.; Ekinci, K. Co-composting of Rose Oil Processing Waste with Caged Layer Manure and Straw or Sawdust: Effects of Carbon Source and C/N Ratio on Decomposition. *Waste Manag. Res.* **2015,** *33*, 332–338.
27. Oviedo-Ocaña, E.; Torres-Lozada, P.; Marmolejo-Rebellon, L.; Hoyos, L.; Gonzales, S.; Barrena, R.; Komilis, D.; Sanchez, A. Stability and Maturity of Biowaste Composts Derived by Small Municipalities: Correlation among Physical, Chemical and Biological Indices. *Waste Manag.* **2015,** *44*, 63–71.
28. Paredes, C.; Bernai, M. P.; Cegarra, J.; Roig, A.; Novarro, A. F. *Nitrogen Transformation During the Composting of Different Organic Wastes*; Kluwer Academic Publishers: Dordrecht, 1996.
29. Petric, I.; Helić, A.; Avdić, E. A. Evolution of Process Parameters and Determination of Kinetics for Co-composting of Organic Fraction of Municipal Solid Waste with Poultry Manure. *Bioresour. Technol.* **2012,** *117*, 107–116.
30. Petric, I.; Mustafić, N. Dynamic Modeling the Composting Process of the Mixture of Poultry Manure and Wheat Straw. *J. Environ. Manag.* **2015,** *161*, 392–401.
31. Poincelot, R.; Day, P. Rates of Cellulose Decomposition During the Composting of Leaves Combined with Several Municipal and Industrial Wastes and Other Additives. *Compost. Sci.* **1973,** *14* (3), 23-25.
32. Raj, D.; Antil, R. S. Evaluation of Maturity and Stability Parameters of Composts Prepared from Agro-industrial Wastes. *Bioresour. Technol.* **2011,** *102*, 2868–2873.
33. Rawoteea, S. A.; Mudhoo, A.; Kumar, S. Co-Composting of Vegetable Wastes and Carton: Effect of Carton Composition and Parameter Variations. *Bioresour. Technol.* **2017,** *227*, 171–178.
34. Saludes, R. B.; Iwabuchi, K.; Miyatake, F.; Abe, Y.; Honda, Y. Characterization of Dairy Cattle Manure/Wallboard Paper Compost Mixture. *Bioresour. Technol.* **2008,** *99*, 7285–7290.
35. Seng, Y. S. M. L. Palm Oil Mill Effluent (POME) from Malaysia Palm Oil Mills: Waste or Resource. *Int. J. Sci. Environ.* **2013,** *2*, 1138–1155.
36. Storino, F.; Arizmendiarrieta, J. S.; Irigoyen, I.; Muro, J.; Aparicio-Tejo, P. M. Meat Waste as Feedstock for Home Composting: Effects on the Process and Quality of Compost. *Waste Manag.* **2016,** *56*, 53–62.
37. Sülük, K.; Tosun, İ.; Ekinci, K. Co-composting of Two-phase Olive-mill Pomace and Poultry Manure with Tomato Harvest Stalks. *Environ. Technol.* **2017,** *38*, 923–932.
38. Sun, Z.-Y.; Zhang, J.; Zhong, X.-Z.; Tan, L.; Tang, Y.-Q.; Kida, K. Production of Nitrate-rich Compost from the Solid Fraction of Dairy Manure by a Lab-scale Composting System. *Waste Manag.* **2016,** *51*, 55–64.
39. Tabassum, S.; Zhang, Y.; Zhang, Z. An Integrated Method for Palm Oil Mill Effluent (POME) Treatment for Achieving Zero Liquid Discharge: A Pilot Study. *J. Clean. Prod.* **2015,** *95*, 148–155.
40. Tiquia, S. M. Reduction of Compost Phytotoxicity during the Process of Decomposition. *Chemosphere* **2010,** *79*, 506–512.
41. Tiquia, S. M.; Tam, N. F. Y.; Hodgkiss, I. J. Effects of Composting on Phytotoxicity of Spent Pig-manure Sawdust Litter. *Environ. Pollut.* **1996,** *93*, 249–256.
42. Unmar, G.; Mohee, R. Assessing the Effect of Biodegradable and Degradable Plastics on the Composting of Green Wastes and Compost Quality. *Bioresour. Technol.* **2008,** *99*, 6738–6744.

43. Villegas, M.; Huiliñir, C. Biodrying of Sewage Sludge: Kinetics of Volatile Solids Degradation Under Different Initial Moisture Contents and Air-flow Rates. *Bioresour. Technol.* **2014,** *174*, 33–41.
44. Wong, J.; Mak, K.; Chan, N.; Lam, A.; Fang, M.; Zhou, L.; Wu, Q.; Liao, X. Co-composting of Soybean Residues and Leaves in Hong Kong. *Bioresour. Technol.* **2001,** *76*, 99–106.
45. Yan, L.; Liu, Y.; Wen, Y.; Ren, Y.; Hao, G.; Zhang, Y. Role and Significance of Extracellular Polymeric Substances from Granular Sludge for Simultaneous Removal of Organic Matter and Ammonia Nitrogen. *Bioresour. Technol.* **2015,** *179*, 460–466.
46. Yang, L.; Zhang, S.; Chen, Z.; Wen, Q.; Wang, Y. Maturity and Security Assessment of Pilot-Scale Aerobic Co-composting of Penicillin Fermentation Dregs (PFDs) with Sewage Sludge. *Bioresour. Technol.* **2016,** *204*, 185–191.
47. Yaser, A. Z.; Rahman, R. A.; Kali, M. S. Co-composting of Palm Oil Mill Sludge-sawdust. **2007,** *10*, 4473–4478.
48. Zahrim, A. Palm Oil Mill Biogas Producing Process Effluent Treatment: A Short Review. *J. Appl. Sci.* **2014,** *14*, 3149–3155.
49. Zahrim, A.; Tahang, T. Production of Non Shredded Empty Fruit Bunch Semi-Compost. *J. Inst. Eng. Malay.* **2010,** *71*, 11–17.
50. Zahrim, A. Y.; Leong, P. S.; Ayisah, S. R.; Janaun, J.; Chong, K. P.; Cooke, F. M.; Haywood, S. K. Composting Paper and Grass Clippings with Anaerobically Treated Palm Oil Mill Effluent. *Int. J. Recycl. Organ. Waste Agric.* **2016,** *5*, 221–230.
51. Zhang, L.; Zeng, G.; Dong, H.; Chen, Y.; Zhang, J.; Yan, M.; Zhu, Y.; Yuan, Y.; Xie, Y.; Huang, Z. The Impact of Silver Nanoparticles on the Co-composting of Sewage Sludge and Agricultural Waste: Evolutions of Organic Matter and Nitrogen. *Bioresour. Technol.* **2017,** *230*, 132–139.
52. Zhou, C.; Liu, Z.; Huang, Z.-L.; Dong, M.; Yu, X.-L.; Ning, P. A New Strategy for Co-composting Dairy Manure with Rice Straw: Addition of Different Inocula at Three Stages of Composting. *Waste Manag.* **2015,** *40*, 38–43.
53. Singh, R. P.; Hakimi Ibrahim, M.; Norizan Esa, M. S. Iliyana. Composting of Waste from Palm Oil Mill: a Sustainable Wastemanagement Practice. *Rev. Environ. Sci. Biotechnol.* **2010,** *9*, 331-344.
54. Wu, T. Y.; Mohammad, A. W.; Jahim, J. M.; Anuar, N. A Holistic Approach to Managing Palm Oil Mill Effluent (POME): Biotechnological Advances in the Sustainable Reuse of POME. **2009,** *27* (1), 40-52
55. Polprasert, C. *Organic Waste Recycling;* Wiley Publishing, 1989.

CHAPTER 15

TAKING INTO ACCOUNT THE FEATURES OF THE WITHIN-FIELD VARIABILITY OF COMPOSITE SOIL FERTILITY IN PRECISION AGROTECHNOLOGIES

RAFAIL A. AFANAS'EV, GENRIETTA E. MERZLAYA, and MICHAIL O. SMIRNOV*

Pryanishnikov All-Russian Scientific Research Institute of Agrochemistry, d. 31A, Pryanishnikova St., Moscow 127550, Russia

**Corresponding author. E-mail: rafail-afanasev@mail.ru; lab.organic@mail.ru; User53530@yandex.ru*

ABSTRACT

The chapter considers the features of the within-field variation of soil fertility agrochemical indices. It is shown that the spatial distribution of within-field areas with different levels of soil fertility is characterized by certain statistical patterns that have a certain effect on the efficiency of mineral fertilizers used to increase crop yields. A description is given of various methods for revealing the within-field heterogeneity of soil fertility and how it is taken into account for the differentiated agrochemical application.

15.1 INTRODUCTION

To meet the demand of the population for high-quality grain products, the most advanced technologies for cultivating grain crops should be applied, including precision agriculture, based, unlike traditional technologies, on the within-field variability of soil fertility. At present, precision agrotechnologies are used in many countries of the world. They make it possible not only to increase crop yields and the quality of plant products but also to exclude environmental contamination by agrochemicals due to more accurate rationing of fertilizer doses to better meet their needs.

15.2 MATERIALS AND METHODOLOGY

Studies on the differentiated fertilizer application, taking into account the variability of soil fertility, were carried out in different soil and climatic conditions of the European part of Russia. To do this, field experiments were conducted with various doses of mineral fertilizers, especially nitrogen, which is usually found in the soils of this region at a minimum. At the same time, the within-field variability of soil fertility agrochemical indices, the patterns of their within-field distribution, and their influence on the efficiency of applied fertilizers were studied. Particular attention was paid in studies specifically to the identification of the features of the within-field distribution of sites characterized by different soil agrochemical properties, which had not previously been given due attention to the science and practice of fertilizer application. Accounting for the within-field variation of soil fertility was aimed at optimizing the plant mineral nutrition through more adequate use of fertilizers.

15.3 RESULTS

Studies on the variability of soil fertility agrochemical indices in soils of different types have shown that despite its seeming chaotic nature, there are a number of patterns that should be taken into account in high-precision agrotechnologies of adaptive landscape agriculture. First, it was found that within the boundaries of a single-crop rotation field (the production site), the soil reaction, the content of humus, and nutrients in the plow layer, to some extent, correspond to the law of normal distribution, that is, the most frequently occurring values of any agrochemical index refer to the average level characteristic for a given field, whereas the minimum and maximum values are in a relative minimum.

The specific features of the spatial distribution of the agrochemical properties of soils are discussed below mainly on the example of the mobile phosphorus content in the plow layer of various soil types.

Thus, on the testing area of the central experiment station (CES) of the Pryanishnikov All-Russian Agrochemistry Research Institute (PARARI) (Moscow region) from 400 plots of a special testing area (each plot had an area of 10 m^2), more than half of these plots had an average mobile phosphorus content, and the number of plots with a minimum$^{6\text{-}15}$ and maximum$^{26\text{-}45}$ content did not exceed four tens.

A similar pattern is characteristic of other types and varieties of soils. At the same time, both within-field and interfield variability do not differ in principle: the distribution of the number of within-field areas (Fig. 15.1) and fields (Fig. 15.2) with different mobile phosphorus contents in the soil basically obeys the law of normal distribution. This feature of the variability of soil fertility should be taken into account when calculating the need for fertilizers and ameliorants, including differential application, bearing in mind that to calculate the total demand for agrochemicals in a certain field area, one can focus on the average agrochemical indicators of these fields, that is, the averaged values will be close to the average values for a given field or crop rotation as a whole.

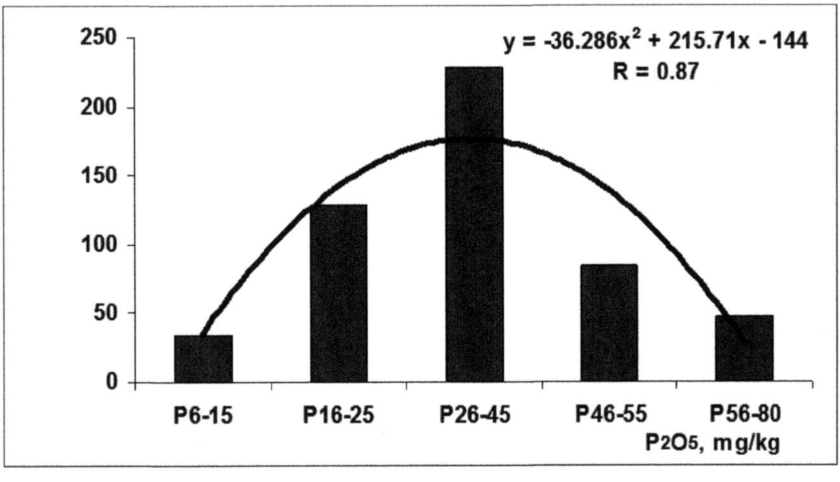

FIGURE 15.1 Distribution curve of the number of 520 elementary sites on the fields of the development farm "Gazyrskoe" with a different content of mobile phosphorus (according to Machigin) in common chernozem, Krasnodar region.

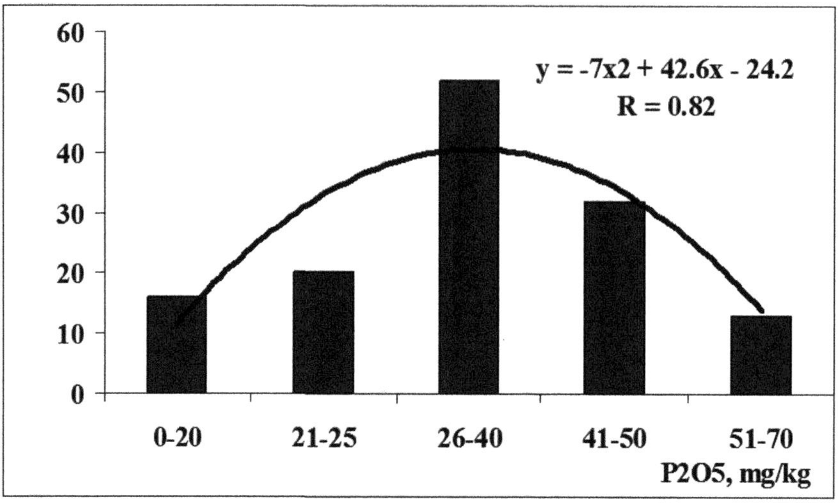

FIGURE 15.2 Distribution curve of the number of 133 fields of the development farm "Gazyrskoe" with a different content of mobile phosphorus (according to Machigin) in ordinary chernozem, Krasnodar region.

The next regularity of the spatial structure of the soil cover is that the greatest variability of agrochemical indices is observed both in regions with relatively smaller and larger values with a decrease in variability in the intermediate interval, that is, on the average, which is indicative, in particular, for variability of phosphorus (Fig. 15.3).

It follows from this regularity of the within-field variation of soil fertility that the agroeconomic effect of a differentiated fertilizer application and ameliorants in each specific case will depend on the ratio of the sites with the marginal values of agrochemical indices to the areas with their average values in this field: the higher this ratio, the higher the efficiency of differentiation of doses, because on a part of the field, whose fertility is leveled, the differentiation of doses is practically not required. However, the number of sites with marginal agrochemical characteristics of the soil, which follows from the first regularity, lies along the edges of the normal distribution curve, that is, their total area is much smaller than the area with average agrochemical parameters for a given field. Therefore, the effectiveness of the differentiated application of agrochemicals is determined by this pattern.

Nevertheless, taking into account these regularities, in particular, the second allows us to choose the optimal strategy for the relative equalization of soil fertility by some index. The most acceptable option is when the average fertilizer dose calculated for the entire field increases on poorly maintained fertility shapes and decreases on highly provided for this element to bring the land to the level for a number of years to an acceptable level. From this same regularity, it follows that a single application of fertilizers with an increased dose will not reduce the variability of soil fertility, since it will not eliminate, but may enhance the above-mentioned uneven fertility of the soil cover. The reason for this may be a lower consumption of phosphorus fertilizers to increase the soil content of mobile phosphorus content per 1 mg/kg in areas with a large initial content compared to sites that have poor soil availability with phosphates.[1]

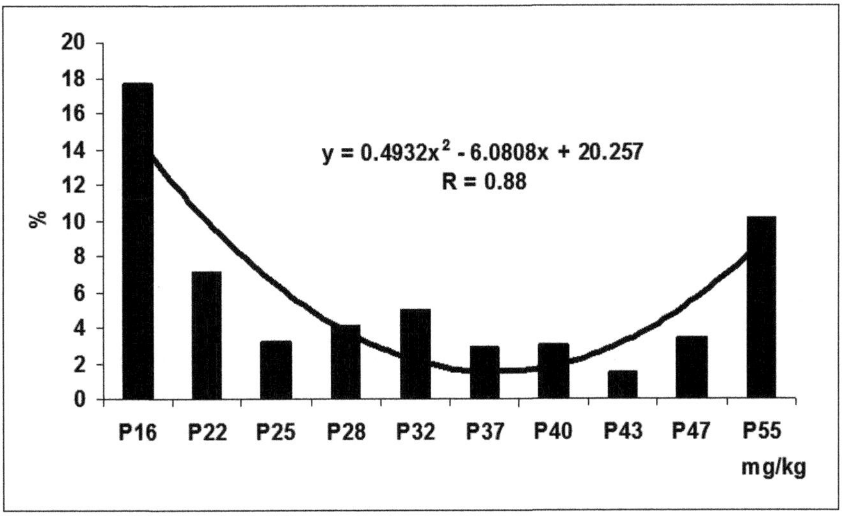

FIGURE 15.3 Variability of the mobile phosphorus content (according to Machigin) in common chernozem of 133 fields of the development farm "Gazyrskoye," Krasnodar region.

The third regularity of the variability of soil fertility lies in the fact that the within-field variability can exceed the interfield variability. This is established by the results of agrochemical survey of fields in the development farm "Gazyrskoye." Thus, the coefficient of variation of mobile phosphorus content for 133 production fields was 34%, and for 520 elementary sections of these fields, it is 42.9%. This property is due to the fact that the increase in the area of arable land, by which averaging of agrochemical indices is carried out, reduces the variability between the contours and increases it within the contours, and vice versa. This was clearly manifested on the testing area of the CES of the PARARI (Table 15.1).

TABLE 15.1 Dependence of Variability of Soil Agrochemical Indices in Arable Layer of Sod-Podzolic Soil on the 4-ha Testing Area of CES PARARI from the Area of Averaged Sites.

Number of plots	Plot area (ha)	Coefficients of variation of agrochemical indices (V%)		
		Humus	P_2O_5	K_2O
40	0.1	0.8	2.1	2.4
20	0.2	1.6	3.5	4.6
8	0.5	3.2	7.3	10.0
4	1.0	5.6	12.6	17.1
2	2.0	10.0	20.8	27.6
1	4.0	18.6	30.7	41.2

It follows from the table that the intragroup variability of humus, mobile forms of phosphorus, and potassium increases significantly, by an order of magnitude or more, with an increase in the area of the elementary section from 0.1 to 4 ha. This regularity should be taken into account when developing high agrotechnologies, finding a compromise between the appropriateness of isolating fertility contours with the least in-loop variability of soil fertility to improve the efficiency of fertilizers, on the one hand, and the smallest number of such contours on the field to reduce the costs of sampling and analysis of soil samples, on the other.

The fourth, according to our accepted order, feature of the spatial heterogeneity of the soil cover consists in smooth, gradual transition from the largest values of agrochemical indices tors to smaller and, conversely, from smaller to larger (Fig. 15.4).

Taking into account this feature of the soil cover structure is important from a practical point of view, since it allows designing and creating machines for the differentiated agrochemical application to provide a relatively smooth change in their doses when moving across the field, which facilitates both the design work and working processes for the fertilizer and ameliorants application in production conditions.

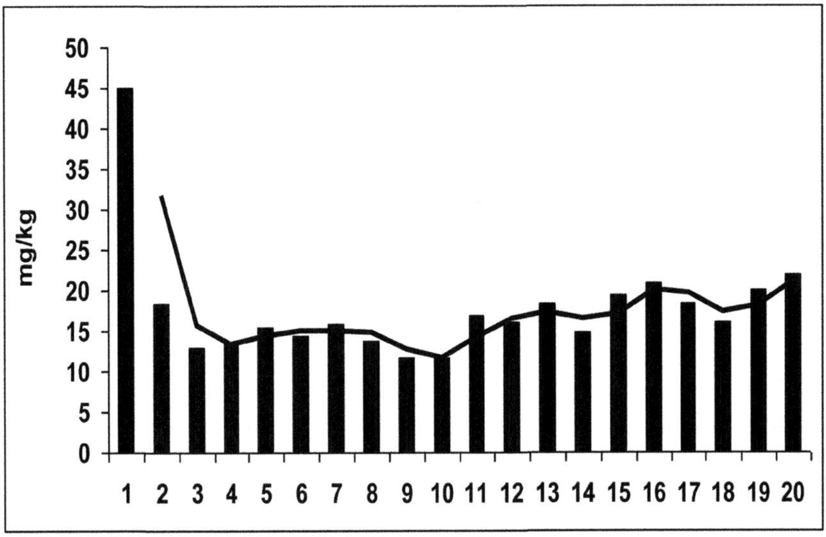

FIGURE 15.4 Change in the content of mobile phosphorus in sod-podzolic soil on a 200-m transect on the plots of the testing area CES PARARI.

The fifth peculiarity of the within-field spatial heterogeneity of soil fertility is the discrepancy between the boundaries of different agrochemical contours.

Statistical analysis of agrochemical data of 48 soil samples taken at the experimental site of the Ol'sha, Smolensk region, established that the correlation coefficients between the main agrochemical indices of soil fertility are invalid, with the exception of the correlation between the humus content and total nitrogen, inasmuch as the total nitrogen in the mass is an integral part of the organic matter of the soil.

On the within-field contours in the fields of the development farm "Gazyrskoye" (Krasnodar region), the agrochemical indices also do not correlate or slightly correlate between themselves, as shown in Table 15.2.

From this, it follows that for each within-field territorial contour, different doses of fertilizers and the ratio of the nutrients in the fertilizer corresponding to the agrochemical characteristics of these contours are needed.

Taking into account this regularity in the technologies of differentiated fertilizer application, it is required either the use of machines with simultaneous sowing of different types of fertilizer or the repeated passage of machines adapted for the introduction of one fertilizer type.

TABLE 15.2 Coefficients of Correlation of Agrochemical Indices of Common Chernozem on the Within-field Contours in the Development Farm "Gazyrskoye" ($n = 40$).

Agrochemical indices	Humus	pH_{water}	N_{min}	P_2O_5	K_2O	Ca
pH_{water}	0.3					
N_{min}	0	−0.1				
P_2O_5	0.1	0	−0.2			
K_2O	0.1	−0.2	−0.1	0.2		
Ca	0.3	0.2	−0.3	0	−0.2	
Mg	−0.5	0.1	−0.1	0.2	−0.2	−0.2

By the next, sixth in the account, the law should be attributed to the greatest variability of the easy-mobile forms of plant nutrients in comparison with the less mobile forms. As shown in Figure 15.5, the variation coefficients of the easy-mobile forms of nitrogen, phosphorus, and potassium significantly, 1.5–2 times, exceed in importance of the coefficients of the less mobile forms of these elements. In this case, the yield of crops depends on the easy-mobile nutrient forms, in particular from phosphorus, to a greater extent, than from less mobile ones (Fig. 15.6).

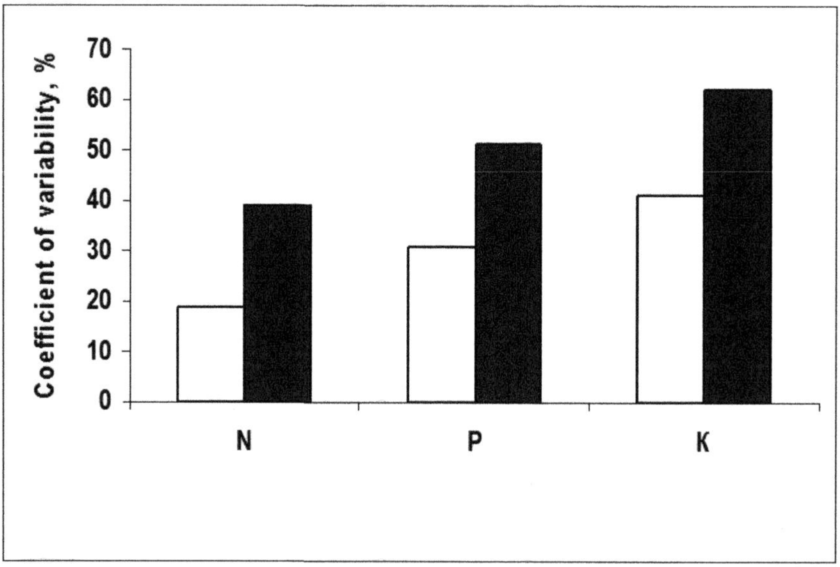

FIGURE 15.5 The variability of the mobile (the first row of columns) and easy-mobile (the second row of columns) NPK forms in the soil of the testing area.

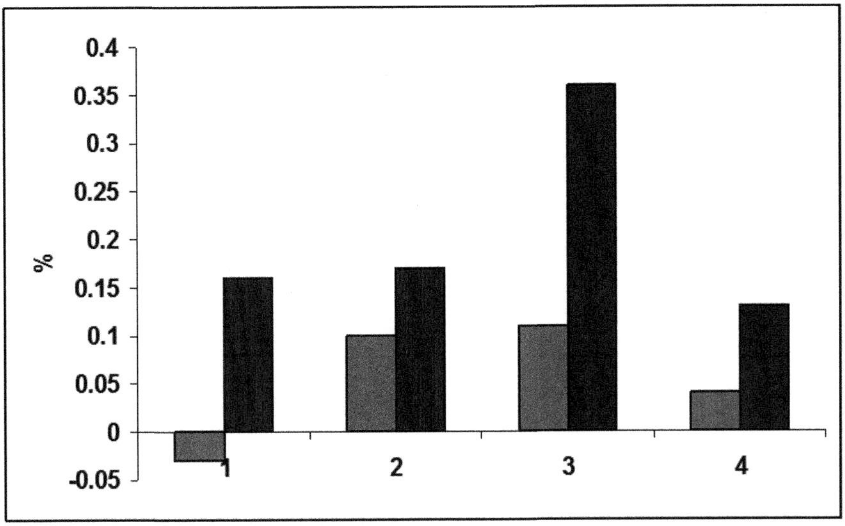

FIGURE 15.6 Influence of mobile (first row of columns) and easy-mobile forms (second row) of phosphorus in the soil of the testing area for crop yields of field crop rotation: (1) oats, (2) perennial grasses, (3) winter wheat, and (4) spring barley.

Accounting for this pattern is very important for increasing the agro-economic efficiency of differentiated fertilizer application, since it allows to more accurately take into account the plants' need for nutrients.

In other words, in agrochemical soil survey, along with the traditional indices of their availability with mobile forms of nitrogen, phosphorus, and potassium, it is expedient to determine in the soil the easy-mobile plant nutrients and to orient oneself to them when calculating the optimal fertilizer doses. This will not only improve the fertilizer effectiveness but also will avoid losing them to the environment.

Next, the seventh, the peculiarity of the within-field variability of soil fertility refers not so much to the actual fertility of soils, but to the use of soil nutrients by plants. It consists in the fact that the yield of crops varies to a lesser degree than the within-field indices of soil fertility. According to Figure 15.7, it can be seen that the variation coefficient in yield of annual grasses hay ($V = 41\%$), which are the precursors of the grain crop—winter wheat, at 400 plots of the testing area of CES PARARI, is significantly lower than the variation coefficients of the easy-mobile forms of phosphorus ($V = 52\%$) and potassium ($V = 62\%$).

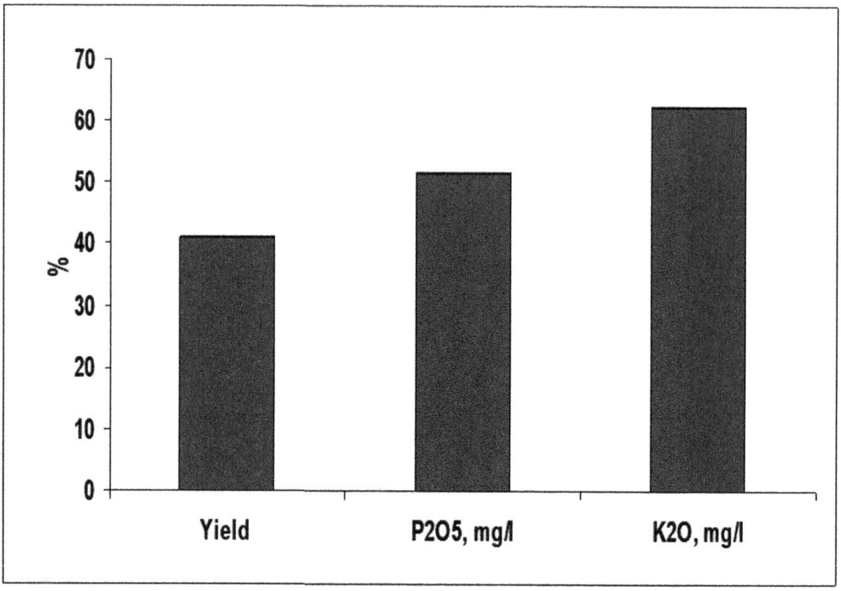

FIGURE 15.7 Variability of the yield of annual grasses hay and the easy-mobile forms of phosphorus and potassium in the soil of the testing area.

This effect is caused by the physiological plasticity of plants, their ability to increase the development of root systems in lack of nutrition and to increase their effect on hardly soluble nutrients of the soil.

Such an adaptive reaction of plants aimed at their normal functioning, however, reduces the differences in crop yields in areas with a different level of soil fertility and, accordingly, the addition yield from differentiated fertilizer application. With a slight difference in the within-field agrochemical parameters or low-differentiated doses of fertilizers, one should not expect a significant effectiveness of their application due to the ecological plasticity of plants.

The final, eighth peculiarity of the within-field variation of agrochemical indices is the dependence of the soil fertility levels on the meso and microrelief of fields, which to a large extent determines the regularities shown above. Soils of one location on the terrain relief affect the surrounding soils, influencing the leaching, transport, and deposition of chemical components, confirmed by many modern studies.[2]

Using the example of the testing area of CES PARARI (Fig. 15.8), it is shown that in the flat areas, the nitrate nitrogen content in the soil prevails over the ammonium nitrogen, while in the depressions, the ammonium nitrogen content is higher than that of nitrate nitrogen because of greater anaerobiosis and reduction of nitrates migrating from intrasoil runoff to lowered locations. In the soil of the lowered parts of the relief, there is also greater mobile phosphorus content.

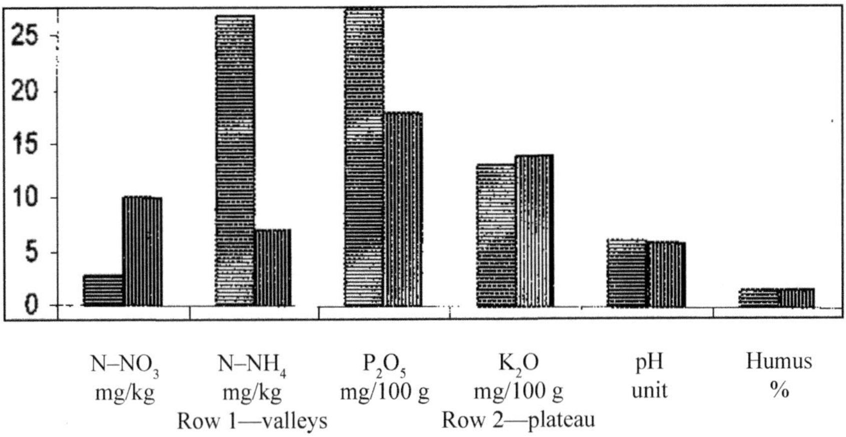

FIGURE 15.8 Dependence of agrochemical indices of sod-podzolic soil from the location on the relief of the testing area CES PARARI.

This pattern makes it possible to use the results of topographical survey of field relief for a priori identification of fertility contours as elementary sites for selection and agrochemical survey of fields in comparison with traditional grid selection. In this case, the difference is important not only in the relative height of the within-field areas but also in the exposure of the slopes.

For example, in the conditions of Krasnodar region during the spring top dressing of winter spike crops, the nitrate nitrogen content in soils located on the southern slopes was 3–4 times higher than on the slopes of the northern exposure.[3]

Taking into account the abovementioned patterns of the within-field variation of soil fertility in conditions of precision agriculture, it will be more rational to use fertilizers to increase the productivity of arable land and reduce the ecological load on the agricultural environment.

One of the primary tasks in this regard is the allocation of agrocontours on the fields, which should serve both for their agrochemical inspection and for the differentiated fertilizer application. Currently, there are several ways to solve this problem.

The research conducted by the All-Union Research Institute of Agrochemistry together with other organizations has shown that aerospace remote sensing of arable lands in the radio and infrared ranges of electromagnetic waves can serve as one of the most promising ways of mapping the within-field variability of soil fertility.

Scanning the surface of the fields with radar is possible at any time of the day and at any cloudiness. It was found that, at a wavelength of 9.8 cm, the fixation of radiosignals reflected from the near-surface layer of the soil with a depth approximately equal to half the wavelength of the radiation, that is, about 5 cm, in the form of a digitized image of the field area allows you to identify areas with relatively homogeneous indices on a visualized image (PC monitor).

The degree of grouping and generalization of signals, that is, the number of allocated contours, differing among themselves by the average digital indices, depends both on the actual heterogeneity of the scanned object (field) and the tasks of the computer operator.

The use of this method creates an objective basis for the rational allocation of soil fertility contours as elementary sites for subsequent land agrochemical survey of fields, that is, mapping of the within-field variability that meets the technological requirements of precision agriculture.

Remote radiolocation and IR spectrometry are also suitable for the rapid diagnosis of nitrogen nutrition of agricultural crops, primarily cereals, for the purpose of vegetative fertilization of crops. In particular, the subsatellite experiments, laid by introducing increasing nitrogen doses from 0 to 120 kg/ha on winter wheat crops in the development farm "Gazyrskoye," in the Krasnodar region, revealed a close relationship between the reflected radiosignals and infrared radiation with the total nitrogen content in the plants.

So, in the conditions of the North Caucasus, the average difference in nitrogen content in plants of winter wheat in the shooting, determined by laboratory and remote methods, was 4.4% (relative).

No less successful was the IR spectrometry of winter wheat crops in conventional field experiments with fertilizers in the development farm "Gazyrskoye" using a helicopter, where the correlation coefficients between the magnitude of the signals and the content of total and nitrate nitrogen in plants were 0.8–0.9.

The use of unmanned aerial vehicles greatly facilitates and cheapens the diagnostic examination of fields for the purpose of carrying out vegetative fertilization of cereal crops, primarily nitrogen fertilizers.

Investigations carried out at the CES PARARI using unmanned aerial vehicle (UAV) for the diagnosis of nitrogen nutrition of winter wheat in the field experiment with increasing doses of nitrogen fertilizers showed high coefficients of pairwise linear correlation of the normalized difference vegetation index (NDVI) parameters obtained from UAVs with the results of the ground diagnostics of the sowing, namely estimating in points photometer "YARA," indices of stem diagnostics, and even with the emission of carbon dioxide from the soil (Fig. 15.9 and Table 15.3).

FIGURE 15.9 Diagnosis of nitrogen nutrition of winter wheat with using a multicopter (UAV).

It was also quite effective to use differentiated organic fertilizer application in particular sewage sludge, taking into account the within-field variability of the provision of sod-podzolic loamy soil with humus. According to the patented method, the differentiated application of the sediment at doses of 50–150 t/ha can increase the yield of cereals to 10–18% compared with the application of their average for the whole field—100 t/ha.[4]

TABLE 15.3 Indicators for the Diagnosis of Nitrogen Nutrition of Winter Wheat, Carried Out by Various Methods During the Shooting, 2017.

NDVI (UAV)	Nitrogen doses (kg/ha N)	"YARA" (points)	Stem diagnosis (indices)	CO_2 emission (g/m^2 24 h)	Yield (t/ha)
0.65	0	356	0	7.7	2.83
0.81	30	511	0.87	8.2	4.41
0.84	60	541	1.4	10.1	4.99
0.85	90	580	2.5	9.7	5.0
0.86	120	620	2.7	12.3	5.1
Correlation coefficient	**0.84**	**0.97**	**0.86**	**0.73**	**0.99**

Based on the results of using the UAV for the nitrogen nutrition diagnosis of winter wheat on the basis of CES PARARI, a method has been developed for using these purposes of conventional commercial unmanned aerial vehicles equipped with digital cameras of general purpose.

The next stage of agrochemical maintenance of grain crop sewing is to create automated interactive programs for calculating fertilizer doses for their planned yields. Its essence is computerization of calculation methods, also based on the results of field experiments.

Calculation of fertilizer doses is carried out in accordance with the received electronic agrochemical cartograms of soil fertility, planned crop yields, taking into account the quality of predecessors, water availability, field relief features, agrophysical soil properties, and a number of other factors.

The development of programs for the automated calculation of mineral fertilizer doses for cereals cultivated in the agricultural regions of our country will serve as a basis for increasing their production, improving the quality of products, ecological safety of the agrochemical application.

The final stage of agrochemical maintenance of precise agriculture is namely a differentiated application. According to the available developments, this is entrusted to robotic units consisting of a tractor, a mounted or trailed fertilizer spreader (Fig. 15.10), electronic equipment including an onboard computer, a global positioning system (GPS)/global national sputnik system (GLONASS) navigator, an automatic driving system (autopilot), and, for differentiated fertilization of vegetative crops with fertilizers, additional sensors (Fig. 15.11).

In general, research on the study of the precise agriculture characteristics in various regions of the European part of our country have shown that a scientifically grounded account of the soil fertility heterogeneity can give an undoubted agroeconomic effect.

FIGURE 15.10 Combined unit for differentiated application of mineral fertilizers.

FIGURE 15.11 Differentiated dressing of winter wheat during shooting in the Voronezh region.

An example is the effective differentiated application of nitrogen fertilizers for winter wheat in the field experiment conducted by the All-Union Research Institute of Agrochemistry (Fig. 15.12).

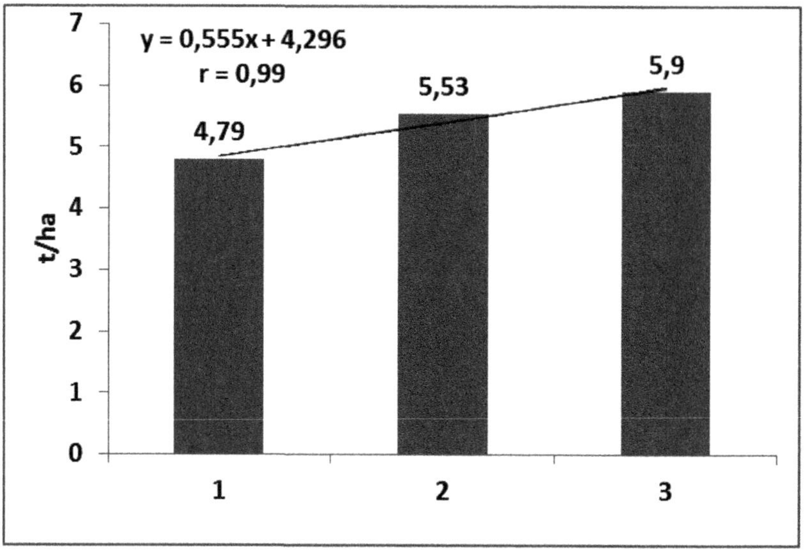

FIGURE 15.12 Yield of winter wheat under differentiated application of nitrogen fertilizers in sod-podzolic soil: (1) control, (2) fixed dose, and (3) differentiated doses.

15.4 CONCLUSIONS

Precise agriculture as a promising stage in the development of world agriculture is largely based on the principles of optimizing the plant mineral nutrition, that is, on the rationally, biologically, economically, and environmentally balanced application of mineral and organic fertilizers and other agrochemicals, taking into account the within-field variability of soil fertility. The use of precise agriculture will increase the provision of the population with grain and, accordingly, high-quality products, aimed at improving its viability and protection from negative environmental influences. It should be taken into account that the rational use of fertilizers, aimed at optimizing the mineral nutrition of plants, promotes the realization in plant organisms of their inherent biochemical processes, the formation of protective reactions and bio-compounds, including antioxidants. At the same time, an increase in the production of grain-based food products of precision agriculture can reduce the danger of negative environmental effects on the population health.

KEYWORDS

- **agricultural crops**
- **soil**
- **procedures of precision agriculture**
- **fertilizers**
- **the environment**

REFERENCES

1. Mineev, V. G.; Kovalenko, A. A.; Vaulin, A. V.; Afanas'ev, R. A. Influence of Phosphorous Agrophones on Fertilizer Efficiency and Crop Rotation Crop Productivity on Sodpodzolic Soil. *Agrochemistry* **2009,** *11*, 22–31.
2. Mineev, V. G. Ecological Functions of Agrochemistry. In *Fertilizers and Chemical Ameliorants in Agroecosystems: Materials of the Fifth Scientific and Practical Conference*; Mineev, V. G., Ed.; Moscow State University Publishing House: Moscow, 1998; pp 6–13.
3. Shirinyan, M. Kh.; Afanas'ev, R. A.; Korobskoy, N. F. Diagnostics and Optimization of Winter Wheat Mineral Nutrition on Common Chernozems of the North Caucasus. In *Scientific Foundations and Recommendations for Diagnostics and Optimization of Mineral Nutrition of Cereals and Other Crops*; Milashchenko, N. S., Ed.; Agroconsult: Moscow, 2000; pp 55–71.
4. Afanas'ev, R. A.; Merzlaya, G. E.; Ladonin, V.; Marchenko, N. M. Method of Introducing Organic Fertilizer Application. Patent RU No. 2260930; 2005.

CHAPTER 16

CHEMISTRY OF MEDICINAL PLANTS, FOODS, AND NATURAL PRODUCTS AND THEIR COMPOSITES

LAURA MARÍA SOLIS-SALAS, LUIS ENRIQUE COBOS-PUC, CRISTÓBAL NOÉ AGUILAR-GONZÁLEZ, CRYSTEL ALEYVICK SIERRA RIVERA, ANNA ILINÁ, and SONIA YESENIA SILVA BELMARES*

Food Research Department, Faculty of Chemistry, Autonomous University of Coahuila, Blvd. Venustiano Carranza, Col. República Oriente, 25280, Saltillo, Coahuila, Mexico

Corresponding author. E-mail: yesenia_silva@hotmail.com

ABSTRACT

This chapter describes the information on medicinal plant uses, the chemistry of medicinal plants, natural products, food chemistry, and natural products, with the purpose of establishing a relationship between the chemistry of medicinal plants, foods, and natural products. According to the information reviewed, the main link shared by natural products that include plants and some foods is the content of phytochemical compounds, and many of them have the capacity to treat or prevent one or more diseases. In addition, the verification of its functionality has been made more and more frequently by the scientific community interested in the development of functional foods; so, the search for new molecules with high biological potential continues to increase.

16.1 INTRODUCTION

At present, people from different places in the world use traditional medicine; consequently, they use plants as a primary treatment for the diseases. The extracts of plants are used more frequently in these countries and reach up to 85%, compared with other remedies used in traditional medicine.[113] On the other hand, 33–47% of cancer patients use alternative treatments is estimated. These treatments are used during and after allopathic treatment to reduce the recurrence risk as well as control side effects, since the phytochemical compounds with antioxidant properties, that is, fruits and vegetables, have been related to the antiproliferative effect.[49, 51]

Alternatively, some herbal compounds work as alternative drugs or adjuvant in the treatment of diabetes, as they are administered together with insulin to reduce some complications derived from this disease.[101] Additionally, some plant extracts, phytochemical compounds, and derivatives are inhibitors of cholinesterase, therefore can be used to develop products focused on Alzheimer's treatment.[63] Other medicinal plants applications are focused on prevention and treatment of inflammatory diseases such as arthritis as well as diarrhea and hypertension.[8,77,119]

The main reasons to use medicinal plants in the treatment of diseases are that they contain bioactive compounds, usually have minimal side effects, and are available at low cost.[3] Bioactive compounds can be found naturally in small amounts in natural products and foods.[74]

Proteins, pigments, phenolic compounds, alkaloids, steroids, and coumarins are some of the bioactive compounds that come from the metabolism of plants. These compounds play a central role in many human diseases.[80,86,105] They are synthesized by the primary and secondary metabolisms through the shikimic acid, malonic acid, and 2*C*-methyl-D-erythritol-4-phosphate (MEP) pathways.[17] The bioactivity of natural products is directly related to the groups of compounds that each plant has. Therefore, the synthesis of compounds in plants linked to factors such as geographical distribution, soil, sunlight, precipitation, and climate, while the processes of collection, storage, transport, and manufacturing influence the conservation and integrity of the compounds.[102]

For this reason, in recent years, there is an increase in the number of investigations focused on the search for new plant species, active compounds, as well as the evaluation of biological activities.[10] Therefore, some studies have focused on properties that favor human health, such as the ability to improve degenerative diseases.[71] According to this, currently consumers are more interested in buying products to improve health; so, they are widely available in the market. This is due to the diffusion that has been given to the properties of medicinal plants, natural products, and some foods. Briefly, some products are known as nutraceuticals or functional foods, as added with nutrients and phytochemical compounds derived from plants or foods.[92]

This chapter focuses on providing information on medical plants uses, chemistry of medicinal plants, natural products, chemistry of foods, and natural products. Additionally, some plants as potential sources of new biocompounds are presented, as they are able to prevent and adjuvate in some diseases.

16.2 MEDICINAL PLANTS USES

Medicinal plants have been used since prehistoric times to treat diseases. At present, according to the use, they are divided into four categories, which include indigenous medicinal herbs, medicinal herbs in systems, modified medicinal herbs, and products imported with herbal medicine.[83,89] In the last 25 years, approximately 63% of anticancer drugs come from natural compounds.[46]

Next, some medicinal plants are described, as well as their use in the treatment of the disease they address. The *Persea americana* leaves decoction is used as a remedy to treat diarrhea, sore throat, diabetes mellitus, and hemorrhage. In addition, infusions are used to treat inflammation, pain, and fever.[6] *Zygophyllum simplex* L. is traditionally used in the Arab and Indian regions to treat gout, asthma, and inflammation.[4] On the other hand, some plants of the Malvaceae family, such as *Abutilon indicum*, *Hibiscus sabdariffa*, *Sida acuta*, and *Sida rombifolia*, have analgesic, anti-inflammatory, antidiabetic, antiobesity, antioxidant, antimicrobial, anxiolytic, cardioprotective, cytotoxic, hepatoprotective, and nephroprotective properties.[1] *Chrozophora tinctoria* traditionally is used as antiwart, emetic, cathartic, and antipyretic.[3] Some species of the Lauraceae family are used in traditional medicine to combat diarrhea, dysentery, toothache, intestinal parasites, hypertension, cancer, menstrual problems, and inflammation.[8] Additionally, the compounds contained in the ethanolic leaf extract and the *P. americana* fruit block the transmission of tumor cell growth signals since they increase the intracellular oxygenated radicals and facilitate the apoptosis of these cells.[76]

Instead, some species of the Apiaceae family, such as cumin, coriander, anise, and parsley, are used as a source of food for their flavoring property, while others such as caraway, fennel, and dill are used to treat diseases of the digestive, endocrine, reproductive, and respiratory systems.[100] In addition, the cardiac glycosides, volatile oil, fat, starch, gum, and sugars of *Digitalis lanata* and *Digitalis purpurea* are traditionally used to treat ulcers, abscesses, headaches, and paralysis.[11]

In most European countries and America, *Passiflora incarnata*, *Passiflora alata* Curtis, *Passiflora coerulea* L., and *Passiflora edulis* Sims are used as sedatives.[57] In addition, in traditional Ethiopian medicine, *Allium sativum*, *Nigella sativum*, *Ruta chalepensis*, and *Moringa stenopetala* are used.[110] One example is the Brazilian plant *Austroplenckia populnea* (Reiss), used to treat dysentery and rheumatism,[14] as well as *Populus tomentosa* Carr. and *Populus canadensis* Moench (Salicaceae) which are used in East Asian medicine to treat inflammatory diseases and diarrhea.[119]

In the countries of the Middle East and the Far East, *Nigella sativa* L. is used as a natural remedy to dissolve kidney stones and has properties such as antimicrobial, anticarcinogenic, cardioprotective, and anti-inflammatory.[23] *Ricinus cmmunis* L. (Euphorbiaceae) is used in traditional medicine to relieve some abdominal disorders, arthritis, back pain, muscle aches, chronic back pain, sciatica, chronic headache, and constipation and expel the placenta, gallbladder pain, menstrual colic, rheumatism, and insomnia, additionally properties such as antidiabetic, contraceptive, anti-inflammatory, antioxidant, and antibacterial attributed to it. This plant contains flavonoids, phenolic compounds, fatty acids, amino acids, terpenoids, and phytosterols. The phytosterols are used to synthesize contraceptives and anti-inflammatories.[81]

Accordingly, the scientific verification of ethnomedicine allows the discovery of pharmaceutical targets and guarantees the safety of the use of medicinal plants.[1]

16.3 NATURAL PRODUCTS

The natural products generally influence some biological systems, since they present pharmacological or toxicological effects on animals or humans.[20] The natural products of plants are considered an incalculable source of new compounds with biological activity, which is why they are used in the pharmaceutical, cosmetic, and food industries.[30] Bioactive molecules derived from natural products have properties similar to those of drugs so that the body can absorb and metabolize them.[72] Most of the active principles of plants are phytochemical compounds derived from secondary metabolism that have an unusual and complex chemical structure, as well as specific characteristics and a restricted botanical distribution.[24] In the world, for centuries, it is known that plants with biomedical properties have a high demand that shows an increase in recent years. At present, this market reaches 60 billion dollars, with annual growth rates between 5% and 15%.[69] Recently, research carried out that includes bioactive compounds from plants and foods focused on reducing blood pressure and lipids as they affect endothelial function and arterial stiffness. Berberine, curcumin, cocoa, lycopene, and green tea are some examples of natural products of plant able to reduce blood pressure.[29]

16.4 CHEMISTRY OF MEDICINAL PLANTS AND NATURAL PRODUCTS

The plants carry out a series of chemical reactions called metabolism that allows the growth of the plant.[96] The plants carry out a primary metabolism as well as a secondary one. In the primary metabolism are synthesized essential molecules for life such as proteins, carbohydrates, lipids, and nucleic acids, while in the secondary metabolism are produced compounds that are not essential for plant life, such as alkaloids, phenolic compounds, polyphenols, polyacetates, coumarins, lignans, quinones, terpenes, sesquiterpenes, steroids, carotenoids, flavonoids, dyes, sapogenins, tannins, and essential oils. Some compounds derived from secondary metabolism have allelopathic activities, defense against pathogens, consumers of plants, and natural enemies.[122]

The medicinal plant substances used to treat diseases are obtained of simple processes.[111] The main categories of bioactive compounds from plants are the terpenes and terpenoids, alkaloids, and the phenolic compounds that include approximately 25,000, 12,000, and 8000 compounds, respectively.[122]

The alkaloids are synthesized from the aromatic amino acids through the shikimic acid pathway as well as from the aliphatic amino acids through the tricarboxylic acid pathway. The phenolic compounds are synthesized through the pathways of shikimic acid and malonic acid in the metabolism of plants, while the terpenes are produced through the mevalonic acid and MEP pathways.[17] Figure 16.1 presents the chemical structure of some plant medicinal compounds, while Figure 16.2 shows the main metabolic pathways.

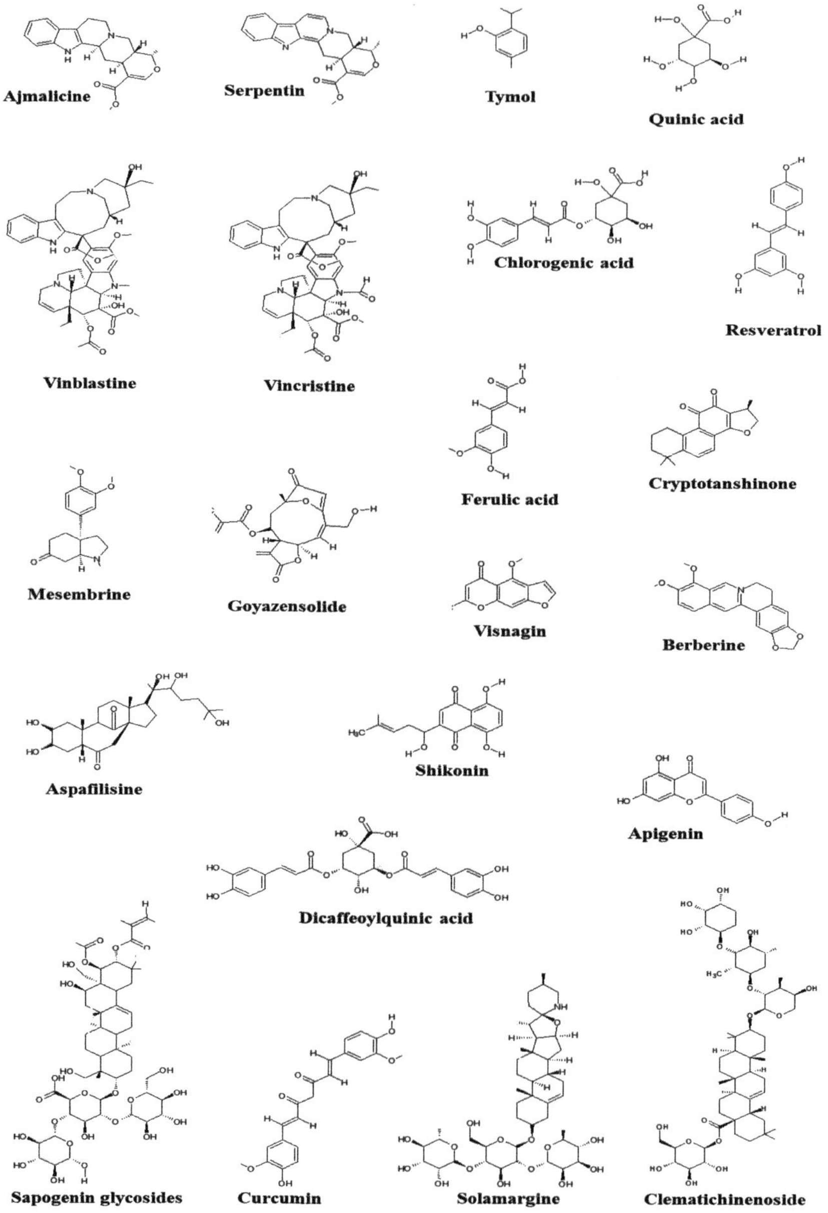

FIGURE 16.1 Chemical structures of compounds isolated from plants.

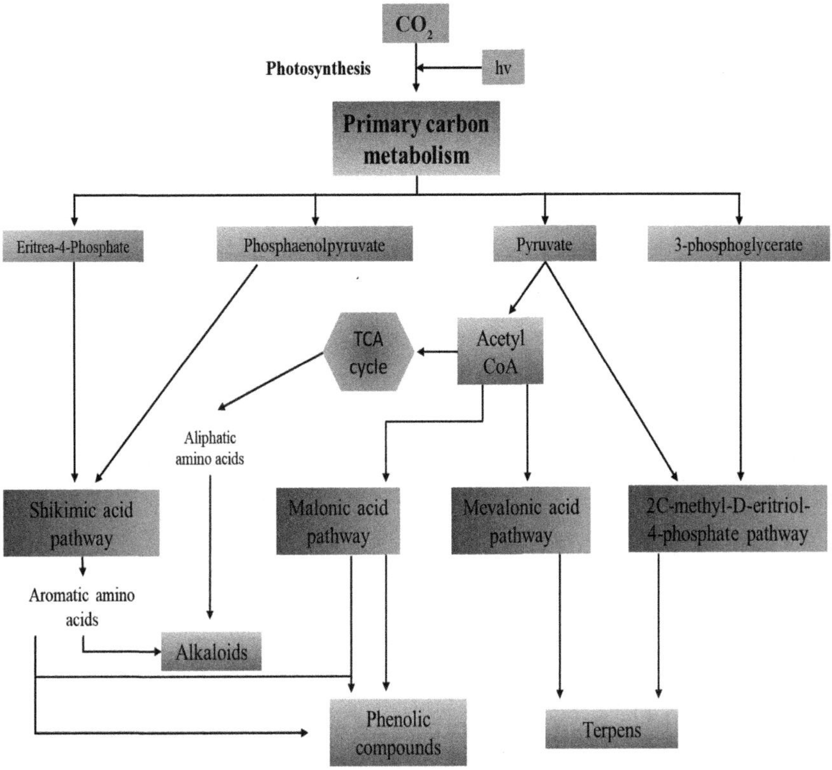

FIGURE 16.2 Main pathways of synthesis of medicinal compounds of plants.

Source: Adapted from Ref. [18].

16.4.1 PEPTIDES AND PROTEINS

The most important bioactive compounds derived from primary plant metabolism are proteins and peptides. The plant proteins are biopolymers that plants synthesize to maintain structural integrity, protection, and physiological functions and to preserve a reserve of energy. The function of each protein is due to the structural configuration that it acquires after the ribosomal synthesis of the amino acid chain. This structural conformation depends on the physical conditions and the biological environment since it is affected by extreme temperatures and reactive molecules. These changes can interrupt the folding of the newly synthesized protein or induce the incorrect folding of existing proteins.[59]

Additionally, some bioactive plant proteins are produced as a subsequent response to infection by fungi and phytopathogenic viruses, since they produce low molecular weight compounds called phytoalexins, peptides, and small proteins with antimicrobial properties.[85,103] An example is the antimicrobial peptide (AMP) production that is induced by cytokines and vitamins when bacteria or microbial debris such as lipopolysaccharides invade eukaryote cells.[60] The AMPs cause the permeation of the bacterial membrane, which leads to filtrations as well as cellular alterations, therefore, this effect is main mechanism of action.[87]

The AMP's present membrane permeation as the main mechanism of action, since it causes filtrations and cellular alterations. However, other bacterial intracellular components may be affected. In both cases, an electrostatic interaction occurs between the phosphate groups of the lipopolysaccharides and the lipoteichoic acids that contain the gram-negative and gram-positive bacteria, respectively.[73]

In addition, it is known that AMPs have an antibacterial and antifungal effect against pathogens of mammals, amphibians, and insects since they regulate the immune system of the host. Defensin, protegrin, histatin, and peptides derived from lactoferrin and filamentous fungal pathogens are some examples since they show an effect on *Candida albicans*.[38,90]

An aspartyl protease (AP) was isolated from potato tubers and leaves, and its antifungal activity was evaluated; finding a direct inhibitory effect on *Phytophthora infestans* cysts and *Fusarium solani* conidia, AP was found in intracellular washing fluids.[56]

On the other hand, there is evidence that AMPs play an important role in human immunity, since almost all tissues and cells that are exposed to microbes can produce AMP. An example is that the AMPs are rich in histidine coming from the saliva since it has an antibacterial effect.[118]

Currently, the search for new AMPs from plant sources has increased, with the aim of discovering molecules that can be used in agriculture and medicine. An example of plant peptides is defensin, which contain 45–54 amino acids.[114] The main reason for using the peptides as a clinical treatment is that they are of low manufacturing cost, which allows the incursion into inaccessible therapeutic areas due to the high cost.[120]

16.4.2 ALKALOIDS

The most important compounds group isolated from medicinal plants is alkaloids that are heterocyclic nitrogenous chemical structures, biosynthetically derived from amino acids and triterpenes.[44] Some of them form salts with acetic, oxalic, lactic, malic, tartaric, and citric acid.[117] This group of compounds includes about 12,000 alkaloids,[42] which are found in several plant families. Apocynaceae, Loganiaceae, and Rubiaceae are some examples.[27]

A large part of the alkaloids are derived from a few amino acids, either open chain or aromatic, and are classified as described below:

1. Alkaloids derived from ornithine and lysine such as alkaloids with tropane, pyrrolizidine, piperidine, and quinolizidine structures.
2. Alkaloids derived from nicotinic acid.
3. Alkaloids derived from phenylalanine and tyrosine such as phenylethylamine and isoquinoline.
4. Alkaloids derived from tryptophan such as indolics and quinoline.
5. Alkaloids derived from histidine such as imidazolic alkaloids
6. Alkaloids derived from anthranilic acid.
7. Alkaloids derived from terpene metabolism such as diterpenes and steroids.
8. Xanthic bases.

The bisyndolic alkaloids known as vinblastine and vincristine are some examples; they are used as a treatment for Hodgkin's lymphoma and leukemia, respectively. Other examples are the monoterpene-indole alkaloids ajmalicine and serpentine used to treat hypertension, cardiac arrhythmias, and improve cerebral circulation.[33, 95]

The steroidal alkaloids β2-solamargine, solamargine, and degalactotigonine, isolated from *Solanum nigrum* L., show cytotoxicity due to the anticancer property.[121] This plant is used in traditional Chinese medicine.[123]

The mesembrine is an isolated alkaloid of *Sceletium tortuosum* L.; this plant is traditionally used to quench thirst, as an analgesic and antidepressant, since it alters the state of mind. Traditionally, dry or fermented aerial material of *S. tortuosum* L. is used, which is known as Kanna.[75]

Bisindolic alkaloids exhibit potent biological, pharmacological, and medicinal properties. The vinblastine and vincristine were isolated for the first time from *Catharanthus roseus* and are used clinically in the treatment of cancer.[91] In Figure 16.3, the structures of these compounds are shown.

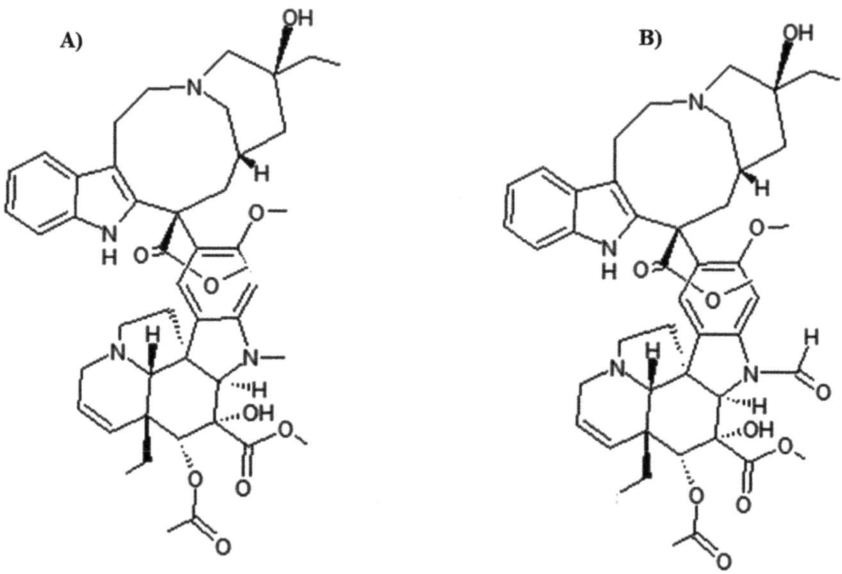

FIGURE 16.3 Structure of alkaloids: (A) vinblastine and (B) vincristine.

16.4.3 POLYPHENOL AND DERIVATIVE COMPOUNDS

The polyphenols isolated from plants function against chronic diseases such as diabetes, cardiovascular diseases, and some cancers due to high antioxidant power.[5,55,65]

Then again, flavonoids show healthy properties as antioxidants, cardioprotectors, anticarcinogens, and hypolipidemics.[108] The quinic, chlorogenic, and dicaffeoylquinic acids isolated from the leaf of *Solanum elaeagnifolium* are some examples of phenolic compounds with anti-cancer property, since they reduce the viability of HeLa (cervical cancer) and MCF-7 (human breast adenocarcinoma) cell lines. In addition, they show effect in human breast cancer explants.[62] In addition, polyphenolic compounds and some derivatives, such as berberine, apigenin, shikonin, ferulic acid, curcumin, and clematichinenoside, show antiarthritic effect.[77]

In addition, some coumarins and polyphenolic compounds come biosynthetically from the phenylpropanoid pathway.[26] Visnagin is a derivative of the furanocoumarins isolated from *Ammi visnaga* L. (Apiaceae); this plant is used in Egypt since ancient times to treat kidney stones.[113] The resveratrol is a polyphenolic compound obtained from grapes, which acts as a modulator of eating behavior and shows the antiobesogenic and antidiabetic properties.[36]

16.4.4 PHENOLIC COMPOUNDS

Phenolic compounds are produced by the vegetal secondary metabolism, and more than 8000 are known. These compounds have different chemical structure and therefore different biological activity.[82] Furthermore, they have at least one hydroxybenzene known as phenol as a functional group which can link to aromatic or aliphatic structures. These are usually found forming polyphenolic structures. However, some compounds derived from phenolic acid are monophenols.[53] The tannins, phytoestrogens, and coumarins are some compounds included in this group.[13,54]

Polyphenols are classified as flavonoids and nonflavonoids based on their chemical structure. Flavonoids contain two benzene groups linked by a tricarbonate bridge; the anthocyanins, flavones, flavanones, flavonoids, flavanonols, flavonoids, condensed tannins, and lignans are some examples.[53] The nonflavonoids are subdivided into noncarboxylic phenols (C_6, C_6–C_1, C_6–C_3) and phenolic acids, derived from benzoic acid (C_6–C_1) and cinnamic acid derivatives (C_6–C_3), respectively.

Flavonoids have a low molecular weight, are part of some plants used as food, and therefore are frequently consumed in the human diet.[46]

Additionally, phenolic compounds have antimicrobial as their main biological activity, since the hydroxyl group causes the inactivation of microbial enzyme,[25] and the hydroxyl group interacts with the cell membrane causing the leakage of cellular components. This causes a change in the fatty acids and phospholipids as well as the deterioration of the energy, which affects the synthesis of genetic materials.[28] In addition, they can penetrate the membranes, causing the inhibition of respiratory enzymes, as well as partial dissipation of the pH gradient and electrical potential.[40]

Phenolic compounds are widely distributed in plants, both in tissues and cells. These compounds offer sensory, nutritional qualities and also provide the color and flavor of some plants.

Thymol, carvacrol, and eugenol represent some examples of the strongest inhibitors of enzymatic processes, while nonphenolic alcohols, such as geraniol and linalool, have a minor effect reduced by the esterification of OH groups.[64,68] Enzymatic inhibition occurs due to the interaction of phenolic compounds with sulfhydryl groups of amino acids. In addition, quinones, flavones, flavonoids, tannins, and flavonols form complexes with the nucleophilic amino acids of proteins, which leads to their inactivation.[84] Then again, it is known that cineole reduces cell division, while limonene and α-pinene inhibit oxygen consumption.[52] Figure 16.4 presents some examples of polyphenolic compounds.

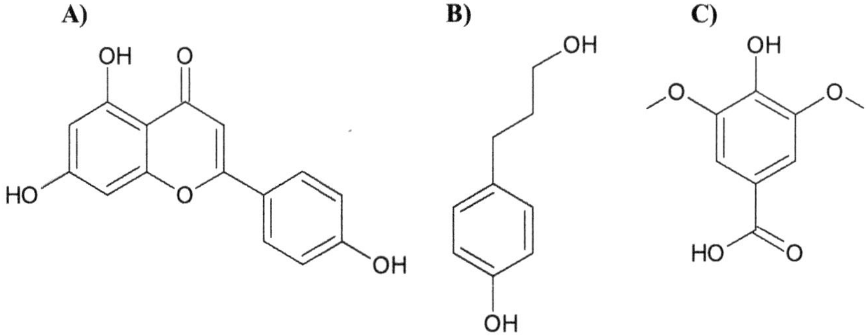

FIGURE 16.4 Chemical structure of (A) apigenin, (B) tyrosol, and (C) syringic acid.
Source: Adapted with permission from Ref. [55].

16.4.5 TERPENES

Terpenes are structurally and biosynthetically fascinating natural products.[124] These compounds derived from isopentenyl diphosphate and dimethyl-allyl-pyrophosphate, which are analogous to isoprene and produced by the mevalonate, methyl eritrol phosphate pathways.[16,43]

Terpenes are found mainly in the essential oils of plants and represent the most abundant group of volatile compounds, covering more than 40,000 compounds.[21] They are also essential for the development of plants as they are part of the structures of the plant cell membrane (sterols C), function as photosynthetic pigments (carotenoids C40), or function as phytohormones [abscisic acid (C15), gibberellins (C20), and the ubiquinone (C 15)].[2]

Terpenes include sesquiterpenlactones, cardiotonic glycosides, and saponins.[54]

The sesquiterpene lactones are compounds that have high toxicity, as well as cytotoxic effect and the ability to generate epidermal allergies.[58] In addition, they attractants pollinators or act as powerful insecticides; the pyrethrin is an example.[58]

The sesquiterpenlactones are found in Asteraceae, Umbeliferaceae, Magnoliaceae, and Lauraceae. These compounds have a fundamental skeleton of 15 carbon atoms, derived from the union of the tail, and the head of three isoprene fragments that give rise to a ring with a lactonic group in its structure and biogenetically come from farnesyl diphosphate.[27] At present, the mechanism of action of sesquiterpene lactones is unknown; however, they account for the capacity of alkylation since they bind to nucleophilic groups (Fig. 16.5).

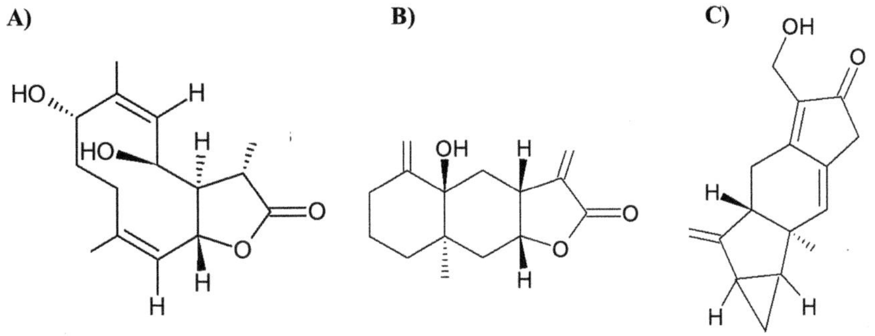

FIGURE 16.5 Chemical structures of some sesquiterpenlactones: (A) germacranolide, (B) eudesmanolide, and (C) onoseriolide.

The terpenes and derivatives are another group of compounds with important biological activities. Consequently, the pharmaceutical industry has exploited them for their potential and effectiveness as medicines.[124] The essential oils of *Carum copticum* (Apiaceae) and *Thymus vulgaris* (Lamiaceae) contain terpenes such as thymol; some of them have effects such as antioxidant, antifungal, antibacterial, and anti-inflammatory.[9,25] Other example is the sesquiterpene lactone goyazensolide, isolated from the ethanolic extract of *Lychnophora passerina*. This plant is known as arnica brasilina and is used in traditional medicine to treat pain, rheumatism, bruises, as well as some inflammatory diseases and insect bites.[9]

16.4.6 STEROIDAL AND TERPENIC STRUCTURE COMPOUNDS

The sapogenins, saponins, steroids, and sterols are some of the compounds contained in agave syrup, used as a sweetener, food, and prebiotic.[106] However, the most studied compound is the steroid sapogenin, known as agavegenin.[104] Another compound with a steroidal structure is aspaphilisin, which has a rearranged seven-member B ring and is uniquely formed by $C_{(7)}$–$C_{(14)}$; this compound was isolated from the roots of *Asparagus filicinus*.[116] The cryptotanshinone is a compound isolated from the *Salvia miltiorrhiza* Bge. (Danshen) roots and commonly used in traditional Chinese medicine to treat high blood pressure.[66]

D. purpurea is another plant that contains carotonic glycosides that have an effect on heart failure. In 1965, these glycosides were discovered that act on the cell membrane at the level of Na^+–K^+–ATPase ion pump. So far, it is unknown whether they act as a protection mechanism or if they are harmful.[48]

Other compounds that contain steroidal structure are steroids that have antidiabetic, antitumor, and antitussive effects and some participate as inhibitors of platelet aggregation (Fig. 16.6).[116]

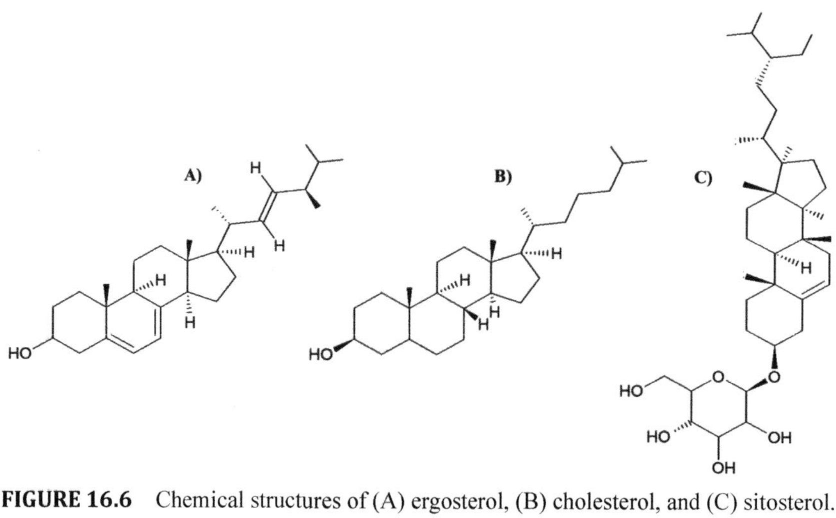

FIGURE 16.6 Chemical structures of (A) ergosterol, (B) cholesterol, and (C) sitosterol.

Saponins with triterpene or steroid structure have different properties such as anticancer, immunostimulant, anti-inflammatory, antimicrobial, hypocholesterolemic, and antioxidant.[104] Saponins have properties like natural detergents because they reduce surface tension and produce a foam on contact with water. Also, it can bind to cholesterol by preventing its absorption.

On the other hand, soy contains a steroidal or triterpenoidal aglycone with the ability to bind to one or more sugar residues through glycosyl linkages.[107] Some plants of Scrophulariaceae that contain saponins are *Verbascum thapsus* and *Scrophularia nodosa*. Another plant that contains these compounds is *Saponaria officinalis*, which is part of the Caryophyllaceae family.[97]

The *Xanthoceras sorbifolium* Bunge seeds contain saponins with a steroidal glycosidic structure; they have a wide variety of physiological effects and high antioxidant potential. For this reason, they are widely used in traditional medicine.[115]

The figure shows the structure of diosgenin, which is found abundantly in legumes and yams (Fig. 16.7).[70]

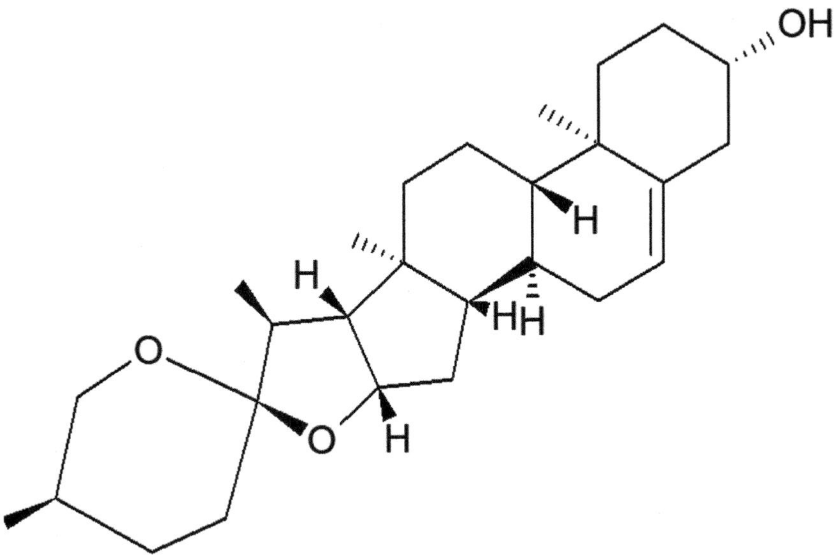

FIGURE 16.7 Chemical structure of diosgenin.

Source: Adapted with permission from Ref. [70].

16.5 CHEMISTRY OF FOODS AND NATURAL PRODUCTS

16.5.1 CHEMISTRY OF FOODS

According to the Alimentarius Codex, "food term" is understood as an elaborated, semiprocessed product intended for human consumption as well as any other substance used in the manufacture, preparation, or processing of food.[31]

Currently, there is an increase in scientific–technological advances as well as changes in the lifestyle of the population who aim to improve and maintain health. For this reason, they acquire the so-called functional foods more frequently.[47]

16.5.2 FOOD WITH MEDICINAL PROPERTIES

Food and health are closely related since they contain beneficial nutrients such as dietary fibers,[61] vitamins, essential minerals, and phytochemicals that promote optimal health during aging, these can obtain through diet, and some examples are isoflavones, anthocyanins, and lycopene contained in soybeans, berries, and tomatoes.[79]

16.5.3 FUNCTIONAL FOOD

The term and the concept of "functional food" were proposed in Japan in the 1980s, when the research changed course, with the idea of preventing diseases through the daily diet of adults. For this reason, scientists raised the concept of "functional food" from the need to prevent diseases and improve health and quality of life.[15]

At present, the Food and Drug Administration does not consider the term "functional food" to be legal, since this institution defines that the compounds contained in foods can promote health as they are part of a varied diet.[19] Until today, there is no official concept of functional food; however, there is a concordance between the definitions established by different authors and organizations. Therefore, a functional food is defined as any natural or processed food, which contains nutritional components as well as additional compounds that provide a favorable effect for health, and the physical and mental capacity of a person.[35]

However, the effects of the food matrix, the preparation processes, as well as the storage conditions can influence the stability and bioactivity of a functional food.[45]

For these reasons, the specific characteristics of a functional food are as follows:

- Represents a conventional food
- Is consumed as part of the daily diet
- Contains natural ingredients in concentrations higher than those of an ordinary food
- Contains ingredients not included in an ordinary food
- Presents a nutritional value as well as positive health effects
- Shows one or more functions that reduce the risk of illness.[26]

Nondigestible fibers, vitamins, proteins, bioactive phytochemical compounds, probiotics, prebiotics, and symbiotics are some examples of functional foods.[99]

These compounds participate in specific functions of the human body to achieve optimal growth, development and metabolism. Additionally, they inhibit oxidative stress and improve the cardiovascular system, intestinal physiology and psychological functions.[47] On the other hand, some compounds, such as milk casein, are used to protect and transport biocompounds with the purpose of incorporating them into food or pharmaceutical products.[94] Additionally, some compounds obtained from vegetable sources give functionality to foods, so they incorporate as showing positive effects on health.[17,88] Table 16.1 shows some examples of functional foods.

TABLE 16.1 Natural and Designed Functional Foods.

Functional food	Compounds	Benefit	References
Human milk	Oligosaccharides and glycoconjugates	Protect the newborn from pathogens	[37, 39]
Mayonnaise	Phenolic compounds	Antioxidant	[98]
Tomato	Lycopene	Protects against oxidation of low density lipoproteins	[93]
Soy	Proteins	Reduces the risk of cardiovascular disease	[78]
Milk and meat of ruminants	Linoleic acid	Anticancer, antilipogenic, and antiteratogenic properties	[41]
Yogurt	Sterols	Weight-loss agents	[99]
Sugar	Fructo-oligosaccharides	Promotes the development of beneficial intestinal flora	[12]

TABLE 16.1 *(Continued)*

Functional food	Compounds	Benefit	References
Fermented lactic foods	Lactobacilli	Gastrointestinal protection	[47]
Tomato puree	Anthocyanin	Anti-inflammatory	[50]
Muffins	Carotenoids microencapsulated	Antioxidant	[112]
Yogurt	Tocotrienol microcapsules	Antioxidant	[109]
Yogurt	Omega-3 fatty acids	Promotes the reduction of cholesterol and triacylglycerol levels	[32]
Carrot juice	Phenolic compounds from orange peel and pulp extracts	Antioxidant	[7]
Linaza	Alfa-linolenic acid	Cardioprotective properties	[45]
Foods and beverages	Polyphenols	Can improve endothelial function, platelet function, insulin sensitivity, and lower blood pressure	[79]
Crab, Shrimp	Chitin, chitosan	Prevention of inflammatory disorders	[67]
Bread	Theanine	Improvement in learning ability and attention, reduction of blood pressure, improvement of immune system, reduction of anxiety	[34]
Strawberry	Ellagic acid, anthocyanins, quercetin, catechin, ascorbic acid, and folic acid	Controlled feeding studies have identified the ability of strawberries to attenuate high-fat diet-induced postprandial oxidative stress and inflammation, or postprandial hyperglycemia, or hyperlipidemia	[18]
Mandarin juice	Probiotic microorganisms	Prevent or improve the treatment of digestive system diseases	[22]

KEYWORDS

- **medicinal plants**
- **natural products**
- **functional foods**
- **synthesis of plant compounds**
- **food chemistry**

REFERENCES

1. Abat, J. K.; Kumar, S.; Mohanty, A. Ethnomedicinal, Phytochemical and Ethnopharmacological Aspects of Four Medicinal Plants of Malvaceae Used in Indian Traditional Medicines: A Review. *Medicines* **2017,** *18*, 1–33.
2. Abbas, F.; Ke, Y. G.; Yu, R. C.; Yue, Y. C.; Amanullah, S.; Jahangir, M. M; Fan, Y. P. Volatile Terpenoids: Multiple Functions, Biosynthesis, Modulation and Manipulation by Genetic Engineering. *Planta* **2017,** *246*, 803–816.
3. Abdallah, H. M.; Almowallad, F. M.; Esmat, A.; Shehata, I. A.; Abdel-Sattar, E. A. Anti-inflammatory Activity of Flavonoids from *Chrozophora tinctorial. Phytochem. Lett.* **2015,** *13, 74–80.*
4. Abdallah, H. M.; Esmat, A. Antioxidant and Anti-inflammatory Activities of the Major Phenolics from *Zygophyllum simplex* L. *J. Ethnopharmacol.* **2017,** *205*, 51–56.
5. Ademosun, O.; Oboh, G. Characterization of the Antioxidant Properties of Phenolic Extracts from Some Citrus Peels. *J. Food Sci. Technol.* **2012,** *49*, 729–736.
6. Adeyemi, O. O.; Okpo; S. O.; Ogunti, O. O. Analgesic and Anti-inflammatory Effects of the Aqueous Extract of Leaves of *Persea americana* mill (Lauraceae). *Fitoterapia* **2002,** *73*, 375–380.
7. Adiamo, O. Q.; Ghafoor, K.; Al-Juhaimi, F.; Babiker, E. E.; Mohamed, A. I. A. Thermosonication Process for Optimal Functional Properties in Carrot Juice Containing Orange Peel and Pulp Extracts. *Food Chem.* **2018,** *245*, 79–88.
8. Agomuo, E.; Amadi, B. D. M. Some Biochemical Studies on the Leaves and Fruits of *Persea americana. IJRRAS* **2012,** *11*, 556–560.
9. Ahmad, K. M. S.; Ahmad, I.; Singh, S. C. *Carum compticum* and *Thymus vulgaris* Oils Inhibit Virulence in *Trichophyton rubrum* and *Aspergillus* spp. *Bras. J. Microbiol.* 2014, *45*, 523–531.
10. Albuquerque, B. C.; de Souza, J.; Ferrari, F. C.; Ferraz Filha, Z. S.; Coelho, G. B.; Saúde-Guimarães, D. A. The Influence of Seasonality on the Content of Goyazensolide and on Anti-inflammatory and Anti-hyperuricemic Effects of the Ethanolic Extract of *Lychnophora passerina* (Brazilian Arnica). *J. Ethnopharmacol.* **2017,** *198*, 444–450.
11. Allkin, B. In *Useful Plants: Medicines: At Least 28,187 Plant Species are Currently Recorded as Being of Medicinal Use*; Willis, K. J., Ed.; Royal Botanic Gardens: London, 2017.

12. Al Snafi, A. E. Phytochemical Constituents and Medicinal Properties of *Digitalis lanata* and *Digitalis purpurea*: A Review. *IAJPR* **2017,** *4*, 225–234.
13. Alvídrez, M. A.; González, B. E.; Jiménez, S. Z. Tendencias en la Producción de Alimentos: Alimentos Funcionales. *RESPYN* **2002,** *3*, 1–6.
14. Andary, C. L.; Boubals, D.; Dai, G. H.; Mondolot, C. L. Involvement of Phenolic Compounds in the Resistance of Grapevine Callus to Downy Mildew (*Plasmopara viticola*). *Eur. J. Plant Pathol.* **1995,** *101*, 541–547.
15. Andrade, S. F.; Cardoso, L. G.; Carvalho, J. C.; Bastos, J. K. Anti-inflammatory and Antinociceptive Activities of Extract, Fractions and Populnoic Acid from Bark Wood of *Austroplenckia populnea. J. Ethnopharmacol.* **2007,** *109*, 464–471.
16. Arai, S. Studies on Functional Foods in Japan. State of the Art. *Biosci. Biotech. Biochem*. **1996,** *60*, 9–15.
17. Ávalos, G. A.; Pérez, U. E. *Metabolismo Secundario de Plantas. Ser. Fisiol. Veg. Red. (Biol.)* **2009,** *3*, 119–145.
18. Azmir, J.; Ghafoor, K.; Jahurul, M. H. A.; Norulaini, N. A. N.; Omar, A. K. M; Sahena, F.; Sharif, K. M.; Rahman, M. M.; Zaidul, I. S. M. Techniques for Extraction of Bioactive Compounds from Plant Materials: A Review. *J. Food Eng.* **2013,** *117*, 426–436.
19. Basu, A.; Nguyen, A.; Betts, N. M.; Lyons, T. J. Strawberry as a Functional Food: An Evidence-based Review. *Crit. Rev. Food Sci. Nutr.* **2014,** *54*, 790–806.
20. Bello, J. Los Alimentos Funcionales o Nutraceuticos. I. Nueva Gama de Productos en la Industria Alimentaria. *Alimentaria* **1995,** *265*, 25–29.
21. Bernhoft, A. *A Bioactive Compounds in Plants: Benefits and Risks for Man and Animals*; The Norwegian Academy of Science and Letters: Oslo, Norway, 2010; p 11.
22. Betts, G.; Di Pasqua, R.; Edwards, M.; Hoskins, N.; Ercolini, D.; Mauriello, G. Membrane Toxicity of Antimicrobial Compounds from Essential Oils. *J. Agric. Food Chem*. **2007,** *55*, 4863–4870.
23. Betoret, E.; Calabuig, J. L.; Patrignani, F.; Lanciotti, R.; Dalla, R. M. Effect of High Pressure Processing and Trehalose Addition on Functional Properties of Mandarin Juice Enriched with Probiotic Microorganisms. *Food Sci. Technol.* **2017,** *85*, 418–422.
24. Bicak, K.; Gulcemal, D.; Demirtas, I.; Alankus, O. Novel Saponins from *Nigella arvensis* var. involucrata. *Phytochem. Lett.* **2017,** *21*, 128–133.
25. Boskabady, M. H.; Alitaneh, S.; Alavinezhad, A. *Carum copticum* L.: A Herbal Medicine with Various Pharmacological Effects. *Biomed. Res. Int.* 2014, *2014*, 1–11.
26. Bourgaud, F.; Gravot, A.; Milesi, S.; Gontier, E. Production of Plant Secondary Metabolites: A Historical Perspective. Review. *J. Plant Sci.* **2001,** *161*, 839–851.
27. Burt, S. Essential Oils: Their Antibacterial Properties and Potential Applications in Foods: A Review. *Int. J. Food Microbiol.* **2004,** *94*, 223–253.
28. Calvo, S.; Gómez, C.; López, C.; Roya, M. *Nutrición, Salud y Alimentos Funcionales*; UNED, Madrid, 2012; p 650.
29. Canto, B.; Galaz, M.; Gutiérrez, C.; Loyola, M.; Moreno, O.; Sánchez, P. Biosíntesis de los Alcaloides Indólicos. Una Revisión Crítica. *Rev. Soc. Quim. Mex.* **2004,** *48*, 1–29.
30. Ceylan, E.; Fung, D. Y. C. Antimicrobial Activity of Spices. *J. Rapid Methods Autom. Microbiol.* **2004,** *12*, 1–55.
31. Cicero, A. F. G.; Fogacci, F.; Colletti, A. Food and Plant Bioactives for Reducing Cardiometabolic Disease Risk: An Evidence Based Approach. *Food Funct.* **2017,** *8*, 6.

32. Compagnone, R.; Suárez, A.; Castillo, A.; Delle Morache, F.; Ferrari, F. Preliminary Phytochemical Study and Bioactivity of the Plant *Melochia villosa* from Amazonas State, Venezuela. *Saber* **1999,** *25*, 382–389.
33. Codex Alimentarius Commission. *Procedural Manual*; Food and Agriculture Organization of the United Nations, 1997. http://www.fao.org/docrep/w5975s/w5975s08.htm (accessed Nov 25, 2017).
34. Comunian, T. A.; Chaves, I. E.; Thomazini, M.; Freitas, M. I. C.; Ferro-Furtado, R.; de Castro, I. A.; Favaro-Trindade, S. F. Development of Functional Yogurt Containing Free and Encapsulate Echium Oil, Phytoesterol and Sinapic Acid. *Food Chem.* **2017,** *237*, 948–956.
35. Cordell, G. A. *The Alkaloids*; Academic Press: San Diego, CA, 1999; pp 261–376.
36. Culetu, A.; Heritier, J.; Andlauer, W. Valorisation of Theanine from Decaffeinated Tea Dust in Bakery Functional Food. *IJFST* **2015,** *50*, 413–420.
37. Chasquibol, N.; Lengua, L.; Delmás, I.; Rivera, D.; Bazán, D.; Aguirre, R.; Bravo, M. Alimentos Funcionales o Fitoquímicos, Clasificación e Importancia. *Rev. Per. Quim. Quim.* **2013,** *5*, 9–20.
38. Chung-Shu, L.; Jia-Ching, W.; Min-Hsiung, P. Molecular Mechanism on Functional Food Bioactives for Anti-obesity. *Curr. Opin. Food Sci.* **2015,** *2*, 9–13.
39. Dai, D.; Nanthkumar, N.; Newburg, D.; Walker, W. Role of Oligosaccharides and Glycoconjugates in Intestinal Host Defense. *J. Pediatr. Gastroenterol. Nutr.* **2000,** *30*, 23–33.
40. Danesi, R.; Senesi, S.; van't Wout, J. W.; van Dissel, J. T.; Lupetti, A.; Nibbering, P. H. Antimicrobial Peptides: Therapeutic Potential for the Treatment of Candida Infections. *Expert Opin. Invest. Drug* **2005,** *11*, 309–318.
41. Debb, D. D.; Parimala, G.; Saravana, D. S.; Chakrabprty, T. Effect of Thymol on Peripheral Blood Mononuclear Cell PBMC and Acute Promyelotic Cancer Cell Line HL-60. *Chem. Biol. Interact.* **2011,** *193*, 97–106.
42. De Bont, J.; Sikkema, J.; Weber, F.; Heipieper, H. Cellular Toxicity of Lipophilic Compounds: Mechanisms, Implications, and Adaptations. *Biocatalysis* **1994,** *10*, 13–122.
43. De La Torre, A.; Debiton, E.; Juaneda, P.; Durand, D.; Chardigny, J. M.; Barthomeuf, C.; Bauchart, D.; Gruffat, D. Beef Conjugated Linoleic Acid Isomers Reduce Human Cancer Cell Growth Even When Associated with Other Beef Fatty Acids. *Br. J. Nutr.* **2006,** *95*, 346–352.
44. De Luca, V.; St Pierre, B. The Cell and Development Biology of Alkaloid Biosynthesis. *Trends Plant Sci.* **2000,** *5*, 168–173.
45. Dickschat, J. S. Modern Aspects of Isotopic Labellings in Terpene Biosynthesis. *Eur. J. Org. Chem.* **2017,** *33*, 4872–4882.
46. Duncan, V. M. S.; O'neil, D. A. Commercialization of Antifungal Peptides. *Fung. Biol. Rev.* **2013,** *26*, 156–165.
47. Edel, A. L.; Aliani, M.; Pierce, G. N. Stability of Bioactives in Flaxseed and Flaxseed-fortified Foods. *Food Res. Int.* **2015,** 140–155.
48. Estévez, S. F.; Said, M.; Brouard, I.; León, F.; García, C.; Quintana, J.; Estévez, F. 30-Hydroxy-3,40-Dimethoxyflavone Blocks Tubulin Polymerization and Is a Potent Apoptotic Inducer in Human SK-MEL-1 Melanoma Cells. *Bioorg. Med. Chem.* **2017,** *25*, 6060–6070.

49. Ferrer, L. B.; Dalmau, S. J. Alimentos Funcionales: Probióticos. *Acta Pediatr. Esp.* **2001,** *59,* 150–155.
50. Furst, R.; Zundorf, I.; Dingermann, T. New Knowledge about Old Drugs: The Anti-inflammatory Properties of Cardiac Glycosides. *Planta Med.* **2017,** *83,* 977–984.
51. Frenkel, M.; Sapire, K. Complementary and Integrative Medicine in Hematologic Malignancies: Questions and Challenges. *Curr. Oncol. Rep.* **2017,** *19,* 79.
52. Gerardi, C.; Albano, C.; Calabriso, N.; Carluccio, M. A.; Durante, M.; Mita, G.; Renna, M.; Serio, F.; Nlando, F. Techno-functional Properties of Tomato Puree Fortified with Anthocyanin Pigments. *Food Chem.* **2017,** *240,* 1184–1192.
53. Grassmann, J.; Hippeli, S.; Elstner, E. F. Plant's Defense and Its Benefits for Animals and Medicine: Role of Phenolics and Terpenoids in Avoiding Oxygen Stress. *Plant Physiol. Biochem.* **2002,** *40,* 471–478.
54. Giles, L.; Peñuelas, J.; Ribas, M. Effects of Allelochemicals on Plant Respiration and Oxygen Isotope Fractionation by the Alternative Oxidase. *J. Chem. Ecol.* **1996,** *22,* 801–805.
55. Gimeno, C. E. *Compuestos fenólicos. Un Análisis de Sus Beneficios Para la Salud*; Elsevier: Amsterdam, 2004; p 23.
56. Giráldez, A.; Frutos, F.; Ramos, G.; Mantecón, A. Plants Secondary Compounds in Herbivores Nutrition. *Arch. Zootec.* **1998,** *47,* 597–620.
57. Gutiérrez, U. J. A.; Chávez, S. A.; Serna, S. S. O. Phenolic Composition, Antioxidant Capacity and In Vitro Cancer Cell Cytotoxicity of Nine Prickly Pear (*Opuntia* spp.) Juices. *Plant Foods Hum. Nutr.* **2009,** *64,* 146–152.
58. Guevara, M. G.; Oliva, C. R.; Huarte, M.; Daleo, G. R. An Aspartic Proteasa with Antimicrobial Activity Is Induced after Infection and Wounding in Intercellular Fluids of Potato Tubers. *Eur. J. Plant Pathol.* **2002,** *108,* 131–137.
59. Hamid, H. A.; Ramli, A. N.; Yusoff, M. M. Indole Alkaloids from Plants as Potential Leads for Antidepressant Drugs: A Mini Review. *Front. Pharmacol.* **2017,** *8, 96.*
60. Harborne, J. B. *Introduction to Ecological Biochemistry*; Academic Press: London, 1993.
61. Hasan, M. K.; Cheng, Y.; Kanwar, M. K.; Chu, X. Y.; Ahammed, G. J.; Qi, Z. Y. Response of Plant Proteins to Heavy Metal Stress: A Review. *Front. Plant Sci.* **2017,** *8,* 1492.
62. Hegedüs, N.; Marx, F. Antifungal Proteins: More than Antimicrobials? *Fungal Biol. Rev.* **2013,** *26,* 132–145.
63. Hernández, C. M.; Quiles, H. A. High Hydrostatic Pressure Treatment as an Alternative to Pasteurization to Maintain Bioactive Compound Content and Texture in Red Sweet Pepper. *IFSET* **2014,** *26,* 76–85.
64. Hernández, O. L.; Carranza, R. P.; Cobos, L. E.; López, L. L. I.; Ascasio, J. A.; Silva, B. S. Y. Bioguided Fractionation from *Solanum elaeagnifolium* to Evaluate Toxicity on Cellular Lines and Breast Tumor Explants. *VITAE* **2017,** *24,* 124–131.
65. Howes, M. J.; Perry, E. The Role of Phytochemicals in the Treatment and Prevention of Dementia. *Drugs Aging* **2011,** *28,* 439–468.
66. Ibler, B.; Knoblock, A.; Pauli, A.; Weigan, H.; Weiss, N. Antibacterial and Antifungal Properties of Essential Oil Components. *J. Essent. Oil Res.* **1988,** *1,* 119–128.
67. Jiang, F.; Dusting, G. J. Natural Phenolic Compounds as Cardiovascular Therapeutics: Potential Role of Their Antiinflammatory Effects. *Curr. Vasc. Pharmacol.* **2003,** *1,* 135–156.

68. Jiang, J.; Liu, J.; Zhang, L.; Owusu, L.; Liu, L.; Zhang, J.; Tang, Y.; Li, W. Anti-tumor and Chemosensitization Effects of Cryptotanshinone Extracted from *Salvia miltiorrhiza* Bge. on Ovarian Cancer Cells In Vitro. *J. Ethnopharmacol.* **2017,** *205*, 33–40.
69. Kadam, S. U.; Prabhasankar, P. Marine Foods as Functional Ingredients in Bakery and Pasta Products. *Food Res. Int.* **2010,** *43*, 1975–1980.
70. Kalodera, Z.; Kustrak, D.; Kosalec, I.; Pepeljnjak, S. *Plant-derived Antimycotics. Current Trends and Future Prospects*; Hartworth Press: New York, 2003; pp 49–79.
71. Kartal, M. Intellectual Property Protection in the Natural Product Drug Discovery, Traditional Herbal Medicine and Herbal Medicine Products. *Phytother. Res.* **2007,** *21*, 113–119.
72. Kim, D. S.; Jeon, B. K.; Lee, Y. E.; Woo, W. H.; Mun, Y. J. Diosgenin Induces Apoptosis in HepG2 Cells through Generation of Reactive Oxygen Species and Mitochondrial Pathway. *Evid. Based Complement. Alterat. Med.* **2010,** *2012*, 8.
73. Kim, G. N.; Shin, J. G.; Jang, H. D. Bioactive Compounds in Foods: Their Role in the Prevention of Cardiovascular Disease and Cancer. *Am. J. Med.* **2002,** *113*, 71–88.
74. Kitazato, K.; Wang, Y.; Kobayashi, N. Viral Infectious Disease and Natural Products with Antiviral Activity. *Drug Discov. Ther.* **2007,** *1*, 14–22.
75. Kosikowska, P.; Lesner, A. Antimicrobial Peptides (AMPs) as Drug Candidates: A Patent Review (2003–2015). *Expert Opin. Ther. Pat.* **2016,** *26*, 689–702.
76. Kris-Etherton, P.; Hecker, K.; Bonanome, A.; Coval, M.; Binkoski, A.; Hilpert, K. Bioactive Compounds in Foods: Their Role in the Prevention of Cardiovascular Disease and Cancer. *Am. J. Med. Sci.* **2002,** *113*, 71–88.
77. Krstenansky, J. L. Mesembrine alkaloids: Review of Their Occurrence, Chemistry, and Pharmacology. *J. Ethnopharmacol.* **2017,** *195*, 10–19.
78. Larijani, L. V.; Ghasem, M.; Abedian Kenari, S.; Naghshvar, F. Evaluating the Effect of Four Extracts of Avocado Fruit on Esophageal Squamous Carcinoma and Colon Adenocarcinoma Cell Lines in Comparison with Peripheral Blood Mononuclear Cells. *Acta Med. Iran.* **2014,** *54*, 201–205.
79. Lü, S.; Wang, Q.; Li, G.; Sun, S.; Guo, Y.; Kuang, H. The Treatment of Rheumatoid Arthritis Using Chinese Medicinal Plants: From Pharmacology to Potential Molecular Mechanisms. *J. Ethnopharmacol.* **2015,** *176*, 177–206.
80. Mahn, K.; Borras, C.; Knock, G.; Taylor, P.; Khan, Y.; Sugden, D.; Poston, L.; Ward, J. P.; Sharpe, R. M.; Vina, J.; Aaronson, P.; Mann, G. Dietary Soy Isoflavone-induced Increases in Antioxidant and eNOS Gene Expression Lead to Improved Endothelial Function and Reduced Blood Pressure In Vivo. *FASEB J.* **2005,** *12*, 1755–1757.
81. Manach, C.; Milenkovic, D.; Van de Wiele, T.; Rodriguez, M. A.; de Roos, B.; Garcia, C. M. T.; Landberg, R.; Gibney, E. R.; Heinonen, M.; Tomaa, B. F. Addressing the Inter-individual Variation in Response to Consumption of Plant Food Bioactives: Towards a Better Understanding of Their Role in Healthy Aging and Cardiometabolic Risk Reduction. *Mol. Nutr. Food Res.* **2017,** *61* 1–16.
82. Martins, S.; Mussato, I. S.; Martínez, A. G.; Montañez, S. J.; Aguilar, C. N.; Texeira, J. A. Bioactive Phenolic Compounds: Production and Extraction by Solid-state Fermentation. A Review. *Biotechnol. Adv.* **2011,** *29*, 365–373.
83. Marwat, S. K.; ur-Rehman, F.; Khan, E. A.; Baloch, M. S.; Sadiq, M.; Ullah, I.; Javaria, S. *Ricinus cmmunis*: Ethnomedicinal Uses and Pharmacological Activities. *Pak. J. Pharm. Sci.* **2017,** *30*, 1815–1827.

84. Metha, P.; Shah, R.; Lohidasan, S.; Mahadik, K. R. Pharmacokinetic Profile of Phytoconctituet(s) Isolated from Medicinal Plants: A Comprehensive Review. *J. Trad. Complement. Med.* **2015,** *5*, 207–227.
85. Modak, B.; Gorai, P.; Dhan, R.; Mukherjee, A. Tradition in Treating Taboo: Folkloric Medicinal Wisdom of the Aboriginals of Purulia District, West Bengal, India Against Sexual, Gynaecological and Related Disorders. *J. Ethnopharmacol.* **2015,** *169*, 370–386.
86. Cowan, M. M. Plant Products as Antimicrobial Agents. *Clin Microbiol Revs.* **1991,** *12*, 564–582.
87. Ng, S. M. S.; Yap, Y. Y. A.; Cheong, J. W. D.; Ng, F. M.; Lau, Q. Y.; Barkham, T.; Teo, J. W. P.; Hill, J.; Chia, C. S. B.; Antifungal Peptides: A Potential New Class of Antifungals for Treating Vulvovaginal Candidiasis caused by Fluconazole-resistant *Candida albicans*. *J. Pept. Sci.* **2017,** *23*, 215–222.
88. Ng, Y. P.; Or, T. C.; Ip, N. Y. Plant Alkaloids as Drug Leads for Alzheimer's Disease. *Neurochem. Int.* **2015,** *89*, 260–270.
89. Odo, Ch. E.; Nwodo, O. F.; Joshua, P.; E., Ugwu, O. P. Acute Toxicity Investigation and Anti-diarrhoeal Effect of the Chloroform–Methanol Extract of the Leaves of *Persea americana*. *Iran J. Pharm. Res.* **2014,** *13*, 651–658.
90. Oomen, R. J.; Séveno-Carpentier, E.; Ricodeau, N.; Bournaud, C.; Conéjéro, G.; Paris, N.; Berthomieu, P.; Marquès, L. Plant defensin AhPDF1.1 is Not Secreted in Leaves But It Accumulates in Intracellular Compartments. *New Phytol.* **2011,** *192:* 140–50.
91. Parveen, A.; Parveen, B.; Parveen, R.; Ahmad, S. Challenges and Guidelines for Clinical Trial of Herbal Drugs. *J. Pharm. Bioallied Sci.* **2015,** *7*, 329–333.
92. Pérez, H. Nutracéuticos: Componente Emergente Para el Beneficio de la Salud. *ICIDCA* **2006,** *3*, 20–28.
93. Rahman, M. T.; Tiruveedhula, V. V.; Cook, J. M. Synthesis of Bisindole Alkaloids from the Apocynaceae which Contain a Macroline or Sarpagine Unit: A Review. *Molecules* **2016,** *14*, 1–67.
94. Ranadheera, C. S.; Liyanaarachchi, W. S.; Chandrapala, J.; Dissanayake, M.; Vasilijvec, T. Utilizing Unique Properties of Caseins and the Casein Micelle for Delivery of Sensitive Food Ingredients and Bioactives. *Trends Food Sci. Technol.* **2016,** *57*, 178–187.
95. Rao, A. Lycopene, Tomatoes, and the Prevention of Coronary Heart Disease. *Exp. Bioland. Med.* **2002,** *227*, 908–913.
96. Ribeiro, M. H. Naringinases: Occurrence, Characteristics, and Applications. *Appl. Microbiol. Biotechnol.* **2011,** *90*, 1883–1895.
97. Roberts, F.; Waterman, G.; Wink, M. *Alkaloids: Biochemistry, Ecology, and Medicinal Applications*; Plenum Press: New York, 1998; pp 87–107.
98. Rost, T.; Barbour, M.; Thornton, R.; Weier, E.; Stocking, R. Metabolismo. In *Botanica: Introducción a la Biología Vegetal*; LIMUSA: México, 1985; Vol. 1, pp 17–32.
99. Ruiz, G.; Price, K.; Rose, M.; Arthur, A.; Petterson, D.; Fenwick, G. The Effect of Cultivar and Environment on Saponin Content of Australian Sweet Lupin Seed. *J. Sci. Food Agric.* **1995,** *69*, 347–351.
100. Saravana, B. P. A.; Aafrin, B. V.; Archana, G.; Sabina, K.; Sudharsan, K.; Krishnan, K. R.; Babuskin, S.; Sivarajan, M.; Sukumar, M. Polyphenolic and Phytochemical Content of *Cucumis sativus* Seeds and Study on Mechanism of Preservation of

Nutritional and Quality Outcomes in Enriched Mayonnaise. *IJFST* **2016,** *51,* 1417–1424.

101. Sarmiento, L. Functional Foods, a New Feeding Alternative. *Rev. Orinoq.* **2006,** *10* (16), 23.
102. Sayed-Ahmed, B.; Talou, T.; Saad, Z.; Hijazi, A.; Merah, O. The Apiaceae: Ethnomedicinal Family as Source for Industrial Uses. *Ind. Crops Prod.* **2017,** *109,* 661–671.
103. Shirpoor, A. Medicinal Plants for Management of Diabetes: Alternative or Adjuvant? *Antol. J. Cardiol.* **2017,** *17,* 460.
104. Shixin, D.; Shao-Nong, Ch.; Jian, Y. Chemistry of Medicinal Plants, Foods and Natural Products. *Hindawi* **2015,** *121849,* 2.
105. Selitrennikoff, C. P. Antifungal Proteins. *Appl. Environ. Microbiol.* **2001,** *67,* 2883–2894.
106. Sidana, J.; Singh, B.; Sharma, O. P. Saponins of Agave: Chemistry and Bioactivity. *Photochemistry* **2016,** *130,* 22–46.
107. Sommela, E.; Pagano, F.; Pepe, G.; Ostacolo, C.; Manfra, M.; Di Sanzo, R.; Carabetta, S.; Campiglia, P.; Russo, M. Flavonoid Composition of Tarocco (*Citrus sinensis* L. Osbeck) Clone "Lempso" and Fast Antioxidant Activity Screening by DPPH-UHPLC-PDA-IT-TOF. *Phytochem. Anal.* **2017,** *28,* 521–528.
108. Srikrishna, D.; Godugu, C.; Dubey, P. K. A Review on Pharmacological Properties of Coumarins. *Mini Rev. Med. Chem.* **2018,** *18,* 113–141.
109. Takahashi, Y.; Li, X. H.; Tsukamoto, C.; Wang, K. J. Categories and Components of Soysaponin in the Chinese Wild Soybean (*Glycine soja*) Genetic Resource Collection. *GRACE* **2017,** *64,* 2161–2171.
110. Tanaka, T.; Makika, H.; Kawabata, K.; Mori, H.; Kakumoto, M.; Satoh, K.; Hara, A; Sumida, T.; Tanaka, T.; Ogawa, H. Chemoprevention of Azoxymethane-induced Rat Colon Carcinogenesis by the Naturally Occurring Flavonoids, Diosmin and Hesperidin. *Carcinogenesis* **1997,** *18,* 957–965.
111. Tan, P. Y.; Tan, T. B.; Chang, H. W.; Tey, B. T.; Chan, E. S.; Lai, O. M.; Baharin, B. S.; Nehdi, I. A.; Tan, C. P. Effects of Storage and Yogurt Matrix on the Stability of Tocotrienols Encapsulated in Chitosan-alginate Microcapsules. *Food Chem.* **2018,** *241,* 79–85.
112. Tamrat, Y.; Nedi, T.; Assefa, S.; Teklehaymanot, T.; Shibeshi, W. Anti-inflammatory and Analgesic Activities of Solvent Fractions of the Leaves of *Moringa stenopetala* Bak. (Moringaceae) in Mice Models. *BMC Complement. Altern. Med.* **2017,** *17,* 1–10.
113. Tilburt, J. C.; Kaptchuk, T. J. *Herbal Medicine Research and Global Health: An Ethical Analysis*; p 577–656 (Expanded Bulletin of the World Health Organization). http://www.who.int/bulletin/volumes/86/8/07-042820/en/ (accessed Aug 8, 2008).
114. Ursache, F. M.; Andronoiu, D. G.; Ghinea, I. O.; Barbu, V.; Ionita, E.; Cotarlet, M.; Dumitrascu, L.; Botez, E.; Rapeanu, G.; Stanciuc, N. Valorizations of Carotenoids from Sea Buckthorn Extract but Microencapsulation and Formulation of Value-added Food Products. *J. Food Eng.* **2018,** *219,* 16–24.
115. Vanachayangkul, P.; Byer, K.; Khan, S.; Butterweck, V. An Aqueous Extract of *Ammi visnaga* Fruits and Its Constituents Khellin and Visnagin Prevent Cell Damage Caused by Oxalate in Renal Epithelial Cells. *Phytomedicine* **2010,** *17,* 653–658.
116. Van Der Weerden, N. L.; Anderson, M. A. Plant Defensins: Common Fold, Multiple Functions. *Fungal Biol. Rev.* **2013,** *26,* 121–131.

117. Venegas, C. M.; Ruiz, M. M. V.; Martinez, F. E.; Garces, R.; Salas, J. J. Characterization of *Xanthoceras sorbifolium* Bunge Seeds: Lipids, Proteins and Saponins Content. *Ind. Crops Prod.* **2017,** *109*, 192–198.
118. Wang, J. P.; Cai, L.; Chen, F. Y.; Li, Y. Y.; Li, Y. Y.; Luo, P.; Ding, Z. T. A New Steroid with Unique Rearranged Seven-membered B Ring Isolated from Roots of *Asparagus filicinus. Tetrahedr. Lett.* **2017,** *58*, 3590–3593.
119. Watson, D. *Natural Toxicants in Food*; CRC: London, 1998.
120. Wiesner, J.; Vilcinskas, A. Antimicrobial Peptides: The Ancient Arm of the Human Immune System. *Virulence* **2010,** *1*, 440–464.
121. Xu, Q.; Wang, Y.; Guo, S.; Shen, Z.; Wang, Y.; Yang, L. Anti-inflammatory and Analgesic Activity of Aqueous Extract of *Flos populi. J. Ethnopharmacol.* **2014,** *152*, 540–545.
122. Zhang, L.; Falla, T. J. Antimicrobial Peptides: Therapeutic Potential. *Expert Opin. Pharmcol.* **2006,** *7*, 653–663.
123. Zhou, X.; He, X.; Wang, G.; Gao, H.; Zhou, G.; Ye, W.; Yao, X. Steroidal Saponins from *Solanum nigrum. J. Nat. Prod.* **2006,** *69*, 1158–1163.
124. Zwenger, S.; Chhandak, B. Plant Terpenoids: Applications and Future Potentials. *Biotechnol. Mol. Biol. Rev.* **2008,** *3*, 001–007.

INDEX

A

Advanced oxidation processes (AOPs), 221–226
Alkaloids, 365–366
  structure, 366
All-atom loop, D
Anaerobically digested POME (AnPOME), 321
  materials and method, 321
  composting, 322–323
  nutrient content of compost, 325
  physiochemical analysis, 323–325
  phytotoxicity assay, 325
  statistical analysis, 326
  weight of compost material, 322
  zeta potential, 325
  results and discussion
  composting nutrient, 333
  degradation of OM, 328–329
  mass reduction, 329–330
  moisture content, 327–328
  pH profile and conductivity, 331
  phytotoxicity, 332–333
  profile of total organic carbon, 329
  temperature profile, 326–327
  zeta potential, 332

B

Backplanes for displays, 143
Bio-inspired computation
  future of engineering science, 307–308
Bipolar junction transistor (BJT), 122

C

Chemical engineering, 306–307
  futuristic vision, 242–243
  science of sustainability, 241–242
Chemists, 153–154
Collision quenching, 182
Composite membranes, 216
Composite science, 264, 292
  applications of CFD, 266–267
  engineering vision for future, 264–265
Computational fluid dynamics (CFD), 259, 260
  advanced fluid dynamics, 278–279
  applications, 266–267
  energy and environmental sustainability, 280
  future trends, 281–282
  literature review, 262–264
  modern society, 279–280
  scientific
  doctrine and definition, 265
  endeavor, visionary, 267
  progress in aerospace engineering, 277–278
  scope, 261–262
  significant scientific research, 268–275
  chemical process engineering, 275–276
  petroleum engineering, 276–277
  vision, 261
Conductors, 155–156
Contact quenching, 182

D

D. purpurea, 370
Dielectrics, 156
Diosgenin
  chemical structure, 371

E

Energy sustainability
  vision for future, 253–254
Engineering science
  applications of fluid dynamics, 266–267
Epoxy-olygomers, 26
  chemical formula, 27
  hardener in commercial tare, 29
  materials and methods, 28–31

morphology, 31–32
resistance to flame/burning, 40–41
strength tests, 32–38
adhesion, 36
compression strength, 37
deformation-load for compression, 38
samples destructed by compression, 37
SEM images, 33–35
wear of composites, 38
thermo-and chemi-resistance, 38–41
acetone swelling, 40
ink-solvent swelling, 39
visual changes, 42
visualization of test methods, 30
Evolutionary computation, 290–291
future research trends, 312–313
Extracts of plants, 358

F

Ferrocene-aminonaphthalate derivatives, 186
Field-effect transistor (FET), 122
Fluid dynamics
visionary scientific endeavor, 267
Fluidization engineering
FCCU, 251
scientific
doctrine, 239–240
research, 248–250
Fluidized catalytic cracking unit (FCCU), 236
scientific advances, 251–253
vision and challenge, 240–241
Fluorescence chemosensors
examples, 184–197
ICT process, 185–190
PET sensors, 190
examples, 190–197
Fluorescence quantum yield ratio, 183
Fluorescence quenching, 181–183
collision, 182
contact, 182
Fluorescence sensing
building blocks, 173–174
concept, 172–174
ideologies, 173
Foods
chemistry, 371–272
functional food, 372–374
medicinal properties, 372
Functional food, 372–374
natural and designed, 373–374

G

Gas sensors, 143–144
Genetic algorithm (GA), 238–239
applications, 243–248
future of engineering science, 307–308
significant scientific research in application, 308–309
Global climate change, 215
Graphene, 4
characterization, 6, 8–9
derivative-based fibers, 9–12
fabrication methods, 10–12
methods of synthesis, 7
oxide and IR active functional groups, 9
preparation, 5–6
real-time use, 5
Graphene-based fibers, 9–10
application, 14–16
actuators, 14
catalysis, 15
energy storage, 15–16
field emitters, 14–15
springs, 15
fabrication methods, 10
dry spinning, 11
dry-jet wet spinning, 12
electrophoretic event, 12
wet spinning, 10–11
properties, 12–14
electrical, 13–14
mechanical, 12–13
thermal, 14
Gypsum, 28
curves of swelling in acetone, A

I

IGFET metal-oxide-semiconductor field-effect transistor (MOSFET), 122
Ink-solvent
swelling in, A

Insulated-gate bipolar transistor (IGBT), 122

L

Logic, 143
Low-density polyethylene (LDPE), 88
experimental, 93–94
luminescence behavior, 92–93
polymer composites, 93
results and discussion
characterization, 95–97
composites characterization, 100–113
contact angle of water, 107
density, 105–107
mechanical properties, 109–111
modified version characterization, 97–100
morphology of fracture surfaces, 100
optical behavior of ZNO and, 111–113
resistivity, 107–109
TG analysis, 104–105
XRD, 100–105

M

Mechanical sensors, 144
Medicinal plants, 359–360
chemistry, 361–371
alkaloids, 365–366
chemical structures of compounds, 362
pathways of synthesis, 363
peptides and proteins, 363–364
phenolic compounds, 367–368
polyphenol and derivative compounds, 366–367
steroidal and terpenic structure compounds, 369–371
terpenes, 368–369
Medicinal properties foods, 372
Membrane science
scientific vision, 217
significant scientific endeavor, 219–221
Microtubule (MT) polymer, 57
bending vibration, 77–82
after loading with solvent, 79
extraction of flexural rigidity and natural frequency, 77–82
flexural stiffness, 80
current study, 66–67
dry system, 77
fluorescence image of accumulation, 59
literature review and previous works, 63–66
mathematical modeling of physical systems, 58–63
mechanical properties, 82–83
longitudinal natural frequency, 83
stress–strain, 82
modeling and coarse graining, 71–74
molecular dynamic modeling and simulation, 68–69
NPT process, 76
NVT, 76
primary requirements for modeling, 70–71
solvated environment, 67–68
solvating the system, 74–75
substructure, B
substructure and dimension, 60
substructure, α- and β-tubulin, 61
Modern science
difficulties and challenges, 254
Modified graphene
structure, 5
Multi-objective optimization, 291
scientific
cognizance behind, 293–294
doctrine behind, 293–294
endeavor in application, 297–302
vision and challenges, 310
Multi-objective-simulated annealing, 291–292
future of computational science, 310–311
scientific
grit, 294–295
profundity, 294–295
vision behind, 294
significant research endeavor, 302–306
technological vision, 294–295

N

Natural products, 360–361
chemistry, 361–371
alkaloids, 365–366

chemical structures of compounds, 362
pathways of synthesis, 363
peptides and proteins, 363–364
phenolic compounds, 367–368
polyphenol and derivative compounds, 366–367
steroidal and terpenic structure compounds, 369–371
terpenes, 368–369
Nontraditional environmental engineering tools, 213
Nontraditional optimization techniques, 288
Novel separation process, 215, 218
composite membranes, 216

O

Optimization, 306–307
Organic electronics, 121, 154
materials used, 154
conductors, 155–156
dielectrics, 156
passivation, 156
semiconductors, 155
substrates, 157
organic electronic devices
LEDS, 157–161
OFETs, 164–167
radical batteries, 161–162
solar cells, 162–164
Organic field-effect transistors (OFETs), 164–167
Organic LEDS, 157
cathode component, 158
device structure, 158
electrons transport layer, 159
substrate, 157–158
technical approaches, 160–161
Organic materials, 153
Organic radical batteries, 161–162
Organic semiconductors, 131, 155
n-type, 131–132
p-type, 132–133
Organic solar cells, 162–164
bilayer heterojunctions, 164
single layer junction, 163–164
single layer organic photovoltaic, 163
Organic thin-film transistors (OTFTs), 121
basic parameters, 123–124
channel conductance, 126
charge carrier mobility, 124
current-gain cutoff frequency, 126
gate leakage, 127
hysteresis/bias-stress, 127
insulator capacitance, 126
ON/OFF current ratio, 124
subthreshold slope and turn-on voltage, 125–126
threshold voltage, 125
transconductance, 125
biomedical field
biological sensors, 144–145
biomedical sterilization, 145
organic bioelectronic tools, 144
current approaches
basic parameters, 123–127
general aspects, 121–123
structures and models, 127–129
factors affecting performance
contact effects and materials features, 141–142
dimensional parameters, 139–141
environmental stability, 141
processing conditions, 139–141
materials used in fabrication, 129–130
dielectric layer, 133
electrode, 130–131
gate dielectric material, 133–136
most used, 138–139
organic semiconductors, 131–133
substrate and passivation materials, 137–139
practical importance
biomedical field, 144–145
electronic industry, 143–144
structures and models, 127
cylindrical gate, 129
dual gate, 128
single-gate, 128
vertical channel, 128–129

P

Palm oil industry, 319–320
Palm oil mill effluent (POME), 320

AnPOME, 321
composting, 321
Passivation materials, 156
Peptides and proteins, 363–364
Petroleum engineering, 306–307
Phenolic compounds, 367–368
chemical structure, 368
Photochemistry
principles, 175–176
fluorescence lifetime, 180–181
fluorescence quenching, 181–183
Jablonski diagram, 176–177, 178
quantum yield of fluorescence, 183–184
Stokes shift, 177, 179–180
structure of PNBD and BPNBD, 181
Polyethylene (PE), 87
AMPTES application methods, 90
luminescence behavior of ZnO, 92–93
polymer composites with ZnO, 93
Polymer matrix
experimental, 47–48
mechanical properties, 46
result and discussion, 48–53
composition and physico-mechanical properties, 50–51
influence of structurants and mineral fillers, 53
Polymer science, 292
Polymeric materials, 88
Polyphenol, 366–367
Precise agriculture, 354
Pristine graphene
structure, 5

Q

Quantized space, 206
Quantum computing, 204
Quantum fluctuations
space–time geometry, 205
Quantum simulation, 204–205

R

Radio-frequency identification (RFID), 143
RBCG method
atomic categorization, C

S

Science and engineering of material science, 239
Sensing, 171
Soil fertility agrochemical, 339
materials and methodology, 340
results, 340–354
change in content of mobile phosphorus, 345
coefficients of correlation of agrochemical, 346
combined unit, 353
dependence of agrochemical indices, 349
dependence of variability of soil, 344
diagnosis of nitrogen nutrition, 351
differentiated dressing, 353
distribution curve, 341–342
indicators for diagnosis, 352
influence of mobile of mobile phosphorus content, 347
remote radiolocation and IR spectrometry, 350
spatial structure, 342
variability of mobile phosphorus content, 343, 346
variability of yield of annual grasses hay and, 348
yield of winter wheat, 354
Speck, 206
Steroidal and terpenic structure
compounds, 369–371
D. purpurea, 370
Substrate, 157
Superconductivity (SC), 203
magnetism coexistence by chemical design, 202
Sustainability
need of mathematical tools, 296–297
success of science, 295–296

T

Technology and engineering science, 237
Terpenes, 368–369
chemical structure, 369

W

Water treatment, 218
  arsenic and heavy metal, 227–229
  sustainable development, 226–227

Z

Zinc oxide (ZnO), 88–89
  experimental, 93–94
  luminescence behavior, 92–93
  polymer composites, 93
  results and discussion
  characterization, 95–97
  composites characterization, 100–113
  cumulative particle size distribution, 95
  FTIR spectrum, 97
  modified version characterization, 97–100
  SEM micrograph, 96
  X-ray diffraction diagram, 96

9781771887403

9781771887403